International Conference on

Nuclear Physics at Border Lines

organized by

University of Messina

Fondazione Bonino-Pulejo Messina

Proceedings of the International Conference

Nuclear Physics at Border Lines

Lipari (Messina), Italy 21–24 May 2001

editors

Giovanni Fazio
University of Messina, Italy

Giorgio Giardina
University of Messina, Italy

Francis Hanappe
Université Libre de Bruxelles, Belgium

Giuseppina Immè
University of Catania, Italy

Neil Rowley
Institut de Recherches Subatomiques, Strasbourg, France

World Scientific
New Jersey • London • Singapore • Hong Kong

Published by

World Scientific Publishing Co. Pte. Ltd.

P O Box 128, Farrer Road, Singapore 912805

USA office: Suite 1B, 1060 Main Street, River Edge, NJ 07661

UK office: 57 Shelton Street, Covent Garden, London WC2H 9HE

British Library Cataloguing-in-Publication Data
A catalogue record for this book is available from the British Library.

Proceedings of the International Conference on
NUCLEAR PHYSICS AT BORDER LINES

ISBN 981-02-4778-8

Printed in Singapore.

CONFERENCE ORGANIZATION AND COMMITTEES

ORGANIZING COMMITTEE:

Chairmen: G. GIARDINA and F. HANAPPE

Scientific Secretaries: G. IMMÈ and N. ROWLEY

INTERNATIONAL ADVISORY COMMITTEE:

M. ARNOULD (Belgium)
B. BORDERIE (France)
A. CUNSOLO (Italy)
M. FREER (United Kingdom)
C.K. GELBKE (USA)
G. GIARDINA (Italy)
W. GREINER (German)
D. GUERREAU (France)
F. HANAPPE (Belgium)
D.C. HOFFMAN (USA)
M.G. ITKIS (Russia)
CH. LECLERCQ-WILLAIN (Belgium)
E. MIGNECO (Italy)
G. MUENZENBERG (Germany)
J.B. NATOWITZ (USA)
YU.TS. OGANESSIAN (Russia)
A. PALMERI (Italy)
R.A. RICCI (Italy)
C. SIGNORINI (Italy)
I. TANIHATA (Japan)
I.N. VISHNEVSKY (Ukraine)

LOCAL ORGANIZING COMMITTEE:

G. GIARDINA (Messina), F. HANAPPE (Bruxelles),
G. IMMÈ (Catania), N. ROWLEY (Strasbourg),
G. FAZIO, A. LAMBERTO and A. TACCONE (Messina),
R. PALAMARA (Reggio Calabria)

ACKNOWLEDGMENTS

The Organizers would like to express their gratitude to the following organizations for their sponsorship and financial supports, through which this Conference was made possible:

University of Messina

Bonino-Pulejo Foundation

Istituto Nazionale di Fisica Nucleare (INFN)

Ente per le Nuove Tecnologie, l'Energia e l'Ambiente (ENEA)

University of Reggio Calabria

Regione Siciliana

Provincia di Messina

Azienda Autonoma Provinciale per l'Incremento Turistico (AAPIT)

Comune di Lipari

PREFACE

The International Conference on *Nuclear Physics at Border Lines* was held from 21st to 24th May 2001 at Lipari (Messina), the main island of the fantastic Eolian Archipelago of Italy.

The aim of the Conference was to emphasize the research fields in which we are extending the border lines of Nuclear Physics by new experimental techniques and presenting theoretical methods and models with important new challenges.

The scientific programme of the Conference was arranged in four major areas:

- Reactions between massive nuclei leading to superheavy element formation.
- Radioactive beams and neutron-rich systems.
- Exotic nuclei and nuclear astrophysics.
- New states of nuclear matter.

These Proceedings is a collection of all invited talks at the plenary sessions and oral contributions given by the speakers at the parallel sessions.

During the meeting, the participants presented new interesting results and discussed widely future plans about exciting new prospects for Nuclear Physics.

Around 150 scientists from about 15 countries, both inside and outside Europe, took part in the Conference.

We would like to thank all participants for their attendance and contributions to the Conference. We are very grateful to all members of the International Advisory Committee, Organizing Committee and Local Organizing Committee for their efforts to make this Conference successful. Special thanks are due to the young members of the Conference Secretariat for their efficient work.

We would like to express our appreciation to World Scientific for publishing this volume.

G. Fazio, G. Giardina, F. Hanappe, G. Immè and N. Rowley

CONTENTS

Preface vii

Opening Talk: Nuclear Structure at Border Lines 1
Yu.Ts. Oganessian

Hard photons: A probe of dynamical effects in heavy ion collisions at intermediate energy 18
R. Alba, C. Agodi, C. Maiolino, A. Del Zoppo, M. Colonna, G. Bellia, R. Coniglione, P. Finocchiaro, K. Loukachine, E. Migneco, P. Piattelli, D. Santonocito, P. Sapienza, M. Bruno, M. D'Agostino, M.L. Fiandri, G. Vannini, N. Colonna, F. Gramegna, P.F. Mastinu, I. Iori, A. Moroni, G.V. Margagliotti, P.M. Milazzo, R. Rui

GDR excitation close to the liquid-gas transition of nuclear matter 26
F. Amorini, G. Cardella, A. Di Pietro, P. Figuera, G. Lanzalone, J. Lu, A. Musumarra, M. Papa, S. Pirrone, F. Rizzo, S. Tudisco

First results with TONNERRE (a new detector for delayed-neutron spectroscopy) 30
J.C. Angélique, N.L. Achouri, S. Grévy, F.R. Lecolley, J.L. Lecouey, E. Liénard, N.A. Orr. J. Péter, S. Pietri, C. Borcea, A. Buta, F. Negoita, C. Timis, F. de Oliveira, M. Lewitowicz, M. Stanoiu, O. Tarasov, P. Baumann, S. Courtin, P. Dessagne, R. Hadeler, A. Knipper, C. Miehe, F. Poirier, G. Walter, D. Guillemaud-Mueller, D.Y. Penionzhkevich, D.Z. Dlouhý, J. Mrazek, J.M. Daugas, W.N. Catford

The S-factor of the ^{3}H(^{3}H,2n)^{4}He and ^{3}He(^{3}He,2p)^{4}He reactions using a three-cluster exit-channel 34
F. Arickx, J. Broeckhove, W. Vanroose, V. Vasilevsky, A. Nesterov

Theoretical estimation for recent experimental results of superheavy elements at Dubna with three-dim Langevin approach 38

Y. Aritomo

From hadronic to deconfinated matter in ultrarelativistic heavy ion experiments 42

A. Badalá

Studying exotic nuclei through direct reactions 46

Y. Blumenfeld

The study of the reaction ^{58}Fe + ^{248}Cm → ^{306}(122) 56

A.A. Bogatchev, M.G. Itkis, Yu.Ts. Oganessian, E.M. Kozulin, I.M. Itkis, M. Jandel, J. Kliman, G.N. Knyajeva, N.A. Kondratiev, I.V. Korzyukov, L. Krupa, I.V. Pokrovski, V.A. Ponomarenko, E.V. Prokhorova, A.Ya. Rusanov, V.M. Voskresenski, F. Hanappe, B. Benoit, T. Materna, N. Rowley, L. Stuttgé, G. Giardina, K.J. Moody

A new shape isomer in the N=Z nucleus ^{72}Kr 60

E. Bouchez, K. Hauschild, A. Hurstel, W. Korten, Y. Le Coz, R. Lucas, M. Rejmund, Ch. Theisen, F. Becker, G. de France, M. Lewitowicz, I. Matea, F. de Oliveira, M. Stanoiu, B. Blank, C. Borcea, A. Buta, F. Negoita, D. Pantelica, A. Emsallem, J. Genevey, J. Pinston, F. Hannachi, P. Rahkila

Spectroscopy of light exotic nuclei by (^{7}Li,^{7}Be) charge exchange reaction 64

F. Cappuzzello, A. Cunsolo, S. Fortier, A. Foti, A. Lazzaro, H. Lenske, A.L. Melita, C. Nociforo, L. Tassan-Got, J.S. Winfield, S. Boninelli, S. Orrigo

Isospin effects on instabilities and fragmentation mechanisms 68

M. Colonna, V. Baran, M. Di Toro, V. Greco

Ground-state properties of neutron-rich nuclei in the sd-shell 77

D. Cortina-Gil, J. Fernandez-Vazquez, K. Markenroth, F. Attallah, T. Baumann, J. Benlliure, M.J.G. Borge, L. Chulkov, C. Forssén, L.M. Fraile, H. Geissel, J. Gerl, K. Ithashi, R. Janik, B. Jonson, S. Karlsson, S. Mandal, M. Meister, X. Mocko, T. Ohtsubo, A. Ozawa, V. Pribora, K. Riisager, G. Schneider, H. Scheit, G. Schrieder, B. Sitar, A. Stolz, P. Strmen, K. Sümmerer, X. Szarka, S. Wan, H. Weick

Cluster states in Be isotopes 81

P. Descouvemont

Changes in neutron magic numbers of neutron-rich nuclei 85

Z. Dlouhý, D. Baiborodin, J. Mrázek, G. Thiamová

Dynamical model of fission fragment angular distribution 89

V.A. Drozdov, D.O. Eremenko, O.V. Fotina, S.Yu. Platonov, O.A. Yuminov, G. Giardina, A. Taccone

The reaction ^{6}He+^{64}Zn around the Coulomb barrier 93

P. Figuera, A. Di Pietro, A. Musumarra, F. Amorini, G. Cardella, J. Lu, M. Papa, M.G. Pellegriti, F. Rizzo, S. Tudisco, T. Davinson, H. Mahmud, C. Ruiz, A.C. Shotter, C. Angulo, S. Cherubini, A. Ninane, M. Milin, N. Soic, R. Raabe, L. Weissman

New opportunities with beams of rare isotopes in the U.S. 97

C.K. Gelbke

Quaternary fission 107

F. Gönnenwein, P. Jesinger, M. Mutterer, A.M. Gagarski, G.A. Petrov, W.H. Trzaska, V. Nesvizhevsky, O. Zimmer

Preliminary results and future activities at the GARFIELD apparatus 112

F. Gramegna, P.F. Mastinu, L. Vannucci, M. Bruno, M. D'Agostino, L. Fiandri, A. Lanchais, G. Vannini, G. Casini, M. Chiari, A. Nannini, A. Bonasera, S. Cavallaro, A. Cosmano, A. Moroni, A. Ordine, U. Abbondanno, P.M. Milazzo, G.M. Margagliotti

Mean-field calculations of super-heavy elements 122

P.H. Heenen

Recent experiments and plans for the synthesis of superheavy elements at the GSI SHIP 132

S. Hofmann

The study of the characteristics of neutron emission in the reactions with ^{48}Ca ions 142

I. Itkis, A.A. Bogatchev, M.G. Itkis, J. Kliman, G.N. Knyazheva, N.A. Kondratiev, E.M. Kozulin, I.V. Korzyukov, L. Krupa, Yu.Ts. Oganessian, I.V. Pokrovski, E.V. Prokhorova, V.M. Voskresenski, N. Amar, J. Péter, G. Giardina, F. Hanappe, T. Materna, P. Desesquelles, O. Dorvaux, N. Rowley, Ch. Schmitt, L. Stuttgé

Nuclear fission at border lines 146

M.G. Itkis, A.A. Bogatchev, I.M. Itkis, M. Jandel, J. Kliman, G.N. Kniajeva, N.A. Kondratiev, I.V. Korzyukov, E.M. Kozulin, L. Krupa, Yu.Ts. Oganessian, I.V. Pokrovski, V.A. Ponomarenko, E.V. Prokhorova, A.Ya. Rusanov, V.M. Voskresenski, F. Hanappe, T. Materna, N. Rowley, L. Stuttgé, G. Giardina, K.J. Moody

Investigation of neutron and gamma multiplicities in reactions with heavy ions leading to the production of superheavy nuclei close to the island of stability 157

E.M. Kozulin, M.G. Itkis, Yu.Ts. Oganessian, A.A. Bogatchev, A.Yu. Chizhov, I.M. Itkis, M. Jandel, J. Kliman, G.N. Kniajeva,

N.A. Kondratiev, I.V. Korzyukov, L. Krupa, I.V. Pokrovski, V.A. Ponomarenko, E.V. Prokhorova, V.M. Voskresenski, F. Hanappe, T. Materna, A. Ninane, N. Rowley, L. Stuttgé, Ch. Schmitt, O. Dorvaux, B. Gall, G. Giardina, J. Péter, N. Amar, J.M. Gauthier, St. Grevy, G. Chubarian, P. Desesquelles, V.A. Rubchenya, W.H. Trzaska, Z. Radivojevic, Ch. Stodel

Charge radius change in the heavy tin isotopes around A=132 from laser spectroscopy 162

F. Le Blanc, E. Cottereau, S. Essabaa, J. Obert, J. Oms, A. Ouchrif, B. Roussière, J. Sauvage, D. Verney, L. Cabaret, J. Pinard, J.E. Crawford, J.K.P. Lee, R. Horn, G. Huber, J. Lassen, ISOLDE Collaboration

Many-body theory at extreme isospin 166

H. Lenske, C.M. Keil, F. Hofmann

Study of $\beta - \nu$ angular correlations in nuclear β decay 176

E. Liénard, G. Ban, G. Darius, P. Delahaye, D. Durand, A.M. Vinodkumar, F. Mauger, O. Naviliat, P. Van Hove

Search for a long-lifetime component in the fission of ^{235}U induced by ~30 MeV alpha-particles 180

F. Malaguti, L. Patrizii, V. Togo, A. Uguzzoni, P. Olivo, D. Eremenko, O. Fotina, S. Platonov, O.A. Yuminov, G. Giardina

Probing correlations in many-body haloes 183

F.M. Marqués Moreno

High spin states populated via projectile fragmentation in very neutron-rich nuclei around mass 180 193

P. Mayet, J. Gerl, Ch. Schlegel, Zs. Podolyák, P.H. Regan, P.M. Walker, M. Caamaño, M. Pfützner, M. Hellström, M.N. Mineva (for the GSI ISOMER Collaboration)

Breakup processes of ^{6,7}Li projectiles from a ^{208}Pb target at Coulomb barrier energies 197

M. Mazzocco, C. Signorini, L. Stroe, A. Andrighetto, A. Vitturi, D. Fabris, M. Lunardon, G. Nebbia, G. Viesti, G.F. Prete, M. Cinausero, E. Fioretto, F. Soramel, A. Brondi, G. La Rana, R. Moro, E. Verdaci, N. Gelli, F. Lucarelli

Theoretical approaches and experimental evidences for liquid-vapor phase transitions in nuclei 202

L.G. Moretto, J.B. Elliott, L. Phair, G.J. Wozniak

Role of the primordial diclustering configuration in the asymmetric mode of fission 213

G. Mouze, R.A. Ricci

The astrophysical S-factors for proton capture on ^{13}C from the asymptotic normalization coefficients for proton removal from ^{14}N 216

A.M. Mukhamedzhanov, A. Azhari, V. Burjan, F. Carstoiu, C.A. Gagliardi, V. Kroha, A. Sattarov, X. Tang, L. Trache, R.E. Tribble

Determination of the Astrophysical Factor and Electron Screening Potential in the ^{2}H(^{6}Li,α)^{4}He reaction by applying the Trojan Horse Method 220

A. Musumarra, R.G. Pizzone, I. Bertuglia, S. Blagus, M. Bogovac, P. Figuera, M. Lattuada, M. Milin, Đ. Miljanić, M.G. Pellegriti, D. Rendić, C. Rolfs, S. Romano, N. Soić, C. Spitaleri, S. Typel, H.H. Wolter, M. Zadro

Study of halo-structure in radioative He and Li nuclei with proton elastic scattering 224

A.V. Dobrovolsky, G.D. Alkhazov, A. Andronenko, A. Bouchet, P. Egelhof, S. Fritz, G.E. Gavrilov, H. Geissel, C. Gross, H. Irnich, A.V. Khanzadeev, G.A. Korolev, G. Kraus, A.A. Lobodenko, G. Münzenberg, M. Mutterer, S.R. Neumaier, T. Schäfer, C. Scheidenberger, W. Schwab, D.M. Seliverstov, T. Suzuki, N.A. Timofeev, A.A. Vorobyov, V.I. Yatsoura

Effects of the shell structure in reactions leading to the same compound nucleus or different isotopes 234

A.K. Nasirov, G. Fazio, G. Giardina, A. Lamberto, R. Ruggeri, A. Taccone, F. Hanappe, R. Palamara, L. Stuttgé

Influence of excited radioactive nuclei for results in large-scale nuclear chronometry 244

V.S. Olkhovsky

New magic number, N=16, near the neutron drip line 248

A. Ozawa

Constrained molecular dynamics approach to fermionic systems 253

M. Papa, T. Maruyama, A. Bonasera

Experiments on super-heavy nuclei at GANIL 257

J. Péter, N. Alamanos, N. Amar, J.C. Angélique, R. Anne, G. Auger, F. Becker, R. Dayras, A. Drouart, J.M. Fontbonne, A. Gillibert, S. Grévy, D. Guerreau, F. Hanappe, R. Hue, A.S. Lalleman, N. Lecesne, T. Legou, M. Lewitowicz, R. Lichtenthäler, E. Liénard, W. Mittig, F. De Oliveira, N. Orr, G. Politi, Z. Sosin, M.G. Saint-Laurent, J.C. Steckmeyer, C. Stodel, J. Tillier, R. De Tourreil, A.C.C. Villari, J.P. Wieleczko, A. Wieloch

Isospin physics with the REVERSE experiment 267

M. Alderighi, A. Anzalone, R. Barnà, V. Baran, I. Berceanu, J. Blicharska, A. Bonasera, B. Borderie, R. Bougault, M. Bruno, G. Cardella, S. Cavallaro, A. Chbihi, M. Colonna, M. D'Agostino, R. Dayras, E. De Filippo, D. De Pasquale, M. Di Toro, E. Geraci, F. Giustolisi, A. Grzeszczuk, P., Guazzoni, D. Guinet, M. Iacono Manno, A. Italiano, S. Kowalski, A. Lanchais, G. Lanzanò, G. Lanzalone, N. Le Neindre, S. Li, S. Lo Nigro, C. Maiolino, Z. Majka, T. Paduszynski, A. Pagano, M. Papa, M. Petrovici, E. Piasecki, S. Pirrone, G. Politi, A. Pop, F. Porto, M.F. Rivet, E. Rosato, S. Sambataro, G. Sechi, V. Simion,

M.L. Sperduto, J.C. Steckmeyer, C. Sutera, A. Trifirò, M. Trimarchi, G. Vannini, M. Vigilante, J.P. Wieleczko, J. Wilczynski, H. Wu, Z. Xiao, L. Zetta, W. Zipper

Collisional relaxation of collective vibration in hot nuclei 271

V.A. Plujko, M.O. Kavatsyuk, O.M. Gorbachenko

Mass-energy distributions of fission fragments of superheavy nuclei produced in the reactions with ^{48}Ca ions 275

E. Prokhorova, M.G. Itkis, Yu.Ts. Oganessian, E.M. Kozulin, A.A. Bogatchev, I.M. Itkis, M. Jandel, J. Kliman, G.N. Kniajeva, N.A. Kondratiev, I.V. Korzyukov, L. Krupa, I.V. Pokrovski, V.A. Ponomarenko, A.Ya. Rusanov, V.M. Voskresenski, F. Hanappe, B. Benoit, T. Materna, N. Rowley, L. Stuttgé, G. Giardina

Thermodynamics of hot nuclei: multifragmentation and phase transition 279

M.F. Rivet

Entrance and exit channels for very heavy and superheavy elements 289

G. Royer, K. Zbiri, R.A. Gherghescu

The effect of the entrance channel in complete fusion reactions leading to the production of heavy nuclei 293

R.N. Sagaidak, V.I. Chepigin, A.P. Kabachenko, O.N. Malyshev, Yu.Ts. Oganessian, A.G. Popeko, A.V. Yeremin, F.P. Heßberger, S. Hofmann, V. Ninov, Ch. Stodel, G. Giardina, A. Lamberto, A.K. Nasirov

Fusion cross section for charged particles by using the optical theorem 297

A. Scalia

Decay of compound nuclei 300

Ch. Schmitt, J. Bartel, A. Surowiec, K. Pomorski

Study of transfer induced fission and fusion-fission reactions for $^{28}Si + ^{232}Th$ system at 340 MeV 305

D. Fabris, M. Lunardon, G. Nebbia, A. Samant, G. Viesti, V. Rizzi, E. Fioretto, G. Prete, M. Cinausero, D.V. Shetty, S. Pesente, A. Brondi, G. La Rana, R. Moro, E. Vardaci, A. Boiano, A. Ordine, N. Gelli, F. Lucarelli, A. Saxena, B.K. Nayak, D.C. Biswas, R.K. Choudhury, S.S. Kapoor

Statistical decay of heavy nucleus-nucleus systems 310

K. Siwek-Wilczyńska

Structure and properties of superheavy nuclei 314

I. Muntian, A. Sobiczewski

Coulomb excitation of Ni and Zn isotopes around the N=40 subshell closure 322

O. Sorlin, S. Leenhardt, C. Donzaud, F. Azaiez, F. Amorini, A. Astier, D. Baiborodin, M. Belleguic, K. Bienczak, C. Bourgeois, C. Borcea, H. Brenning, D.M. Cullen, I. Deloncle, Z. Dlouhý, Zs. Dombrádi, E. Dragulescu, J. Duprat, M. Gorska, H. Grawe, S. Grévy, D. Guillcmaud-Mueller, G. Hagemann, B. Herskind, J. Kiener, R. Lemmon, M. Lewitowicz, D. Pantalica, S.M. Lukyanov, Yu.-E. Penionzhkevich, M.G. Porquet, P. Mayet, F. de Oliveira Santos, L. Petizon, F. Pougheon, N. Redon, M.G. Saint-Laurent, J.A. Scarpaci, G. Sletten, M. Stanoiu, O. Tarasov, Ch. Theisen

Backtraced neutron multiplicities and capture dynamics in the superheavy region 326

L. Stuttgé, O. Dorvaux, C. Schmitt, F. Hanappe, B. Benoit, E. De Góes Brennand, T. Materna, K. Siwek-Wylczynska, J. Wylczynki, L. Donadille, P. Désesquelles, E. Liatard, M.G. Itkis, Yu.M. Itkis, N.A. Kondratiev, E.M. Kozulin

Microscopic clusters in the ground states of lithium and beryllium isotopes 336

M. Tomaselli, T. Kühl, P. Egelhof, W. Nörtershäuser, C. Kozhuharov, A. Dax, H. Wang, S.R. Neumaier, D. Marx, H.-J. Kluge, I. Tanihata, S. Fritzsche, M. Mutterer

Astrophysical S factors from asymptotic normalization coefficients 340

R.E. Tribble, A. Azhari, P. Bem, V. Burjan, F. Carstoiu, J. Cejpek, H.L. Clark, C.A. Gagliardi, V. Kroha, Y.-W. Lui, A.M. Mukhamedzhanov, J. Novak, S. Piskor, A. Sattarov, E. Simeckova, X. Tang, L. Trache, J. Vincour

Neutron induced reactions on radioactive beryllium and argon isotopes 350

C. Wagemans, G. Goeminne, J. Wagemans, U. Köster, M. Loiselet, M. Gaelens, P. Geltenbort

Pre-fusion capture cross sections for heavy nucleus-nucleus systems 354

J. Wilczyński, K. Siwek-Wilczyńska

MAGNEX: A large-acceptance spectrometer for radioactive nuclear beam experiments 364

A. Cunsolo, F. Cappuzzello, A. Foti, A. Lazzaro, A.L. Melita, C. Nociforo, V. Shchepunov, J.S. Winfield

What we have learnt from study of the direct reactions with 25 MeV/n ^{6}He and ^{8}He secondary beams on hydrogen and helium targets? 368

R. Wolski, A.S. Fomichev, A.M. Rodin, S.V. Stepantsov, S.I. Sidorchuk, G.M. Ter-Akopian, Yu.Ts. Oganessian

Hydrogen-4 and hydrogen-5 from transfer reactions induced by a 57.5-MeV triton beam on deuterium and tritium targets 372

Yu.Ts. Oganessian, G.M. Ter-Akopian, D.D. Bogdanov, A.S. Fomichev, M.S. Golovkov, A.M. Rodin, S.I. Sidorchuk, R.S. Slepnev, S.V. Stepantsov, R. Wolski,

V.A. Gorshkov, M.L. Chelnokov, M.G. Itkis, E.M. Kozulin, A.A. Bogatchev, N.A. Kondratiev, I.V. Korzyukov, A.A. Yukhimchuk, V.V. Perevozchikov, Yu.I. Vinogradov, S.K. Grishechkin, A.M. Demin, S.V. Zlatoustovsky, A.V. Kuryakin, S.V. Fil'chagin, R.I. Il'kaev, F. Hanappe, T. Materna, L.Stuttgé, A.H. Ninane, A.A. Korsheninnikov, E.Yu. Nikolski, I. Tanihata, P. Roussel-Chomaz, W. Mittig, N. Alamanos, V. Lapoux, E.C. Pollacco, L. Nalpas

Kinematic separator and magnetic analyzer for the identification of recoiling nuclei 376

Yu.Ts. Oganessian, A.V. Yeremin, A.G. Popeko, O.N. Malyshev, A.V. Belozerov, V.I. Chepigin, M.L. Chelnokov, V.A. Gorshkov, A.P. Kabachenko, R.N. Sagaidak, A.I. Svirikhin

Decay time characteristics of the heavy excited nuclei 385

D.O. Eremenko, O.V. Fotina, S.Yu. Platonov, O.A. Yuminov, G. Giardina, A. Lamberto, F. Malaguti

Nucleon collectivization as the main mechanism of the fusion-fission dynamics 389

V.I. Zagrebaev

Author Index 393

NUCLEAR STRUCTURE AT BORDER LINES

Yu.Ts.OGANESSIAN

Flerov Laboratory of Nuclear Reaction, Joint Institute for Nuclear Research
141980 Dubna, Moscow Region, Russian federation

E-mail: oganessian@flnr.jinr.ru

The properties of neutron-rich nuclei near the drip-line are considered for two limiting cases: for the lightest elements, namely isotopes of hydrogen with A=4-5 and superheavy elements with A=288 and 292. The isotopes of ^{4}H and ^{5}H have been investigated using the 58 MeV triton beam in the reactions t+d and t+t with registering charged particles and neutrons. In the spectra of the excitation energies certain states have been found which determine the structural properties of these nuclei. A programme of further investigations with the use of the accelerator complex of radioactive beams DRIBs, which is being created at the moment, is discussed. Characteristics of beams of radioactive ions with light and medium masses, which are expected to be yielded using DRIBs after its commissioning, are presented.
Results on the synthesis of even-even isotopes with Z=110-116 in reactions with ^{48}Ca ions are discussed. It is shown that an increase in the neutron numbers of heavy nuclei greatly enhances their stability in respect to α-decay and spontaneous fission. Radioactive properties of the new nuclides confirm theoretical predictions concerning the existence of the "islands of stability" of superheavy nuclei.

1 Introduction

According to the topics of the conference the present talk refers to the properties of nuclei near the drip-line. It includes the data of a number of works carried out within the past year at the Flerov Laboratory of Nuclear Reactions (FLNR, JINR) and devoted to the study of the lightest and extremely heavy nuclei, in the properties of which neutron correlations play a decisive role. It is hardly possible that the lightest and the heaviest nuclei can be theoretically described using a unified approach. However, intensive development within the past years of microscopic models has been provided by one and the same basis and the choice of one or other nucleon-nucleon interaction can be applied to the description of the structure of a wide variety of nuclei.

According to theoretical predictions, in a number of cases slight changes in the nucleon composition lead to substantial changes in their properties which is extremely attractive for experimental verification.

Extension of experimental possibilities connected with the use of beams of radioactive nuclei is an important step on this way.

2 Properties of heavy isotopes of hydrogen

As is known, changes in the nuclear matter density at approaching the neutron drip-line were observed for light nuclei with 2≤Z≤ 6. Formation of unusual (exotic) structures in the form of the "neutron halo" (^{6}He, ^{11}Li, ^{11}Be etc) or the "neutron skin" (^{8}He) shows that at a high neutron excess neutron correlations play an important role. From this point of view, the study of the systems, situated beyond the drip-line, which are even greater enriched with neutrons is of certain interest.

The highest concentration of neutrons can be achieved in the isotopes of the lightest elements – helium and hydrogen.

Fig. 1 shows possible structures originating in highly neutron-rich nuclei. As will follow from the later chapters, these systems can be investigated using beams

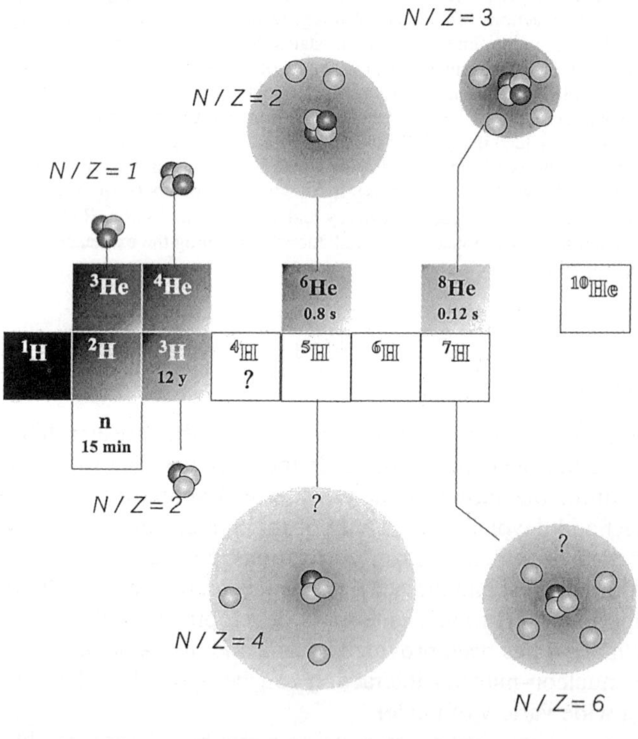

Fig. 1. Isotopes of hydrogen and helium. Protons are designated in black color, neutrons - in white. For the hydrogen isotopes with A= 5 and 7 a supposed nuclear structure is shown.

of radioactive nuclei – ^{6}He and ^{8}He. It is not excluded that the structures of the "neutron halo" or the "neutron skin" type observed in He isotopes will manifest themselves also in hydrogen isotopes located beyond the drip-line.

First attempts to study the structure of the ^{5}H nucleus was made in 2000 at the FLNR using the reaction ^{6}He(p,2p)^{5}H induced by the 25 MeV/n ^{6}He beam ($3\cdot10^4$/s in intensity). The beam was produced at a tandem of U-400M + separator ACCULINA in the reaction ^{7}Li+^{9}Be [1]. Analyzing combined spectra of two protons one can make a conclusion that in the system ^{5}H the low-lying state is manifested in the form of the resonance at the energy E_r=1.7±0.3 MeV and the width Γ_{obs}=1.9±0.4 MeV.

Further experiments were continued in another reaction: ^{3}H+^{3}H using more sophisticated experimental techniques.

In this reaction a transfer of one neutron leads the formation of the ^{4}H nucleus and a transfer of two neutrons - to the formation of ^{5}H. A triton beam with an even higher energy resolution (ΔE/E~0.5%) can be produced as a result of a direct acceleration of tritium ions at the cyclotron, and the intensity of the beam can be increased up to 10^{10}/s.

The scheme of the experiment is presented in Fig. 2.

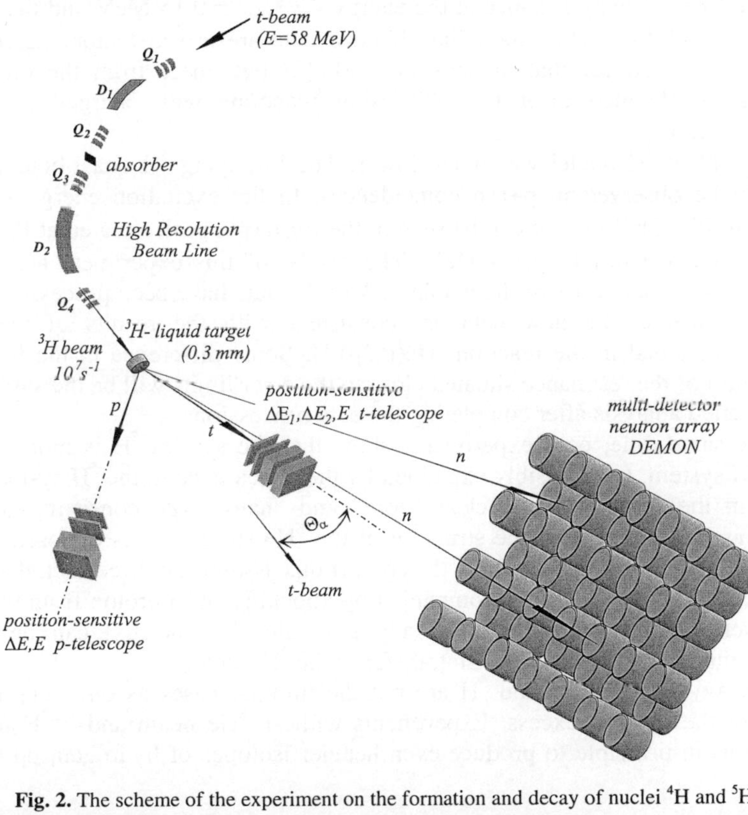

Fig. 2. The scheme of the experiment on the formation and decay of nuclei ^{4}H and ^{5}H.

A beam of molecular ions DT^+(m/e=5) accelerated up to an energy of 19.3 MeV/n was extracted from the cyclotron chamber in the form of a triton beam (E_L=58.0 MeV) by means of stripping on a thin carbon foil. The separator ACCULINA was switched into the regime of a high resolution beam-line aiming at decreasing the energy spread down to $\Delta E \sim 300$ keV. Liquid deuterium and tritium enclosed into a miniature container 10 mm in diameter and 10μ inox entrance windows were used as targets. The thickness of the target layer was 0.3 mm ($1.5 \cdot 10^{21}$ atoms/cm^2).

The intensity of the ion beam used in the experiments was $2 \cdot 10^7$/s. For the registration of charged particles (p,d,t,He) position-sensitive telescopes were used. Detection of neutrons and determining their energies by the time-of-flight measurements was performed with the use of multi-detector system DEMON (41 modules).

Using the arrangement as described above, it was possible to carry out the missing mass as well as invariant mass experiments.

Fig. 3 shows the spectra of ^{4}H and ^{5}H excitation energies. The data for ^{4}H were obtained in the reaction ^{3}H(^{2}H,p)^{4}H→ t+n by measuring the p+t and p+t+n coincidences. In both cases the location of the first level of that nucleus is clearly seen in the form of the resonance at the energy $E_r = 3.22 \pm 0.15$ MeV and the width $\Gamma_{obs} = 3.33 \pm 0.25$ MeV (Fig. 3a). The obtained data are in satisfactory agreement with those obtained for that nucleus in work [2] but differ from the other 11 measurements obtained prior to 1992 using reactions with charged particles, neutrons and π^--mesons.

The yield of ^{5}H nuclei was much lower. The low-lying ^{5}H quasi-bound state could only be observed in p+t+n coincidences. In the excitation energy spectra presented in Fig. 3b this state can be seen in the form of a peak situated at $E_r = 2.5 \pm 0.3$ MeV with a width $\Gamma_{obs} \leq 1.0$ MeV. The results of this experiment are so far preliminary since only 60% of the whole bulk of the data have been processed.

On the whole, the new data are consistent with the results of the first experiment obtained in the reaction ^{6}He(p,2p)^{5}H. Some difference in the location and the width of the resonance situated close to the error limits will be the subject of a more detailed analysis after completing the data procession.

At the same time, both experiments show that the system ^{5}H is more bound than the ^{4}H system. It is possibly explained by the appearance in the ^{5}H system of a structure in the form of the nuclear "core" and "halo" type consisting of two neutrons, as is the case with the structure of the ^{6}He structure. As opposed to the ^{6}He system, here ^{3}H plays the role of the core. If this assumption is correct, than it is not excluded that the reaction accompanied by knocking-off a proton from the ^{6}He core (which by the definition is already the halo-like nucleus) can be more successful than the reaction of the 2n transfer to the ^{3}H nucleus.

The heavy isotopes ^{4}H and ^{5}H are not the limiting cases as one goes in the direction of the neutron excess. Experiments with an ^{8}He beam and ^{1,2,3}H targets will allow us in principle to produce even heavier isotopes of hydrogen, up to ^{7}H.

Similarly to ^{8}He, this nucleus may have the “neutron skin” and can be more bound than the system ^{5}H. In the decay $^7H \rightarrow t+n+n+n+n$ one can investigate neutron correlations, formation of neutron clusters etc.

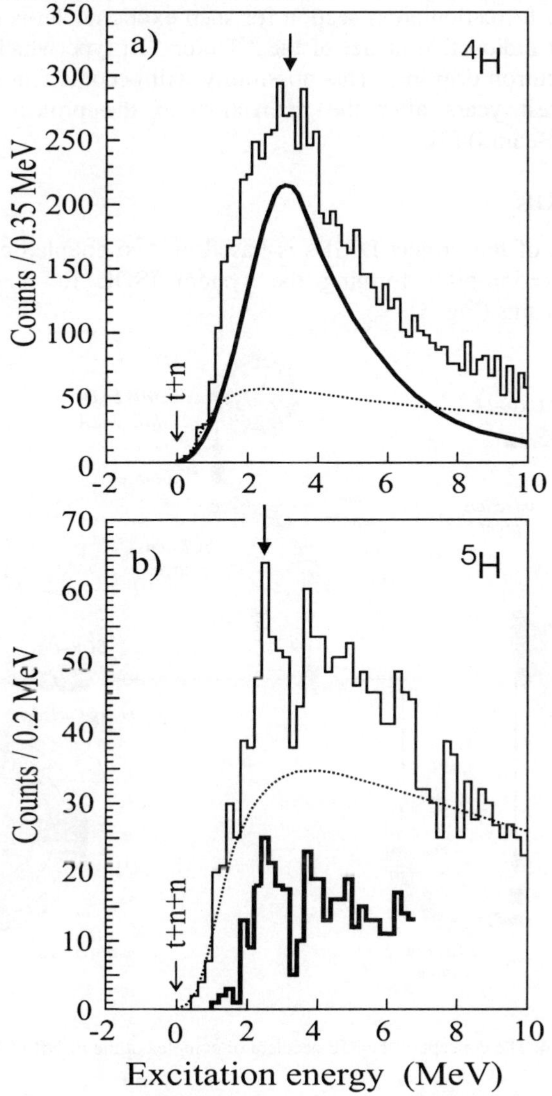

Fig. 3. Spectra of excitation energies of nuclei:
a) ^{4}H in the reaction t(d,p), b) ^{5}H in the reaction t(t,p).
The thin lines are the experimental spectra. The dotted curves are the estimated continium. The thick lines are the difference between the first and the second curves. The arrows point to the location of the maximums in the observed resonances.

In other words, without having at one's disposal a method for the study of purely neutron matter, here it becomes possible to study the micro-systems of the nuclear scale consisting up to 86% of neutrons.

However the formation cross section for such exotic nuclei is low and requires intense beams of radioactive nuclei of the ^{6,8}He or ^{11}Li type which themselves are situated at the neutron drip line. This possibility will become feasible at the FLNR within the nearest years after the realization of the project DRIBs (Dubna Radioactive Ion BeamS) [3].

3 Project DRIBs

The concept of the project DRIBs is based on two coupled cyclotrons U-400 and U-400M (k=550-600) forming the system ISOL for the production of radioactive ion beams (Fig. 4).

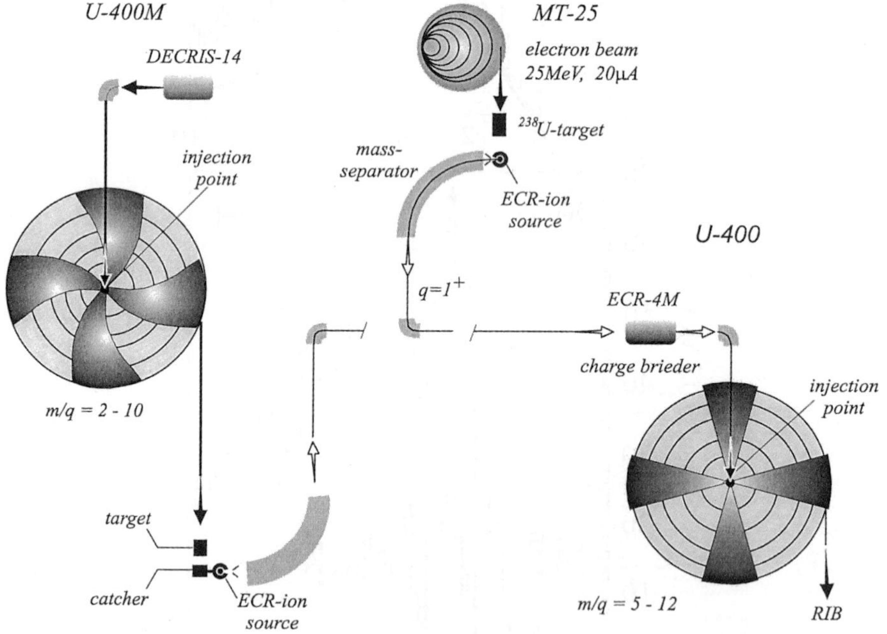

Fig. 4. The concept of the RIB accelerator complex at the FLNR (JINR).

For producing ^{6}He and ^{8}He nuclei at a ^{9}Be target (~1 g/cm^2) ion beams of ^{7}Li and ^{11}B will be used. The intensity of Li and B beams at present amounts to about 10 ρμA. Nuclei of ^{6}He and ^{8}He knocked out of the target are stopped in a porous graphite catcher heated up to T~2000K. The stopped atoms due to thermodiffusion move to the ECR ion source (f_0=2.45 Hz). The beam of single-charged He ions with low energies is transported at a distance of ~130 m and is injected into the second cyclotron U-400M. The extraction of the ion beam is realized by means of charge exchange on the graphite foil.

Expected energies and intensities of the beams are presented in Table 1.

Table I. Intensities and energies of the radioactive ion beams produced at the on-line separator and expected in the ISOL-system of the DRIBs complex.

RIB	On- line ACCULINA separator	ISOL DRIBs
^{6}He (0.81s)	$1.5\cdot10^{6}$pps 25 MeV·A	$9\cdot10^{9}$pps 8 - 13 MeV·A
^{8}He (0.12s)	$2\cdot10^{4}$pps 28 MeV·A	$2\cdot10^{7}$pps 6 - 8 MeV·A
^{11}Be (13.8s)	$9\cdot10^{4}$pps 36 MeV·A	$2\cdot10^{8}$pps 4 - 16 MeV·A

The above mentioned investigations devoted to the formation of ^{4}H and ^{5}H nuclei in the reactions ^{3}H(^{2}H,p)^{4}H and ^{3}H(^{3}H,p)^{5}H can be considered as the first pre-DRIBs experiments. As it follows from Fig. 5, it will be possible to study a wider range of neutron- and proton-rich nuclei within the same experimental setting using RIB and radioactive targets of ^{3}H, ^{7}Be, ^{10}B etc.

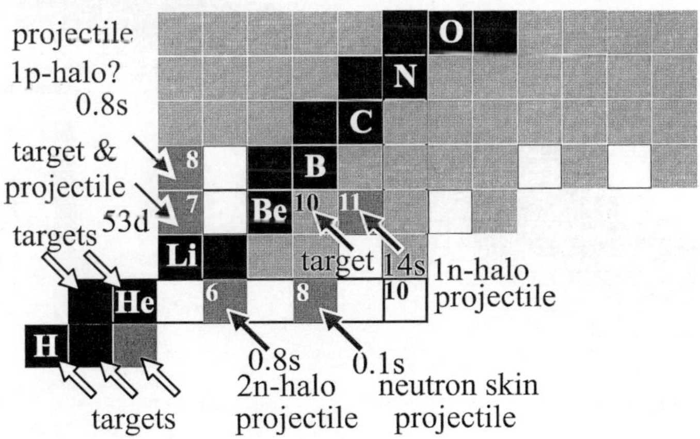

Fig. 5. Isotopes of light elements which may be used as particles and targets for the study of the exotic nuclear structure.

In the first experiments with RIB and ^{6,8}He ions it is supposed to investigate:

- the structure of extremely neutron-rich nuclei of the lightest elements with Z=1-3;
- elastic and inelastic scattering of exotic nuclei; exotic cluster transfer;
- fusion and fission with the use of targets of different isotopic composition;

The second stage of the Project DRIBs is connected with the production of beams of neutron-rich nuclei of medium masses – uranium fission fragments [4].

It is planned to use the reaction of ^{238}U photofission in the region of the giant dipole resonance (E_γ=11-15 MeV) for obtaining the high yield of neutron-rich nuclei. Similarly to U fission induced by low-energy neutrons (E≤1.5 MeV) in photofission the maximal yield of neutron-rich fragments is achieved. The yields of the ^{238}U fission fragments different in masses and charges are presented in Fig. 6. They have a character typical for the low-energy ^{238}U fission, with a clearly expressed asymmetry of the fission fragment mass distribution. It is supposed to use the acting electron accelerator of the microtrone type MT-25 (FLNR). It produces beams of electrons 25 MeV in energy and 20 μA in intensity (the beam power W=0.5 kW).

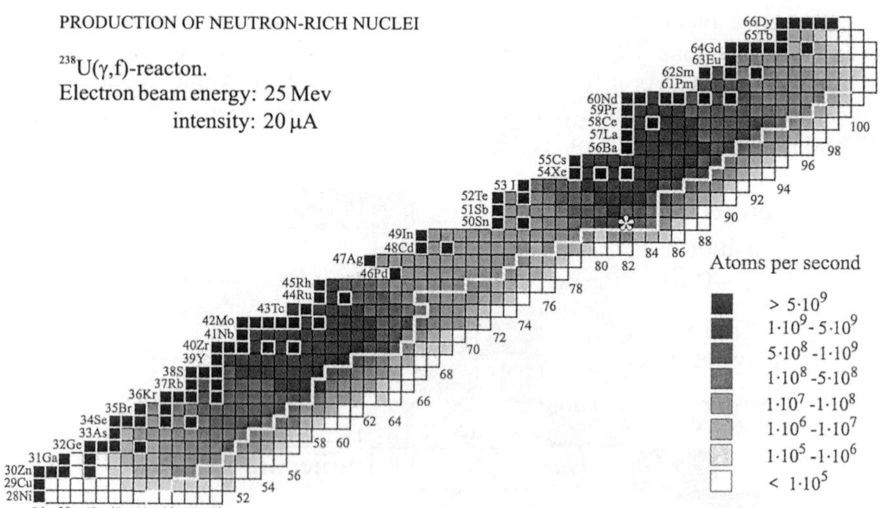

Fig. 6. The yields of isotopes of different elements in the photofission of ^{238}U. The calculations are made for the 0.5 kW electron beam produced at the MT-25 microtron. The white line separates the isotopes with $T_{1/2}$>1s. The isotope ^{132}Sn is indicated by the asterisk.

Calculations and direct experiments carried out using the above mentioned set-up show that with a uranium target 35 g/cm^2 in thickness (the total weight of metallic uranium – about 10g) we may have $1.3\cdot10^{11}$ fission events per second. At such a fission rate the yields of neutron-rich nuclei ^{142}Xe(1.24s), ^{144}Xe(1.25s) and ^{132}Sn(39.7) are $2\cdot10^9$/s, $5\cdot10^7$/s and $1.5\cdot10^9$/s, respectively.

In our opinion the isotope ^{132}Sn stands out among the others. Similarly to the hypothetical superheavy nucleus ^{298}114 (N/Z=1.61), it is a doubly magic nucleus by the neutron concentration N/Z=1.64 and it is second only to ^{78}Ni(N/Z=1.79). That is why the study of the nuclear structure in the region of ^{132}Sn as well as reactions induced by the beam of these nuclei are of great interest. In view of spectroscopic investigations, the intensity of low-energy ^{132}Sn beams at the DRIBs complex may reach 10^7-10^8 atoms/s, and according to the estimations, the intensity of the beams of these ions accelerated up to an energy of 5÷18 MeV·A will amount to $5\cdot10^6$/s.

In the case of strongly asymmetric photofission neutron-rich isotopes of more light elements with Z≥28 are also formed with a relatively high probability. According to the estimations, the yield of extremely neutron-rich nuclei of the ^{78}Ni type, for example, will reach about 10/s.

The problem of separation of the fission fragments in accordance with their charges and masses is one of the key tasks in the production of radioactive beams of medium mass ions. This problem is now in the stage of R&D so that the optimal variant could be chosen as a result of it.

Note that the realization of stage II of the project DRIBs offers possibilities for the realization of future RIB projects. Modern superconducting electron accelerators are capable of providing beams of electrons with E_e=50 MeV and intensity of up to 1mA (the beam power - 50 kW). Under such conditions the rate of fission in the U target increases up to ~10^{14} fission/s, thus increasing the yield of the ^{132}Sn type up to ~10^{12}/s. Hence when using the electron driver one can in principle produce a beam of ^{132}Sn nuclei with an intensity of up to 1pnA. A similar variant of producing RIB of medium mass nuclei is now being considered in view of realizing the second stage of acceleration using the cyclotron CIME (GANIL) [5].

The major part of the work with beams of neutron-rich nuclear fragments is aimed at the study of collective processes in the vicinity of the nuclear fusion threshold. This area of research includes nuclear fusion (inverse fission), investigation of collective dynamics for the massive nuclei with different N/Z ratios etc.

Another area for the investigation at borderlines of nuclear stability refers to extremely heavy nuclei.

4 Synthesis of superheavy nuclei and their properties

It is known that one of the fundamental consequences of nuclear theory is the prediction of the "island of stability" of superheavy elements in the region of hypothetical superheavy elements. This intriguing hypothesis suggested more than 35 years ago has been developed lately and now seems to find its experimental confirmation in the currently conducted experiments.

A considerable increase in the nuclear stability at approaching closed spherical shells Z=114 (and possibly 120-126) and N=184 which follow the doubly magic nucleus ^{208}Pb (Z=82, N=126) is expected for the isotopes of superheavy elements with a high neutron excess (Fig.7). That is why for the synthesis of nuclei with Z=114 and 116 we chose the fusion reactions ^{244}Pu, ^{248}Cm + ^{48}Ca in which the reaction products after the evaporation of neutrons have the maximal neutron excess*). In the fusion reactions with a doubly magic nucleus ^{48}Ca the compound nuclei ^{292}114 and ^{296}116 formed at the Coulomb barrier have the excitation energies of 31 and 33 MeV, respectively. It can be assumed that at these energies shell effects are still present in the excited nucleus which enhances the survival probability of evaporation residues as compared with typical "hot fusion" reactions ($E_x \geq 45$ MeV) used by us earlier for the synthesis of heavy isotopes with Z=106,108 and 110 [6].

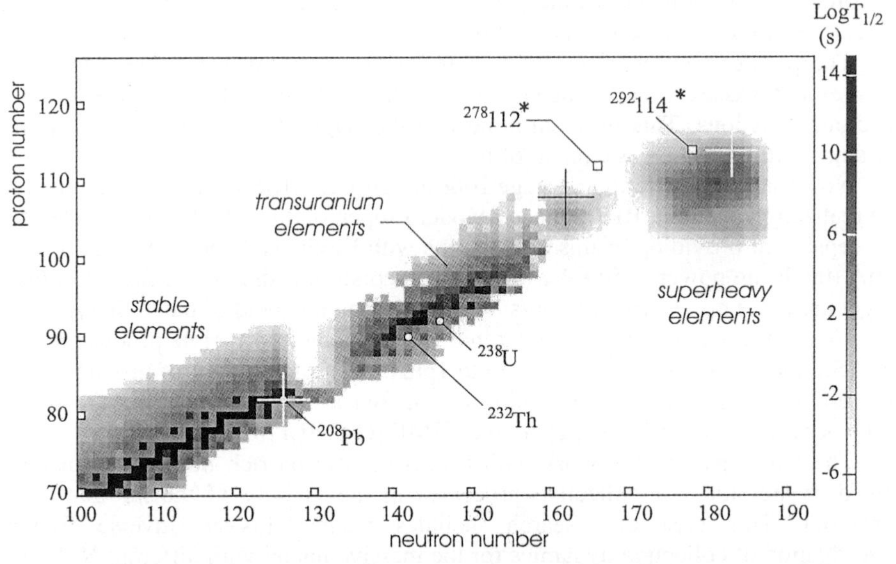

Fig. 7. The map of nuclides in the region of heavy elements. The intensity of the color reflects the half-lives of the nuclei (the right-hand scale). The crosses indicate the location of the doubly magic nuclei with closed spherical shells Z=82, N=126 and Z=114, N=184 and with closed deformed shells Z=108, N=162. The white squares indicate the compound nuclei with Z=112 and 114 formed in the reactions of cold fusion ^{70}Zn+^{208}Pb and hot fusion ^{48}Ca+^{244}Pu, respectively.

Despite of these advantages, all previous attempts to synthesize new elements in reactions with ^{48}Ca ions and actinide targets only yielded the upper limit of their

*) As is seen from Fig.7, in the region of deformed shells Z=108 and N=162 an increase in the stability of heavy nuclei is also expected. The properties of nuclei in this region will be considered in the talk by Dr. S.Hofmann (GSI).

production cross section [7]. It was vital to increase the sensitivity of the experiment by three orders of magnitude to go down to the level of 0.5 pb where formation of superheavy nuclides was expected in the 3n and 4n evaporation channels[**)]. An increase in the sensitivity of the experiments could be achieved first of all due to increasing the intensity of the ^{48}Ca ion beam. For this purpose a new ion source ECR-4M operating on metallic Ca vapors was created. At a consumption of about 0.3 mg/h, a ^{48}Ca ion beam with the energy E_L=6 MeV/A and intensity of about 0.5-0.8 pμA was produced [8].

Enriched ^{242,244}Pu and ^{248}Cm isotopes were used as a target matter. Rotating targets (about 0.3 mg/cm^2 in thickness) deposited on a Ti (1.5 μm) backing having a total area of about 30 cm^2 were used. Recoil nuclei knocked out of the target layer were separated in-flight from ^{48}Ca ions and other products of incomplete fusion reactions by kinematic separators. Two types of recoil separators were used in the experiments: VASSILISSA (energy selector) and a gas-filled separator DGFRS (Dubna Gas-Filled Recoil Separator).

Now briefly about the conditions of the experiment.

Separated heavy atoms are implanted into a strip position-sensitive detector (~50 cm^2 in area) situated in the focal plane. The front detector is surrounded with side detectors so that the whole assembling looks like a box with an open front wall. It increases the efficiency of detection of α -particles from the decay of implanted nuclei up to 87% of 4π. For every implanted atom velocity is measured (by TOF detectors) as well as energy and location of implants on a sensitive surface of the front detector. In the case the nuclei of implanted atoms emit α-particles or fission fragments, the last-mentioned will be registered in a strict correlation with the implant.

Experiments have been carried out in such setting since late 1998 and up to date. Without mentioning the results of the first experiments with ^{238}U, ^{242}Pu and ^{244}Pu targets which were published in works [9], we will only present the data obtained within the past year.

In May-November 2000, the second irradiation of the ^{244}Pu target was performed with ^{48}Ca ions, the beam dose was $1\cdot10^{19}$ ions [10].

The highly enriched (98.5%) target matter was provided by our colleagues from the Livermore National Laboratory. In that experiment two more identical decay chains were observed. Each of them consisted of two sequential α-decays and was terminated with spontaneous fission accompanied by a high energy release in the detectors (Fig. 8a). The life-time of the new nuclei was about 0.5 min. The probability that the observed decays were the result of incidental coincidences of the signals in the front detector was less than $5\cdot10^{-13}$. Note that both events were observed at an energy of the ^{48}Ca ion beam corresponding to the excitation energy

---

[**)] Some aspects of fusion and fission of compound nuclei formed in the reactions of cold and hot fusion will be considered in the talk by Dr. M.Itkis (FLNR, Dubna).

of the compound nucleus E_x=36±2 MeV. At this energy the most probable channel of the ^{292}114 nucleus de-excitation corresponds to the emission of 4 neutrons and γ-rays. Proceeding from this, new decay chains may be attributed to the decay of the even-even isotope of element 114 with mass 288.

The experiment that followed was devoted to the study of the reaction ^{48}Ca+^{248}Cm and was aimed at testing this conclusion.

The target matter – enriched isotope of ^{248}Cm (Z=96) – was produced at the high-flux reactor in Dimitrovgrad (Russia) in the quantity of 10 mg. The other target from ^{248}Cm was provided by the Livermore National Laboratory (USA).

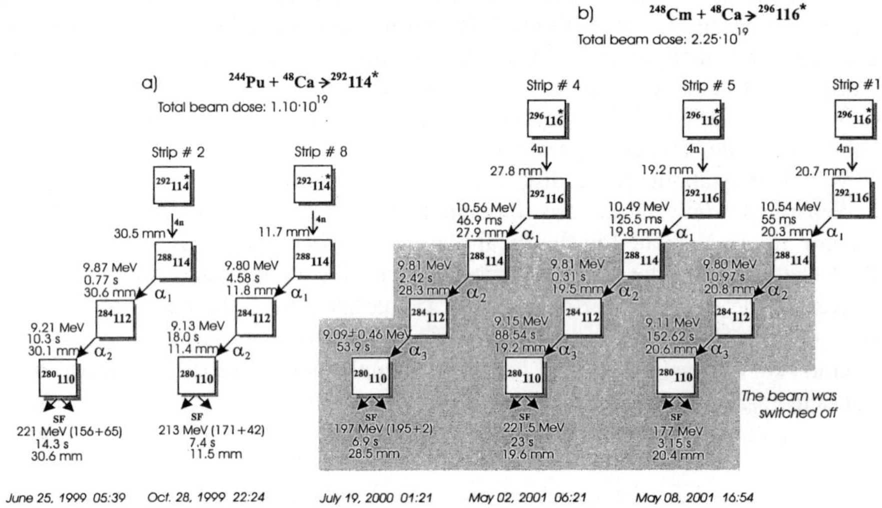

Fig. 8. Chains of radioactive decay of nuclei synthesized in the reactions: a) ^{48}Ca+^{244}Pu, b) ^{48}Ca+^{248}Cm. The shaded area corresponds to the switched-off beam.

Changing the ^{244}Pu target to the ^{248}Cm one, all other conditions of the experiment being kept, should lead to the formation of the new heavy nucleus with Z=116 and mass 292 in the 4n evaporation channel of the fusion reaction ^{48}Ca+^{248}Cm. As a result of the α-decay of this nucleus, expected with a high probability, we should obtain the daughter nucleus-isotope ^{288}114 earlier produced in the reaction ^{48}Ca+^{244}Pu. That is why after the decay of the ^{292}116 nucleus the whole chain of the daughter nucleus decay $^{288}114 \to ^{284}112 \to ^{280}110$ should also be observed in the experiment.

Usually the separator operates with a continuous ^{48}Ca beam. In the experiment on the synthesis of element 116 this regime was changed.

After the implantation into the detector focal plane of the heavy nucleus with the expected parameters (energy and velocity) and its decay with emission of an α-particle with $E_\alpha \geq 10$ MeV (two signals are strictly position-correlated) the beam was switched off.

Measurements made straight after switching off the beam showed that the rate of the α-particles ($E_\alpha \geq 9$ MeV) and spontaneous fission fragments in any strip within $\Delta x = 0.8$ mm, which is determined by the position resolution of the detector, is 0.45/year and 0.1/year, correspondingly. Incidental coincidences of the signals simulating the 3-step 1.5 min.-chain of the nucleus decay $^{288}114$ (α-α-SF) are practically excluded even for a single event [11].

In these conditions, at a beam dose of $2.25 \cdot 10^{19}$ ions three decay chains of element 116 were registered (Fig.8b). After the emission of the first α-particle ($E_\alpha = 10.53 \pm 0.06$ MeV) the following decay occurred in the absence of the beam (see the gray area). As is seen from Fig. 8, two decay chains of the $^{288}114$ nucleus produced in the reaction $^{48}Ca + ^{244}Pu$ and three new decay chains observed in the reaction $^{48}Ca + ^{248}Cm$ are strictly correlated: five signals arising in the front detector, i.e., the recoil nucleus, three α-particles and fission fragments (in the case of the $^{288}114$ nucleus decay: R-α-α-SF), differ in their position by no more than 0.6 mm (Fig.9a).

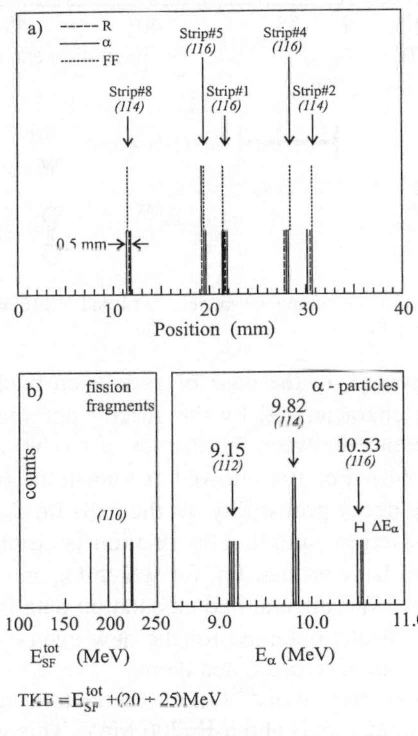

Fig. 9. a) position and b) energy correlations in the decay chains shown in Fig. 8.

Energies of α-particles and nuclear half-lives obtained in the reaction ^{48}Ca+^{248}Cm coincide with each other within the limits of the detector energy resolution (ΔE_{α}~60 keV) and statistical fluctuations determined by the nuclear decay in the chains. Fig. 9b shows α-particle energy spectra for the three events corresponding to the ^{292}116 α-decay and the five events corresponding to the ^{288}114 and ^{284}112 decay, as well as the spectrum of combined energies of fission fragments from the five events of the ^{280}110 spontaneous fission obtained in the experiments with ^{244}Pu and ^{248}Cm targets.

Isotopes of elements 114 and 116 produced in the reactions ^{244}Pu, ^{248}Cm + ^{48}Ca are most likely formed in the 4n evaporation channels. This conclusion follows from the excitation energy values for compound nuclei which after the emission of neutrons and gamma-rays led to the observed evaporation products (Fig. 10).

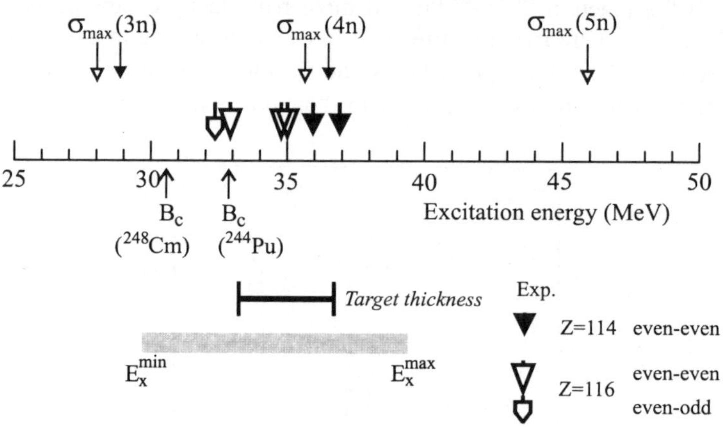

Fig. 10. Excitation energies of the compound nuclei ^{292}114 and ^{296}116 at which the decay chains were observed (shown by Fig. 8).

As it had been expected, in the case of even-even nuclei the experimentally observed α-radiation is characterized by the strictly definite decay energy which corresponds to the difference between the masses of mother and daughter nuclei in the ground states (Fig. 9b). For the allowed α-transitions (even-even nuclei) the decay energy Q_{α} and the decay probability (or the half-life T_{α}) are connected by the well-known relation of Geiger-Nuttall. This relation is strictly fulfilled for all the known by now 60 nuclei heavier than Pb, for which Q_{α} and T_{α} have already been measured. Fig. 11 shows experimental and calculation data for the region of heavy nuclei with Z≥100. The results obtained for the new even-even superheavy nuclei with Z=112, 114 and 116 are also presented there.

Finally, in spontaneous fission of ^{280}110 nuclei the total energy release from the fission fragments in the detectors is about E=206 MeV. This value with a correction to the energy losses in the dead layers of the focal plane and side detectors corresponds to the mean total kinetic energy of the fission fragments TKE~230 MeV. Such a high value of the fission fragment energy points to the spontaneous fission of a rather heavy nucleus. Note that in the fission of ^{235}U induced by neutrons TKE =168 MeV.

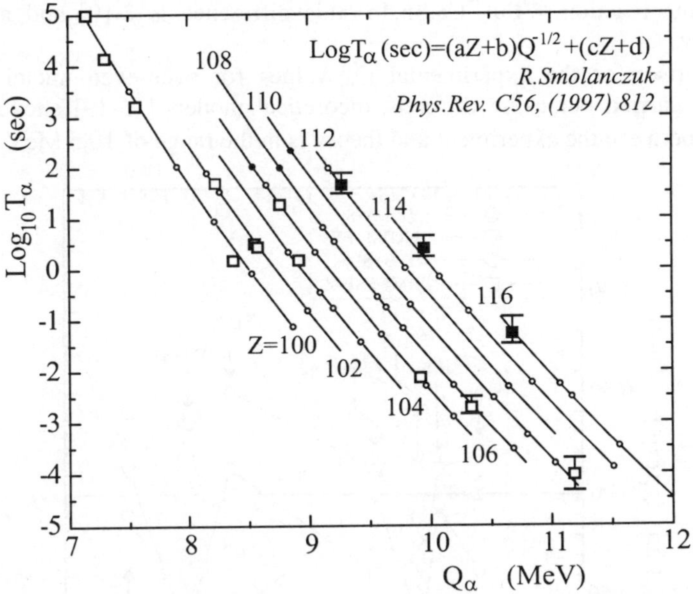

Fig. 11. Calculated [12] and the experimental T_α vs.Q_α relationship for even-even isotopes of transfermium elements.

Radioactive properties of the new nuclides are presented in Table II. Comparison of the partial half lives T_{sf} and T_α of the synthesized nuclei with Z=110 and 112 with the data obtained earlier for more light isotopes of these elements [13]

Table II. Radioactive properties of even-even isotopes of the superheavy elements with Z=110-116.

Isotope	Decay mode	Q_α (MeV)	T_α
^{292}116	α	10.68 ± 0.05	53^{+62}_{-19} ms
^{288}114	α	9.96 ± 0.05	$2.6^{+2.0}_{-0.8}$ s
^{284}112	α	9.28 ± 0.05	$0.75^{+0.57}_{-0.23}$ min
^{280}110	SF	TKE ~ 230	$7.6^{+5.8}_{-2.3}$ s

point to a considerable increase in the stability of heavy nuclei with increasing the neutron number. Addition of 10 neutrons to the $^{270}110$ nucleus (T_α~100 μs) leads to an increase in the half-life almost by 10^5 times. For the nuclei of element 112 with A=277 and the isotopes with A=284 and 285 formed in the synthesis of element 114 nuclei in the reaction ^{244}Pu(^{48}Ca;3n,4n) this difference is $2\cdot10^5$ and about 10^6, respectively.

Comparison of the experimental Q_α values for even-even nuclei with the calculation carried out using different theoretical models [16-19] shows that the difference between the experiment and theory is in the range of ±0.5 MeV (Fig. 12).

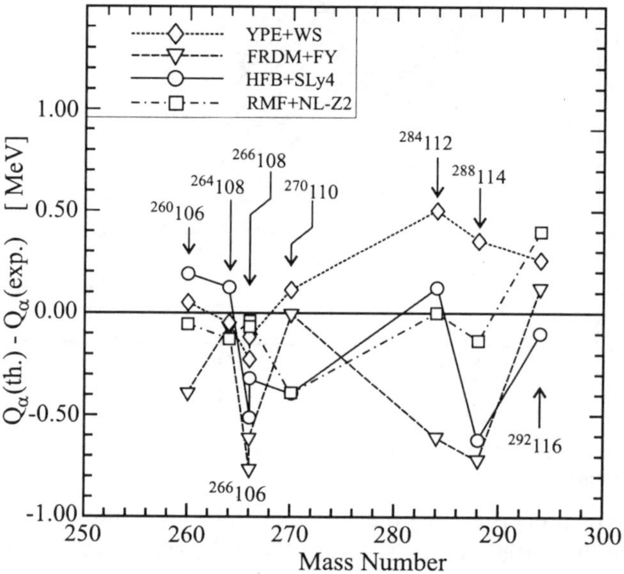

Fig. 12. The difference between calculated and experimental values of Q_α for even-even isotopes with Z=106-116 and N=154-176.

Without making so far a more detailed analysis, one can conclude that theoretical predictions about the decisive influence of the nuclear structure on the stability of superheavy nuclides has been confirmed not only qualitatively but also to some extent quantitatively.

An increase in the half-lives of the most heavy nuclei up to tens of seconds and minutes substantially extends the area for the investigation of superheaviers, including the study of their chemical properties, measuring atomic masses etc. On the other hand, development of acceleration and experimental technique will allow us to make an advent into the region if even heavier nuclei in the future.

5 Acknowledgements

Experiments described in the present talk were carried out at the heavy ion accelerators of the JINR in collaboration with LLNL (Livermore), GSI (Darmstadt), RIKEN (Wako-shi), GANIL (Caen), Kurchatov Institute (Moscow), IReS (Strasbourg) as well as universities of Brussels, Messina and Bratislava.

Taking this opportunity, I would like to express gratitude to my colleagues who took part in the experiments and to Profs. G.Muenzenberg, W.Greiner, S.Hofmann, A.Sobiszewski, M.Itkis, A.Korsheninnikov and M.Zhukov for their interesting discussions and valuable remarks.

References

1. Korsheninnikov A.A. et al. Phys. Rev. Lett. **87** (2001) 092501-1.
2. Blagus S. et al., Phys. Rev. **C44** (1991) 325.
3. Gulbekian G.G. In the Proceeding.
4. Oganessian Yu.Ts. et al. In the Proc. of the V Int. Conf. on Radioactive Nuclear Beams, Apr. 2-8, 2000, Divone, France (to be published). Preprint JINR, E7-2000-83, Dubna 2000.
5. Goutte D. In the Proceedings.
6. Lazarev Yu.A. et al. Phys. Rev. Lett. **33** (1994) 624, **75** (1995) 1903; Phys. Rev. **C54** (1996) 620.
7. Hulet E.K. et al. Phys. Rev. Lett. **89** (1977) 385, Illige J.D. et al. Phys. Lett. **B78** (1978) 209, Otto R.J. et al. J. Irors. Nucl. Chem. **40** (1978) 509. Ghiorso A. et al. LBL Report, LBL-6575 (1977); Oganessian Yu.Ts. et al. Nucl. Phys. **A294** (1978) 213, Armbruster P. et al. Phys. Rev. Lett. **54** (1985) 406.
8. Oganessian Yu.Ts. Phys. of Atomic Nucl. **63** (2000) 1315.
9. Oganessian Yu.Ts. et al. Eur. Phys. J. **A5** (1999) 63, Phys. Rev. Lett. **83** (1999) 3154, Nature **400** (2000) 242.
10. Yu.Ts. Oganessian et al. Phys. Rev. **C62** (2000) 041604 (R), Phys. of Atomic Nucl. **63** (2000) 1769.
11. Oganessian Yu.Ts. et al. Phys. Rev. **c63** (2000) 011301 (R).
12. Smolanczuk R. Phys. Rev. **C56** (1997) 812.
13. Hofmann S. Rep. Prog. Phys. **61** (1998) 639.
14. Smolanczuk R., Skalski J., and Sobiczewski A., in Proceeding of the International Workshop XXIV on Cross Properties of Nuclei and Nuclear Excitations "”Extremes of Nuclear Structure", Hirschegg, Austria, 1996 (GSI, Darmstadt, 1996), 35.
15. Möller P. et al. At. Data Nucl. Data Tables **66** (1997) 131.
16. Čwiok S. et al. Phys. Rev. Lett. **83** (1999) 1108.
17. Bender M. Phys. Rev. **C61** (2000) R031302.
18. Berger J.F. et al. Nucl. Phys. **A685** (2001) 1c.

HARD PHOTONS: A PROBE OF DYNAMICAL EFFECTS IN HEAVY ION COLLISIONS AT INTERMEDIATE ENERGY

R.ALBA, C.AGODI, C.MAIOLINO, A.DEL ZOPPO, M.COLONNA, G.BELLIA, R.CONIGLIONE, P.FINOCCHIARO, K.LOUKACHINE, E.MIGNECO, P.PIATTELLI, D.SANTONOCITO, P.SAPIENZA

INFN-Laboratori Nazionali del Sud and Dipartimento di Fisica dell'Università , Catania - Italy; E-mail: alba@lns.infn.it

M.BRUNO, M.D'AGOSTINO,M.L.FIANDRI, G.VANNINI.

Dipartimento di Fisica dell'Università and INFN Bologna (Italy)

N.COLONNA

INFN Bari (Italy)

F.GRAMEGNA, P.F.MASTINU,

INFN-Laboratori Nazionali di Legnaro (Italy)

I.IORI, A.MORONI

Dipartimento di Fisica dell'Università and INFN Milano (Italy)

G.V.MARGAGLIOTTI, P.M.MILAZZO, R.RUI

Dipartimento di Fisica dell'Università and INFN Trieste (Italy)

Thermal photons have been used as a clock to determine the time of IMF emission during the evolution of the nuclear reaction. The method has been applied to the reaction ^{58}Ni + ^{197}Au at 30 and 45 MeV/amu incident energy. The results put in evidence that the relative contribution of the two possible production mechanisms (dynamical and statistical) is quite different in the two cases. A comparison with theoretical calculations strongly supports the experimental findings.

1 Introduction

Heavy ion reactions at intermediate energy are known to proceed through density oscillations (Fig.1) from an early pre-equilibrium phase, where nucleon-nucleon (n-n) collisions play the dominating role, to an equilibrium phase, where the de-excitation of the system is purely statistical. At the beginning of the reaction the system experiences a strong compression during which nucleons collide and this leads the system to a following expansion. During this expansion phase, if the system is driven into the spinodal region, instabilities can start and disassemble it. If the system survives this phase, it will undergo a new compression to restore the normal density and, at the end, a heavy and hot system (HHS) will be produced. During this time evolution several interesting processes can occur, that are typical of the intermediate energies. Let's focus on some of them. During the first compression, as a consequence of first chance n-n collisions, high energy ($E_\gamma \geq 25$ MeV) bremsstrahlung photons ("direct") are produced [1]. Their production stops as soon as the system starts to expand and it is not influenced by the following evolution of the reaction. If, while the system is expanding, very low densities are reached and the system enters the spinodal region (as at $t = t_0$ in Fig.1), another interesting phenomenon can occur, i.e. the dynamical production of Intermediate Mass Fragments (IMF's) that destroys the original composite system and interrupts the path towards the formation of a HHS. If this doesn't happen, the second compression forces the n-n collisions to start again and, with them, the bremsstrahlung photon production, with the difference that now their energy spectrum reflects the energy distribution of the nucleons in the equilibrating system. It's important to note that this contribution, called "thermal"[2], is observed only if the dynamical evolution of the reaction leads to the formation of a HHS and, in this sense, it is a signature of its formation. The hot HHS can later loose its energy by statistical decay, including statistical IMF emission.

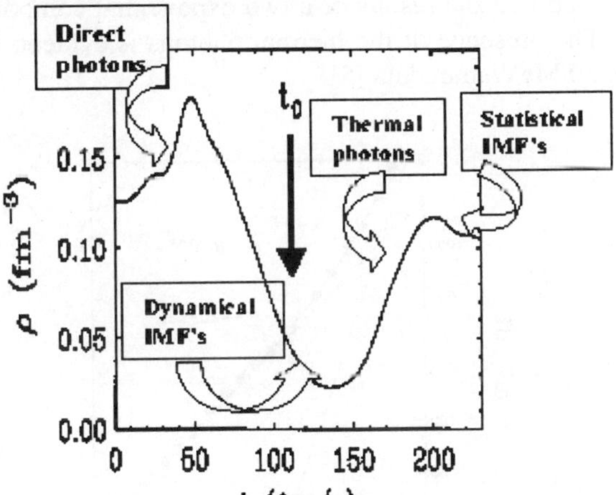

Figure 1: Example of density oscillation in central collisions (BNV) and related phenomena.

2 The experiment

With the aim of investigating the mechanisms of IMF production, an experimental program has been undertaken at Laboratori Nazionali del Sud with the beams of the

Superconducting Cyclotron. The idea was to use thermal photons to tag the presence of a HHS surviving the density oscillation. To observe IMF's in coincidence with such photons is an evidence of their statistical origin, while their dynamical origin will be signaled by an anticoincidence.

Since these studies require a very performant apparatus, able to detect and identify simultaneously hard photons and IMF's, we have used the two multidetector arrays MEDEA[3] and MULTICS[4], measuring with good efficiency and granularity hard photons in MEDEA, light charged particles (LCP's) in both and IMF's in MULTICS. The system chosen for this study had to be heavy enough to experience a significant compression and following expansion and, also, two different bombarding energies, where the two possible production mechanisms could be expected to show up with different probabilities, were necessary to make a comparison. For these reasons the system ^{58}Ni + ^{197}Au at 30 and 45 MeV/amu incident energy was chosen.

3 Experimental results

First, a detailed analysis of the hard photons has been performed. Fig.2 shows the experimental hard photon energy spectrum for the system ^{58}Ni + ^{197}Au at 45 MeV/amu. Superimposed the results of a two exponential component fit to the data are also shown. The presence of the thermal photons is evident. Similar results are observed for the 30 MeV/amu data [5].

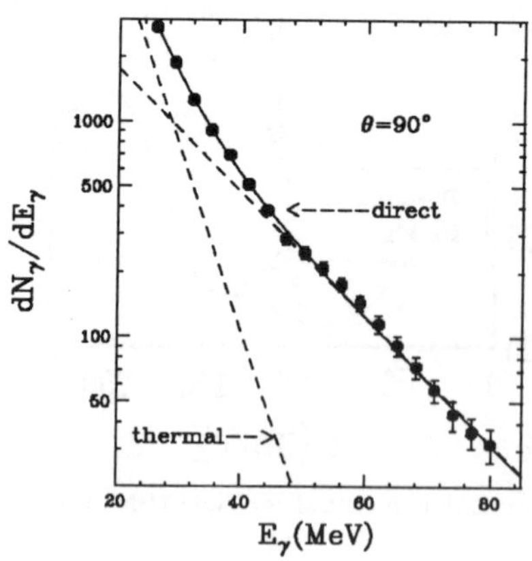

Figure 2: Experimental hard photon energy spectrum. Dashed lines represent the contribution of direct and thermal photons. It can be seen that thermal photons mainly affect the lower energy part of the spectrum.

Using LCP's to select the impact parameter b the hard photon energy spectra have been sorted in several b bins as a function of the detection angle. A simultaneous fit to the energy and angular distributions for each b interval allowed us to determine some characteristics of the thermal photon source. A source velocity approaching that of the nucleus-nucleus center of mass with increasing centrality and a very small anisotropy have been deduced, supporting the interpretation [2] of a late production mechanism. Thermal photons are always present in the spectra stating that a HHS is formed in a significant fraction of events. The ratio of the thermal component to the direct one, however, is observed to decrease with increasing bombarding energy, qualitatively indicating the onset of a competing mechanism.

The experimental tool chosen to put in evidence the main source of IMF's was the study of the thermal photon-IMF correlation factors, defined as

$$1+R_{\gamma\text{-IMF}} = \frac{\langle m_{IMF}\rangle^{\gamma}}{\langle m_{IMF}\rangle}$$

where $\langle m_{IMF}\rangle^{\gamma}$ and $\langle m_{IMF}\rangle$ are the mean values of the multiplicity distribution of IMF's in events gated by thermal photons and not gated, respectively. Values of this quantity smaller than one signal that in the events in which a hard photon is emitted a decrease in IMF production is observed. In the framework of Fig.1 this can be easily understood if a relevant fraction of IMF's is of dynamical origin. Indeed, due to the large excitation energies expected mainly in central collisions (of the order of GeV) the emission of a thermal photon is not expected to affect the emission of statistical IMF's. For the gamma's gating the IMF multiplicities an energy between 25 and 40 MeV was selected to maximize the ratio of thermal to direct photons without loosing too much in counting statistics. In this energy range this ratio is almost 1. When the correlation factors are deduced in such a way, what is observed, in effect, is the sum of the thermal and direct correlation factors wheigted by the fraction in which each source contributes to the total photon yield in the selected photon energy range:

$$1+R_{\gamma\text{-IMF}} = \sum \frac{N^{\gamma}_{\text{therm,direct}}}{N^{\gamma}} \left(1+R_{\gamma\text{-IMF}}\right)_{\text{therm,direct}} \qquad (1)$$

The photon and IMF data have been sorted as a function of the impact parameter for both the beam energies. For each impact parameter bin, the correlation factors have been determined in three IMF parallel velocity windows: the window W_1 centered at the nucleus-nucleus center of mass, the window W_2 centered at half-beam velocity and the window W_3 around the projectile-like velocity.

First, we have checked in the data that the direct photon-IMF correlation factor is always very close to one and this is consistent with the production of IMF's occurring later and independently of the direct photons. Therefore, the effect of the

$(1+R_{\gamma\text{-IMF}})_{direct}$ contamination in (1) is that it tends to mantain the correlation factors close to one. Then we have considered the experimental correlation factor (1). We have found that it is always close to 1 with the only exception of the IMF velocity window W_1 in central collisions at 45 MeV/amu, where it is significantly smaller than one. In other words photons and IMF are always uncorrelated except one case, at 45 MeV/amu, where they are anticorrelated. Fig.3 shows the results for the window W_1 for both the incident energies.

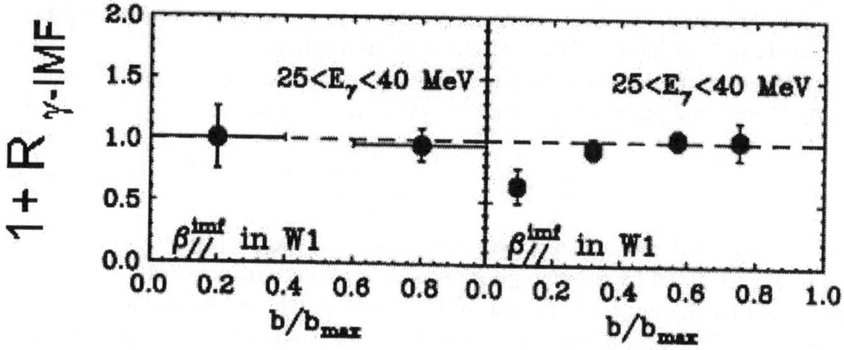

Figure 3: Experimental correlation factors as a function of the collision centrality. Left: 30 MeV/amu; right: 45 MeV/amu.

We can distinguish two different reasons for the absence of correlation. First, IMF's and photon's are emitted by independent sources moving with different velocities: this is likely the case of the correlation factors with IMF's located in W_2 and W_3. Second, IMF's and thermal photons are produced independently within the same source: this is the case of the 30 MeV/amu central collisions with IMF's in W_1. This is a case of particular interest in this work since it signals a dominance of statistical fragmentation. The observation of a correlation factor smaller than 1 at 45 MeV/amu for IMF's in W_1 is also of very special interest for the aim of this work, since it is the type of signal expected when dynamical fragmentation dominates. Moreover, in the correlation factors shown in Fig.3 the true signal of dynamical fragmentation is partially dimmed by the contamination of the direct photons. Then, with the caution due to the experimental error bars, from the correlation data in Fig.3 and using (1) we can deduce that the **thermal** photon - IMF correlation factor in central Ni + Au collisions decreases from 1 at 30 MeV/amu to 0.25 at 45 MeV/amu. The indication of the whole set of data is that at 30 MeV/amu the IMF production is dominated by statistical decay while at 45 MeV/amu, at least in central collisions, it is strongly affected by the dynamical evolution of the collision.

4 Theoretical calculations

In order to corroborate the interpretation of these data with a theoretical model, we have performed a three steps simulation of the IMF production in the reactions studied in this work. In the first step, for each bombarding energy, we have calculated the density profile versus time of Fig.1 by means of dynamical BNV simulations of central collision events. In the second step, the BNV configuration at the time t_0 at which, during the expansion phase, the system enters the spinodal region was used as the starting point to introduce density fluctuations, that could lead to fragmentation, and solve the Boltzmann-Langevin equation [6]. Although the BNV density profiles at the two energies are very similar, the fragment charge distributions predicted by the BNV+Boltzmann-Langevin calculations are completely different (Fig.4). At 30 MeV/amu in most of the events the system experiences a second compression resulting in a HHS formation. On the other hand, at 45 MeV/amu this happens in only a small fraction (about 20%) of the total number of events.

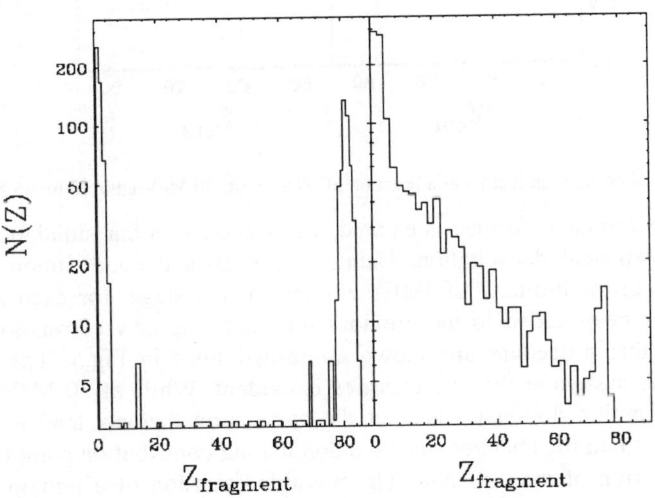

Figure 4: Z-distribution of fragments after BNV + Boltzmann-Langevin. Left: 30 MeV/amu; right: 45 MeV/amu.

At this stage of the simulation we have calculated the ratio

$$1+R=\frac{\langle m_{IMF}\rangle^{HHS}}{\langle m_{IMF}\rangle}$$

of the mean multiplicity $<m_{IMF}>^{HHS}$ of the IMF production which is followed by the transit through a HHS to the mean inclusive multiplicity $<m_{IMF}>$. Similarly to (1) this ratio expresses the correlation factor between dynamical IMF production and heavy remnant formation so that 1+R results to coincide with $(1+R_{\gamma\text{-}IMF})_{therm}$.
Full lines in Fig.5 show this theoretical correlation factors for each bombarding energy and for central collisions, calculated for heavy remnant charge $Z>Z_{cut}$ versus Z_{cut}. For both energies the calculations predict a negative correlation which becomes stronger with increasing Z_{cut} and reflects mainly charge conservation: the larger the charge of the heavy remnant the smaller the multiplicity of fragments.

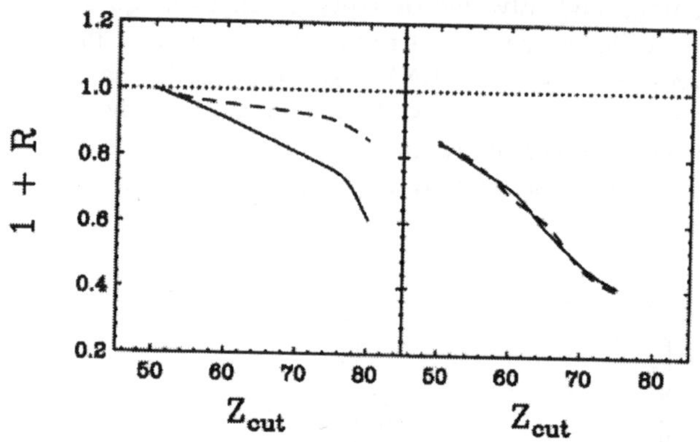

Figure 5: Theoretical correlation factors as a function of Zcut. Left: 30 MeV/amu; right: 45 MeV/amu.

Since the heavy remnant formed is excited, we introduce in our simulation a third step, i.e. the statistical deexcitation. Using the statistical code Simon we have deduced the mean multiplicity of IMF's emitted at this stage, for each Z_{cut}. This contribution has been added to the previous one and the new correlation factors resulting from this procedure are shown as dashed lines in Fig.5. The different behaviour of the system at the two energies is evident. While at 30 MeV/amu the statistical contribution dominates and let the correlation factors tend to 1, at 45 MeV/amu almost nothing changes since the dominating contribution comes from the dynamical evolution of the collision. The possible emission of a photon bringing away about 40 MeV excitation energy before starting the deexcitation was checked to be ininfluent at 45 MeV/amu because of the large total excitation energy and to affect only slightly the results at 30 MeV/amu. This seems to exclude the observed anticorrelation to be a trivial energy balance effect.

5 Conclusions

Using thermal photons as a probe, the main source of IMF emission in Ni + Au reactions at 30 and 45 MeV/amu incident energy has been deduced exploiting the hard photon - IMF correlation function versus the reaction centrality and IMF velocity. IMF's produced by statistical deexcitation of a heavy system formed **at the end** of the dynamical evolution of the collision have been found to dominate at 30 MeV/amu while IMF's emitted **during** the dynamical evolution dominate at 45 MeV/amu.

The set of calculations we have performed largely agrees with our experimental results and supports our interpretation and, implicitly, the use of thermal photons as a signature of the formation of a HHS.

References

1. H.Nifenecker and J.A.Pinston, Prog.Part.Nucl. Phys. 23 (1989)271; E.Migneco et al. Phys.Lett. B 298 (1993)46; Martinez et al. Phys.Lett. 334 (1994)23; P.Sapienza et al. Phys.Rev.Lett. 73 (1994)1769; H.Nifenecker and J.P.Bondorf, Nucl.Phys. A 442 (1985) 478; W.Cassing et al., Phys.Rep. 188 (1990)363.
2. A.Schubert et al., Phys.Rev.Lett. 72(1994)1608; G. Martinez et al., Phys. Lett. B349 (1995) 23; F.M.Marques et al., Phys. Lett. B 349 (1995) 30.
3. Migneco et al., Nucl.Inst.Methods A 314 (1992) 31.
4. I.Iori et al., Nucl.Instrum.Methods A 325 (1993) 458.
5. R.Alba et al., Nucl. Phys. A654 (1999)761c and Nucl. Phys. A681(2001)339c.
6. M.Colonna et al., Nucl.Phys. A 613(1997)165.

GDR EXCITATION CLOSE TO THE LIQUID-GAS TRANSITION OF NUCLEAR MATTER

AMORINI F.[1,2], CARDELLA G.[3], DI PIETRO A.[1], FIGUERA P.[1], LANZALONE G.[1,2], LU J.[1], MUSUMARRA A.[1], PAPA M.[3], PIRRONE S.[3], RIZZO F.[1,2], TUDISCO S.[1,2]

[1]*INFN – Laboratorio Nazionale del Sud, Catania, Italy*

[2]*Dipartimento di Fisica, Università di Catania, Italy*

[3]*INFN, Sezione di Catania, Italy*

γ-ray spectra in coincidence with charged particles for the reactions ^{40}Ca+ ^{48}Ca,^{46}Ti at E_{inc}=25 A MeV are presented. The spectra statistical analysis shows the survival of the GDR up to excitation energy around 4 MeV/A for a system with mass of 50-60 amu. This result was obtained by improving statistical calculations by including the deuteron decay channel, the mass dependence of the GDR parameters in the decay chain, and the GDR isospin splitting.

1 Introduction

One of the most challenging subjects in nuclear physics concerns the evolution of GDR parameters with increasing energy in the nucleus. From previous experiments [1,2] performed on nuclei with mass around 100-110, the GDR yield exhibits a tendency to saturation, explained with the hypothesis of a limiting excitation energy for the GDR existence. To go deeper in the study of this subject, we tried to improve the reliability of the statistical calculations at high excitation energy in order to extract more precise information on the maximum excitation energy at which we still observe collective excitations in nuclear matter. We studied the systems ^{40}Ca+^{48}Ca,^{46}Ti at 25 A MeV. These systems were used also to evaluate the presence of a GDR yield dependence from the N/Z ratio of projectile and target nuclei [N/Z(^{40}Ca)=1, N/Z(^{48}Ca)=1.4, N/Z(^{46}Ti)=1.09] and its influence on the evaluation of a limiting excitation energy [3,4].

2 Experimental setup and data analysis

The experiment was performed at the Superconducting Cyclotron of the Laboratori Nazionali del Sud (LNS) Catania. A 25 A MeV beam of ^{40}Ca was used impinging on 3 mg/cm^2 thick, self-supporting targets of ^{48}Ca,^{46}Ti 95% isotopically enriched. We used the 63 BaF_2 crystals of the multidetector system TRASMA described in ref.[5].

Charged particles in coincidence with γ-rays were detected by using a ΔE-E hodoscope placed at forward angles (θ_{Lab} = 3°- 6°). Fragments stopped in the fitst stage (300 μm annular Si detector) were charge identified by using a pulse shape analysis of the signal[6,7]. In fig.1a typical Energy-"Rise-Time" (rise time in the silicon detector signal) scatter plot is reported showing the charge identification. A concentration of events can be observed in the low energy region (ER).

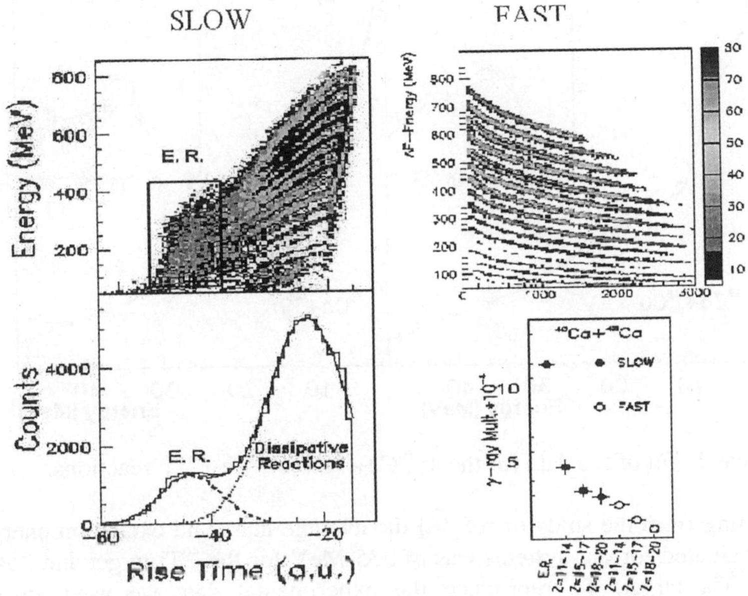

Figure 1. Particle identification: a) slow particles ; b) fast particles passing through the Si detector; c) projection of a) on rise time axis; d) high energy γ-ray yield.

The corresponding energy spectrum (fig.1c) is peaked around 200 MeV. From previous results [8] and GEMINI calculation these events can be attributed to evaporation residues in fusion-like reactions. This is confirmed by the observation of high energy γ-rays detected in coincidence of such residues. In fact, bremsstrahlung γ-ray yield is connected to the number of n-p first chance collisions[9,10]. If we scale the yield of high energy γ-rays ($E\gamma \geq 30$ MeV) corresponding to the various reaction products (fig.1d), we can deduce an average impact parameter for the ER selection of about 2-3 fm. This confirms that these events come from fusion like reactions.

3 Results and discussion

In fig. 2 the obtained γ-ray spectra for the two reactions in coincidence with ER residues are showed, together with the statistical model calculation.

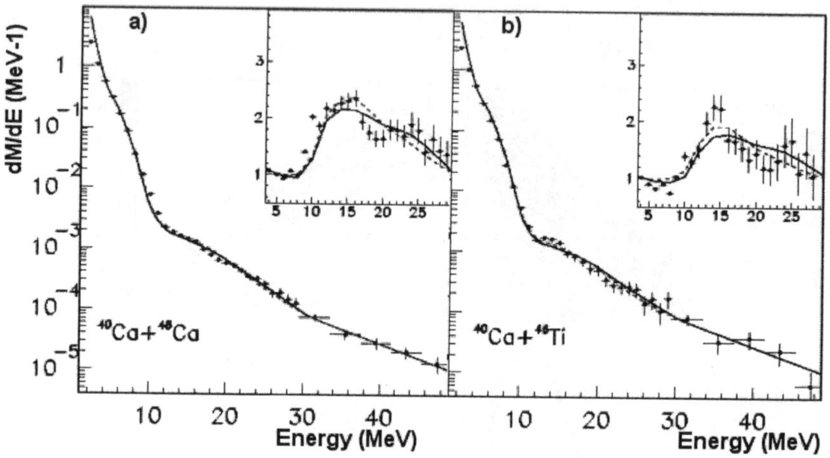

Figure 2. Fit of the data for the a) ^{40}Ca+^{48}Ca, b)^{40}Ca+^{46}Ti reactions.

Starting from the study of ref. [8] the average mass and excitation energy that can be evaluated for our systems was of 335 MeV for the ^{46}Ti target and 354 MeV for the ^{48}Ca target. To reproduce the experimental data we used an isospin dependent[11] version of the code CASCADE with the Reisdorf parameterization of the level density and zero isospin mixing parameters. We added the deuteron decay channel to the CASCADE code because such particles have, for the investigated systems, an emission probability similar than α-particles. Also the mass dependence of the GDR centroid[12] was included due to the large mass variation of the emitting nuclei (from 60 to 40 amu) along the decay path of the excited nucleus. Moreover the even larger isospin splitting of the GDR into two components $T_<=T$ and $T_>=T+1$ has been considered[13,14]. In fig.2a we report the obtained results for the reaction ^{40}Ca+^{48}Ca. The agreement of calculations with experimental data can be better observed in the inset, where calculation and data are reported as divided by a curve calculated with zero GDR strength. The dashed line is obtained assuming Γ_{GDR}=15 MeV (the largest Γ_{GDR} reported around mass 60) in order to test the constant Γ_{GDR} hypothesis assumed at high excitation energy in the standard systematic [11]. In this case to reproduce the data it is necessary to assume an energy cut-off $E^*_{cut\text{-}off}$=260 MeV. However one can obtain a better χ^2 value (0.20±0.10 against 0.37±0.15) using a Γ_{GDR}=26 MeV without any cut-off, full line fig.2a. The differences between the two calculations are mainly in the region above 20 MeV better reproduced using Γ_{GDR} =26 MeV. For ^{46}Ti data the best χ^2 (0.88±0.4) is obtained for $E^*_{cut\text{-}off}$ =200 MeV (dashed line). Assuming no cut-off we get the best fit with Γ_{GDR}=25 MeV (fig.2b).

On the basis of the obtained results it is difficult to decide between the hypotheses of constant Γ_{GDR} or increasing Γ_{GDR} with excitation energy. From the point of view of the limiting excitation energy at which GDR is still surviving, the first hypothesis (Γ_{GDR} =15 MeV) gives a smaller value. Therefore we can assume this value to be the lower limit. From the average excitation energy carried away by each evaporated mass unit (about 13 MeV), a mass of 56 amu can be evaluated for ^{48}Ca data at the $E^*_{cut\text{-}off}$, and the corresponding excitation energy is of 4.6±0.5 MeV/A. For the ^{46}Ti target the $E^*_{cut\text{-}off}$ extracted is slightly smaller and a mass of 53 amu can be evaluated for the system with 200 MeV average excitation energy i.e. 3.8±0.6 MeV/A. This difference can be connected with the difference in charge asymmetry of the two reactions. In conclusion we extracted excitation energy per nucleon at which we still have GDR emission. The values obtained are close to the one where disassembling of nuclear matter is expected[15]. This could be a signature of the connection between the disappearance of the GDR and the start of such disassembling.

References

1. P.Piattelli et al., Nuclear Physics A599(1996) 63c and ref. therein.
2. T. Suomijarvi at al., Phys. Rev C 53(1996)2258.
3. S.Flibotte et al; Phys.Rev.Lett. 77(1996)1448.
4. M.Cinausero et al; Nuovo Cimento 111(1998)613.
5. A.Musumarra et al; NIM A370(1996)558.
6. G.Paush et al; NIM A365(1995)176 and references therein.
7. J.Lu et al; NIM. in print.
8. T.M.V. Bootsma et al ; Z.Phys.A359(1997)391.
9. H.Nifenecker, J.A.Pinston Ann.Rev.Nucl.Part.Sci. 40(1990)113.
10. P.Sapienza et al; Il Nuovo Cimento IIIA(1998)999.
11. M.N.Harakeh et al; Phys.Lett B176(1986)279
12. J.J.Gaardhoje, Ann. Rev. Nucl. Part. Sci. 42(1992)483; A.van der Woude, Prog. Part. Nucl. Phys. 18(1987)217; K.A.Snover, Ann. Rev. Nucl. Part. Sci. 36(1986)586.
13. R.O.Akyuz and S.Fallieros; Phys.Rev.Lett. 27(1971)1016.
14. T.J.Bowles et al; Phys.Rev.Lett. 48(1982)986.
15. A.Bonasera et al; La rivista del Nuovo Cimento. 23.2(2000)1.

FIRST RESULTS WITH TONNERRE (A NEW DETECTOR FOR DELAYED-NEUTRON SPECTROSCOPY)

J.C. ANGELIQUE, N.L. ACHOURI, S. GREVY, F.R. LECOLLEY, J.L. LECOUEY, E. LIENARD, N.A. ORR, J. PETER, AND S. PIETRI.
Laboratoire de Physique Corpusculaire F-14050 Caen Cedex, France.
E-mail:angelique@caelav.in2p3.fr

C. BORCEA, A. BUTA, F. NEGOITA AND C. TIMIS.
Institute of Atomic Physics, Bucharest-Magurele P.O Box MG6, Rumania.

F. DE OLIVEIRA, M. LEWITOWICZ, M. STANOIU AND O. TARASOV.
Grand Accélérateur National d'Ions Lourds, F-14021 Caen Cedex, France.

P. BAUMANN, S. COURTIN, P. DESSAGNE, R. HADELER, A. KNIPPER, C. MIEHE, E. POIRIER AND G.WALTER.
Institut de Recherche Subatomique, BP20, F-67037 Strasbourg Cedex, France.

D. GUILLEMAUD-MUELLER.
Institut de Physique Nucléaire, F-91406 Orsay Cedex, France.

D. Y. PENIONZHKEVICH.
Flerov Laboratory of Nuclear Reactions, JINR, 141980 Dubna, Russia.

D. Z. DLOUHY AND J. MRAZEK.
Nuclear Physics Institute, 250 68 Rez, Czech Republic.

J.M. DAUGAS.
IKS, KU Leuven, Celestijnenlaan 200D, 3001 Leuven, Belgium.

W. N. CATFORD.
Department of Physics, University of Surrey, Guildford GU2 5XH, UK.

The experimental spectroscopy of very-neutron-rich nuclei adds a new dimension to our understanding of nuclear structure. In this contribution, a description of the multi-detector for β-delayed neutron spectroscopy TONNERRE is presented. Following a brief discussion of the motivations behind the project and the design of the array, some perspectives concerning future physics directions are proposed. An example which illustrates the physics capabilities of TONNERRE, namely the neutron energy spectrum in the β-decay of ^{34}Al obtained with this array, is shown.

1 Introduction

During recent years our knowledge concerning the structure of medium-light nuclei has been greatly improved due to the study of new isotopes far from stability and particularly of very neutron-rich nuclei. Decay studies provide one of the primary spectroscopic tools for the investigation of such nuclei. For very exotic nuclei, high decay energies (Q_β >10 MeV) along with relatively low proton or neutron separation energies in the daughter nuclei are encountered. As a result β-delayed particle emission often becomes the dominant decay mode. Access to the location and structure of unbound levels requires the measurement of the delayed particles – in particular the energies. For very neutron-rich nuclei, delayed neutron emission dominates. In the decay of ^{11}Li, for example, the delayed neutron emission probability, Pn, is 92%, with P1n = 88% and P2n = 4%.

Clearly then, the decay spectroscopy of very neutron-rich nuclei requires a neutron detection system of high efficiency (particularly for energies below ~ 5 MeV), with good energy and angular resolution, and a granularity sufficient to provide for multi-neutron decay measurements. Guided by these requirements the TONNERRE ("TONneau pour NEutrons REtardés") array was constructed in 1998 by LPC-Caen in collaboration with IFIN-Bucharest [1].

2 Design

The final design chosen for the TONNERRE array is shown in figure 1. It consists in 32 elements, each comprising a 160 cm long scintillator bar (BC400), 20 cm wide and of thickness 4 cm. The radius of curvature and neutron flight path is 120 cm. Each element is viewed by two photomultiplier tubes (PMT). The start for the TOF measurement is provided by the detection of the decay electrons in an implantation detector located at the centre of the array [2].

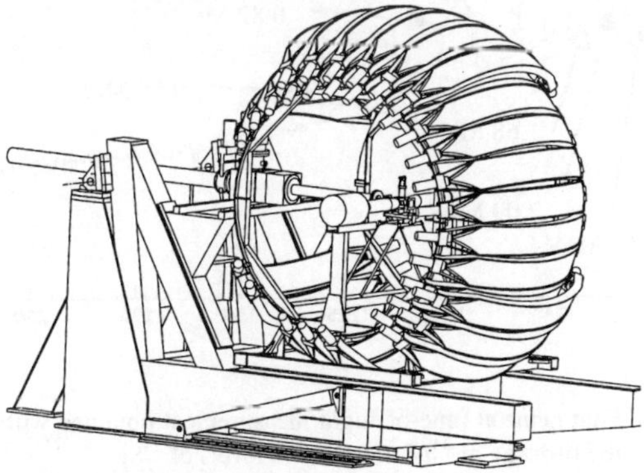

Figure 1. General view of TONNERRE installed on the LISE3 beam-line.

3 Performance

The characteristics of the array were first determined experimentally via an in-beam test performed in October 1998 on the LISE3 separator at GANIL. The nuclei ^{15}B, ^{16}C, ^{17}N were selected for measuring the intrinsic detection efficiency in the range En = 0.8 – 5 MeV. The energy resolution is about 90 keV at 1 MeV.

It should be noted that the β efficiency in the implantation plastic detector was more than 90%. With a high overall detection efficiency of more than 10%, TONNERRE is well placed to undertake a competitive program of delayed-neutron decay spectroscopy. Indeed, besides the neutrons of known energies, a new neutron peak at 3.3 MeV was observed following the β-decay of ^{16}C [3].

4 Physics with TONNERRE

4.1 Spectroscopy around the N=20 shell closure

The experiment using for the first time TONNERRE, concerns an investigation of the collapse in the N=20 shell model closure through a study of the β-n decay of ^{32,33}Mg and ^{34,35}Al [4]. This experiment was performed at GANIL in April 2000. In conjunction with TONNERRE, we used two clover detectors from EXOGAM, one LEPS of Surrey and eight low energy neutron detectors from IReS Strasbourg. From these measurements we have obtained the associated neutron energies. For example, figure 2 below shows the original time-of-flight spectrum for the delayed neutrons from ^{34}Al.

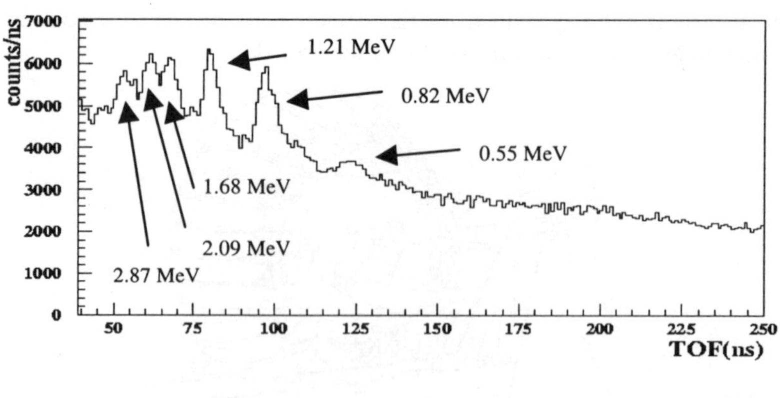

Figure 2. First neutron time-of-flight measurement obtained with TONNERRE in the β-decay of ^{34}Al to unbound levels of ^{34}Si.

The preliminary results demonstrates that the array performed up to the design specifications. Further analysis is underway to investigate delayed neutrons and γ-rays data in order to construct the detailed decay schemes. The resulting spectroscopic information will provide stringent tests of large scale shell model calculations including np – nh excitations aimed at understanding the structural effects occurring in this region .

4.2 *Deformation in the region of N=28*

A comprehensive β-decay study around N=28 shell closure is proposed through the complete spectroscopy of ^{43}P and ^{44}S [5]. The results of the measurements will complement the information from COULEX experiments and will be compared with the predicted deformations in this region. This experiment is planned for in June 2001 at GANIL.

4.3 *The Physics cases around Z=20*

In order to answer the unresolved questions on the structure of the low lying states in the daughter Ca isotopes and of the GT distribution over the full Q_β window, we propose to study the decay $^{51\text{-}53}$K [6]. This experiment proposed by an IReS Strasbourg collaboration will be performed at ISOLDE in 2002.

5 Conclusion and future programs

Clearly the aforegoing experimental programs far from exhaust the possibilities offered by TONNERRE. The potential of a low energy ISOL facility associated with SPIRAL for producing much more intense secondary beams than currently available significantly enhances not only the details with which spectroscopy might be performed but also the limits which may be reached. In addition, other more speculative projects, such as the search for n-n correlations in decays, also remain to be explored.

References

1. Martin T., *Ph.D. Thesis*, **LPCC T98-03**, LPC Caen, (1998).
2. Buta A. et al: *NIMA* **455** (2000) pp. 412-423.
3. Angélique J.C. et al: *GANIL Proposal* **E333** *and references therein,* (1999).
4. Grévy S. et al: *Physical Review C,* **Vol.63 Number 3** , (2001) pp. 1-4.
5. Grévy S. et al: *GANIL Proposal* **E377** *and references therein,* (2000).
6. Miehé C. et al: INTC Proposal **P120** and references therein, (2000).

THE S-FACTOR OF THE $^3H(^3H,2N)^4HE$ AND $^3HE(^3HE,2P)^4HE$ REACTIONS USING A THREE-CLUSTER EXIT-CHANNEL

F. ARICKX, J. BROECKHOVE AND W. VANROOSE
Universieit Antwerpen (RUCA), Groenenborgerlaan 171, B-2020 Antwerpen, Belgium
E-mail: arickx@ruca.ua.ac.be

V. VASILEVSKY AND A. NESTEROV
Bogoliubov Institute for Theoretical Physics, Metrolohichna st. 14-b, Kiev 252143, Ukraine

We study the reactions $^3H(^3H,2n)^4He$ and $^3He(^3He,2p)^4He$ within a fully microscopic cluster model coupling two- and three-cluster configurations and featuring a three-cluster exit channel. The three-cluster continuum is described by Hyperspherical Harmonics. The dynamical equations are solved by the Modified J-Matrix Method, which includes potential asymptotic information for fast convergence. The resulting astrophysical S-factor of both reactions is in good agreement with experimental data.

1 The Modified J-Matrix Method

Several methods in quantum scattering theory reduce the problem to a set of algebraic equations by an introduction of a square integrable (L^2) basis. We consider the *J-matrix method*[2], which has been reformulated in a nuclear physics context as the *algebraic method*[1].

We introduce modifications to the J-matrix method in the oscillator basis that significantly improve the convergence. That is to say, only very small interaction matrices are needed to calculate converged phase shifts. The approach is based on asymptotic approximations for the matrix elements of the potential energy operator in the oscillator basis[2].

2 Model description for the reactions

The six-nucleon wave functions are built by using both two- and three-cluster configurations, each one fully antisymmetrized:

$$\Psi_L = \mathcal{A}\{\Phi_{3N}\Phi_{3N}\ f_L(\mathbf{q}_0)\} + \mathcal{A}\{\Phi_\alpha\Phi_N\Phi_N\ g_L(\mathbf{q}_1,\mathbf{q}_2)\} \tag{1}$$

where the Φ_A (N stands for either nucleon, α for ^4He, and $3N$ for ^3H or ^3He) represent cluster component wave functions, f_L and g_L respectively refer to the wave functions of relative motion for the two-, and three-cluster

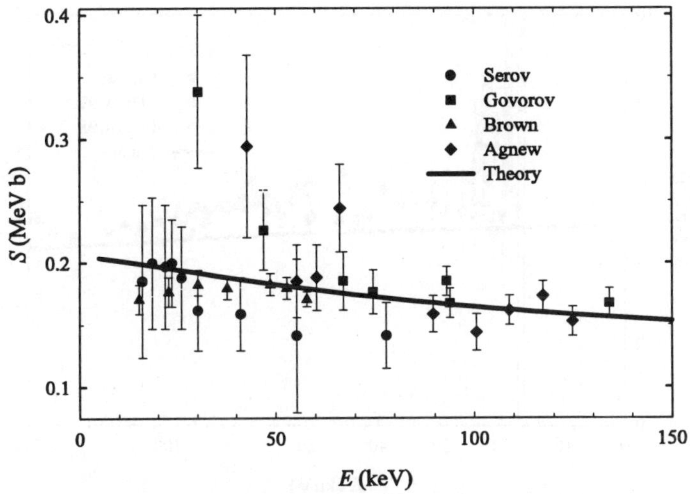

Figure 1. S-factor of the reaction $^3H\left(^3H, 2n\right)^4He$. Experimental refs. see [4].

system, and $\mathbf{q}_i$ are a choice of Jacobi coordinates describing the configuration of relative motion of the clusters.

One of the basic assumptions of the model is that the internal cluster states are defined by having all their nucleons frozen in a (0s)-oscillator wave-function. Consequently all of the reaction dynamics, in particular the behavior of the S-matrix elements, is concentrated in the functions describing the relative motion, i.e. f_L and g_L.

For the two-cluster configuration (in $^3H,^3H$ resp. $^3He,^3He$) we use spherical coordinates $\mathbf{q}_0 = \{q_0, \widehat{\mathbf{q}}_0\}$ and take the quantum numbers $\mu = \{n, L, M\}$ to classify the basis states where n is the familiar radial oscillator quantum number.

For the three-cluster configurations (in $^4He, p, p$ resp. $^4He, n, n$) we use Hyperspherical coordinates $\{\mathbf{q}_1, \mathbf{q}_2\} = \{\rho,\ \Omega\}^3$. This choice is consistent with a set of quantum numbers $\nu = \{N, K, (l_1 l_2)LM\}$, in which $N = 2n_\rho + K$ is the total number of oscillator quanta, K is the hypermomentum, and l_1 and l_2 are the partial angular momenta associated with the Jacobi vectors $\mathbf{q}_1$ and $\mathbf{q}_2$. As discussed in reference[3] K is not a good quantum number and a coupled K-channel calculation has to be performed to solve the dynamical equations.

In the current study we consider both incoming and outgoing waves for the two-cluster configurations, and only an exit-channel for the three-cluster

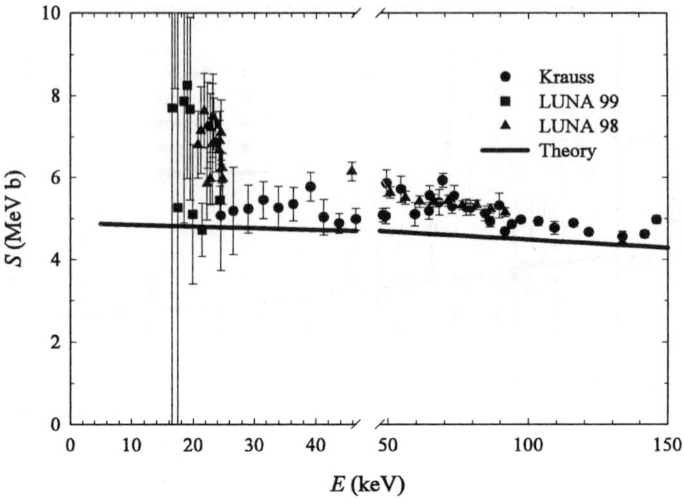

Figure 2. S-factor of the reaction $^3He\left(^3He,2p\right)^4He$. Experimental refs. see[4].

configuration. This is sufficient to obtain the S-matrix elements describing the reaction initiated in the two-cluster channel leading to the three-cluster breakup.

The matching of internal and asymptotic regions is equivalent to the one in the traditional Resonating Group Method (RGM), but now in terms of the oscillator quantum number rather than the coordinate[3].

3 Results and conclusion

We use the Volkov $NN-$interaction[4] in our calculations. It is known to provide an acceptable description for the bound state of ^6He, as well as for the low energy range in the three-cluster continuum of ^6He and ^6Be.

In figures 1 and 2 we show the total S-factor for the reaction $^3H\left(^3H,2n\right)^4He$ respectively $^3He\left(^3He,2p\right)^4He$ in the energy range $0 \leq E \leq$ 150 keV. One notices that the theoretical curve is very close to the available experimental data (see [4] for references to these data). The S-factor for both reactions is seen to be a monotonic function of energy, and does not exhibit any irregularities to be ascribed to a hidden resonance. Thus no indications are found towards explaining the solar neutrino problem.

The present model allows to study the dynamics of the reactions in de-

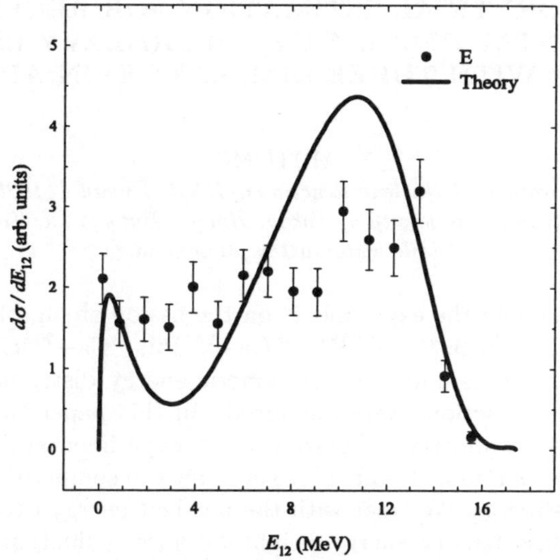

Figure 3. Calculated and experimental total proton yield for the reaction $^3He(^3He, 2p)^4He$. Ref. to experimental data see[4].

tail, and to analyze available experimental data. In figure 3 we show the calculated total proton yield, which has a good qualitative agreement with the experimental data.

Acknowledgments

The authors wish to acknowledge the "Fonds voor Wetenschappelijk Onderzoek Vlaanderen (FWO)" and the UA-RUCA-Rafo fund for financial support.

References

1. V. S. Vasilevsky, F. Arickx, *Phys. Rev.* A **55**, 265 (1997).
2. W. Vanroose, PhD thesis (2001).
3. V. S. Vasilevsky, A. V. Nesterov, F. Arickx, and J. Broeckhove, *Phys. Rev.* C **63**, 034606 (2001).
4. V. S. Vasilevsky, A. V. Nesterov, F. Arickx, and J. Broeckhove, *Phys. Rev.* C **63**, 064604 (2001).

THEORETICAL ESTIMATION FOR RECENT EXPERIMENTAL RESULTS OF SUPERHEAVY ELEMENTS AT DUBNA WITH THREE-DIM. LANGEVIN APPROACH

Y. ARITOMO
Flerov Laboratory of Nuclear Reactions, JINR, Dubna 141980, Russia
Department of Physics, University of Tokyo, Hongo, Bunkyo-ku, Tokyo 113, Japan
E-mail: aritomo@cv.jinr.dubna.su

Recently at Dubna the experiments on the fission of superheavy nuclei in the reactions ^{48}Ca+^{208}Pb, ^{48}Ca+^{238}U, ^{48}Ca+^{244}Pu, ^{48}Ca+^{248}Cm and ^{58}Fe+^{248}Cm carried out [1]. The mass and the total kinetic energy distributions of fission fragments of these reactions were measured. In this paper, we focus on the fusion-fission process and try to reproduce the experimental data by using a fluctuation-dissipation model taking into account the competition between the fusion and quasi-fission. We treat with the incident energy over the Coulomb barrier or the Bass barrier energy height because a fluctuation-dissipation model is a classical one. We estimate the fusion-fission cross section σ_{CN} as;

$$\sigma_{CN} = \frac{\pi\hbar^2}{2\mu_0 E_{cm}} \sum_{l=0}^{\infty}(2l+1)T_l P_{CN}, \tag{1}$$

where μ_0 denotes the reduced mass in the entrance channel and E_{cm} denotes the incident energy in center-of-mass frame. T_l is the barrier penetration coefficient of the lth partial wave through the potential barrier. T_l is calculated with parabolic approximation of the combined Coulomb potential and proximity potential. P_{CN} is the probability of forming a compound nucleus in the competition with quasi-fission.

In this work, we employ the Langevin equation [2]. We adopt the three-dimensional nuclear deformation space with the two-center parametrization. As the three collective parameters to be described by the Langevin equation, we treat z_0 (distance between two potential centers), δ (deformation) and α (mass asymmetry of the colliding partner); $\alpha = (A_1 - A_2)/(A_1 + A_2)$, where A_1 and A_2 denote the mass number of target and projectile, respectively. Hydrodynamical inertia tensor is adopted with the Werner-Wheeler approximation for the velocity field, and the wall-and-window one-body dissipation is adopted for the dissipation tensor.

For the purpose of the calibration of our model calculation, firstly we analyze the fusion-fission cross section for the ^{48}Ca+^{208}Pb reaction, where we can utilize the enough data of fusion-fission cross section [1]. The calculation

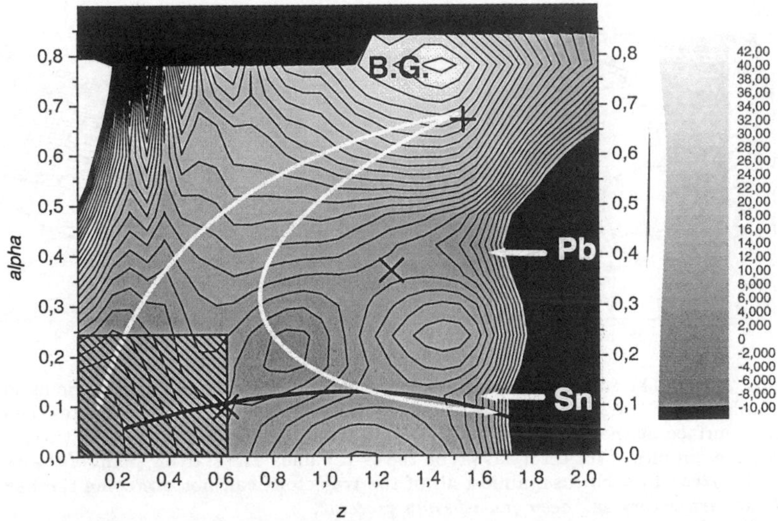

Figure 1: The potential energy surface of liquid drop model with shell correction energy in nuclear deformation space for $^{292}114$. The abscissa and the ordinate denote the separation between two potential center and the mass asymmetry, respectively. Symbols are given in text. The allow indicate the fusion and fission trajectory.

results of the mass distribution of the fission fragments and the excitation function of the fusion-fission cross section give a good agreement with the experimental data beyond the Bass barrier region [3].

Next, we present the analysis of the ^{48}Ca+^{244}Pu reaction. We assume that both shapes of the target and the projectile are spherical at touching point of the colliding system. We also take into account the temperature dependent shell correction energy for the potential energy surface [2,3]. For example, at $T = 0$, the potential energy surface in the reaction ^{48}Ca+^{244}Pu is shown in Fig. 1. $z = \delta = \alpha = 0$ corresponds to a spherical compound nucleus. The contact point in the reaction and saddle points are denoted by (+) point and ($\times$) points, respectively. The shadow box denotes the fusion box which is defined as the inside of the fission saddle point,$\{z < 0.6, \delta < 0.2, |\alpha| < 0.25\}$. We can see that the fission fragment of Pb and Sn corresponds to $\alpha \sim 0.4$ and $\alpha \sim 0.12$, respectively, which are indicated in Fig. 1. In the calculation, the mean trajectory at $E^* = 33$ MeV goes to quasi-fission at $\alpha \sim 0.4$ through the path passing the outer-saddle point, and the fission fragment Pb is produced. Actually, we can find such Pb fragments in the experimental data at the excitation energy [1].

At Dubna the experiments on the fission of superheavy nuclei in the reaction ^{48}Ca+^{244}Pu carried out and they present the fusion-fission cross section of compound nuclei which is derived from the mass symmetric fission fragments $(A/2 \pm 20)$ [1]. The subsequent important question is whether all of the mass symmetric fission fragments come from the compound nuclei or not. As

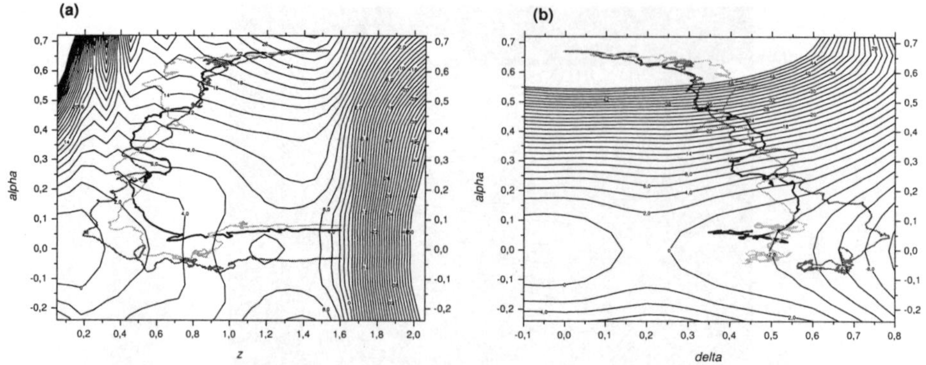

Figure 2: (a): The samples of the trajectory which go to the mass symmetric fission on the $z-\alpha$ plane at $E^* = 33$ MeV in the reaction ^{48}Ca+^{244}Pu. The contour denotes the potential energy surface at the excitation energy. Such trajectory takes 6.8 % of all trajectory. (b): The same samples of trajectory as (a) on the $\delta-\alpha$ plane. The starting point of the calculation is (0.0,0.676). In such case, almost all of the trajectory can not overcome the barrier. We call such trajectory as *"deep quasi-fission process"*.

the final results of the experiments the mass symmetric fission fragments are detected, but there exists two possibilities where it comes from. One is that the mass symmetric fission fragments come from the compound nuclei and the other is that they come from the quasi-fission. In Fig. 1, paths are presented. We try to check them by using three-dimensional trajectory calculation with Langevin equation.

Figure 2 (a) shows the samples of the trajectory which go to the mass symmetric fission on the $z - \alpha$ plane at $E^* = 33$ MeV. The contour denotes the potential energy surface at this excitation energy. We can see clearly that the trajectories enter the pocket around the spherical region and go back to the mass symmetric fission. Such a trajectory occupies 6.8 % of all trajectory.

However, all of such trajectory does not enter the fusion box defined above. Figure 2 (b) shows the same samples of trajectories as the previous one, but they are on the $\delta - \alpha$ plane. In Fig. 2 (b), the starting point of the calculation is (0.0,0.676) and trajectories go to the direction of the large deformation of fragments and of the small mass asymmetry. In such case, almost all of the trajectory can not overcome the barrier on the $\delta - \alpha$ plane and can not reach to the region of the compound nucleus. We call such trajectory as *"deep quasi-fission process"*[3]. That is to say, the deep quasi-fission process contributes to the yield of the mass symmetric fission fragments. In fact, at $E^* = 33$ MeV, only 0.08 % of all trajectory can overcome the barrier and enter to the spherical region or fusion box. As a result, we can say that 0.08 % of all trajectory enter the three-dimensional fusion box in the reaction.

Figure 3 shows the cross section by the calculation and the experiment in the reactions ^{48}Ca+^{244}Pu. The open and closed scatters denote the capture cross section σ_{cap} and the cross section $\sigma_{A/2\pm20}$ which derived by the yield of the mass symmetric fission fragments with $\frac{A}{2} \pm 20$ in the experiments, re-

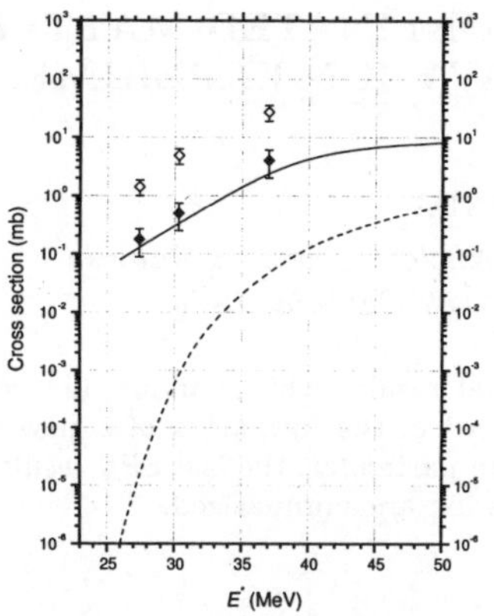

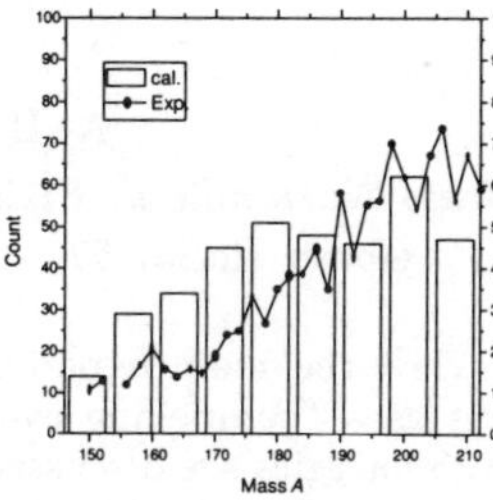

Figure 3: (left) The cross section by the calculation and the experiment in the reactions ^{48}Ca+^{244}Pu. Lines and scatters are given in the text. Figure. 4: (right) The mass distribution of the fission fragments in the reaction ^{48}Ca+^{244}Pu.

spectively [1]. The $\sigma_{A/2\pm20}$ by the calculation is denoted by the solid line. It presents very good agreement with the experimental data. The fusion-fission cross section σ_{CN} by the calculation is denoted by the dashed line. It is calculated by the trajectory crossing the three-dimensional fusion box mentioned above. The fusion-fission cross section σ_{CN} by the calculation is one or two order magnitude smaller than the cross section $\sigma_{A/2\pm20}$. We see that the cross section $\sigma_{A/2\pm20}$ includes the deep quasi-fission events. Such an information is very important to estimate the evaporation residue cross section.

Furthermore we try to calculate the another systems, the reactions ^{48}Ca+^{238}U and ^{48}Ca+^{248}Cm. Also in these case, the calculation for the cross section $\sigma_{A/2\pm20}$ agrees with experimental data well. And we can see that the calculation for the fusion-fission cross section σ_{CN} also is one or two order magnitude smaller than the cross section $\sigma_{A/2\pm20}$ [3]. Figure 4 shows the mass distribution of the fission fragments in the reaction ^{48}Ca+^{244}Pu. The calculation result gives the good agreement with the experimental data.

As the next step, we take into account the decay process which is described by the statistical model and try to estimate the evaporation residue cross section of the superheavy elements. The author is grateful to Prof. M. Ohta, Prof. Zagrebaev and Prof. Yu. Ts. Oganessian for their helpful suggestion and valuable discussion through the present works.

References

1. M.G. Itkis et al., Proceeding of *Fusion Dynamics at the Extremes* , Dubna, May 25 -27, 2000 (World Scientific, Singapore, 2001) p93.
2. Y.Aritomo et al., Proceeding of *Fusion Dynamics at the Extremes* , Dubna, May 25 -27, 2000 (World Scientific, Singapore, 2001) p123.
3. Y. Aritomo, paper submitted to Phys. Rev. C.

FROM HADRONIC TO DECONFINATED MATTER IN ULTRARELATIVISTIC HEAVY ION EXPERIMENTS

A. BADALÁ

Istituto Nazionale di Fisica Nucleare, Sez. di Catania
Corso Italia, 57 – I 95129 Catania, Italy

In this short review the main experimental results which, about one year ago, were pointed out as a "compelling evidence" of the formation of a new state of matter at the SPS energies are discussed. In particular, the last SPS results about strangeness enhancement and J/ψ suppression are summarized.

1 Introduction

Lattice QCD calculations predict that at a transition temperature of 170-180 MeV [1] (corresponding to an energy density of 1-3 GeV/fm^3) a new state of matter, the quark gluon plasma (QGP), will be produced. This plasma exhibits the so called "chiral symmetry", which in normal nuclear matter is spontaneously broken. Moreover, in this plasma quarks and gluons are no longer confined within the hadron dimensions.

Simple estimates indicate that at the AGS beam energy one can, at best, barely reach the initial energy density required for QGP formation. Therefore, it is from the CERN-SPS experiments that evidences of a possible onset of deconfinement could come. In particular evidences can come from central Pb+Pb collisions at 158 A GeV/c ($\sqrt{s} \sim 17$ GeV/fm^3). For these reactions an energy density of 3.2 GeV/fm^3, which is well above the characteristic QGP-phase value, is reached [2].

2 Thermal and chemical equilibrium

A general feature emerging from many measurements at SPS energies is that the invariant distribution $d^2N/m_t dm_t dy$ has an exponential shape and the inverse slope T increases linearly with the mass of the particle[3]. This fact can be explained as arising from the superposition of the thermal motion of the expanding system plus a radial collective flow. The most accurate analysis to disentangle thermal motion and radial flow was done by NA49 experiment, which found a value of T$\sim$ 120 MeV and $\beta_T \sim 0.55$[4]. It is just for this large expansion velocity that the term "Little Bang" was used to describe the expansion of the hadronic phase[5].

The chemical freeze-out temperature and the baryon chemical potential determined in the frame of statistical models of the hadronization process[6] place the system very close to where the phase boundary between plasma and hadrons is expected. Thus there is not enough time for a kinetic equilibration of the hadron abundances by inelastic rescattering and the observed hadronic chemical equilibrium is pre-established in the QGP phase[7]. This will have a confirmation in the strangeness enhancements of WA97 experiment.

3 Strangeness enhancement

The idea of strangeness enhancement as a QGP signal was formulated a long time ago[8]. Moreover, if the hadron formation proceeds by statistical hadronization from a color-deconfined quark state the enhancement with respect to p-A interaction is expected to follow a hierarchical behaviour, i.e. $E_\Omega > E_\Xi > E_\Lambda$ [5]. Such a pattern contradicts the expectations from rescattering in a hadronic fireball since the multi-strange production is suppressed by high thresholds and low cross sections.

The above-mentioned hierarchical behaviour was reported from the CERN experiment WA97 which studied strange particle production at central rapidity in Pb+Pb, p+Pb, p+Be collisions at 158 A GeV/c [9]. In Fig. 1 the hyperon yields normalized to the p-Be data and to the number of wounded

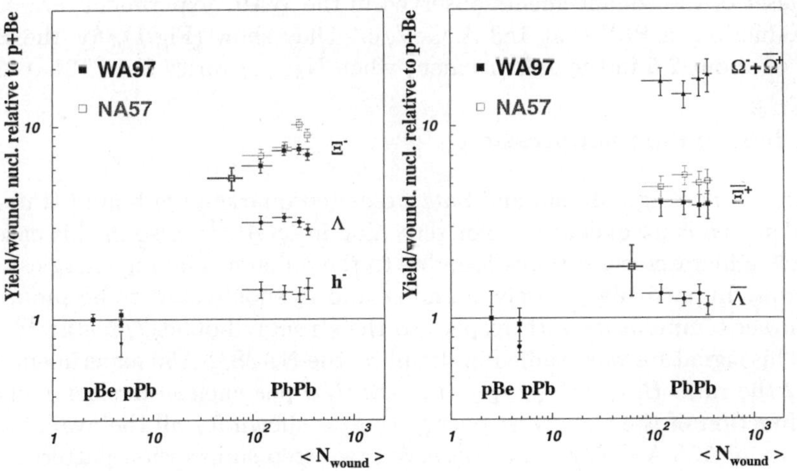

Figure 1. Yields per participant relative to the p-Be yields (from ref. 10).

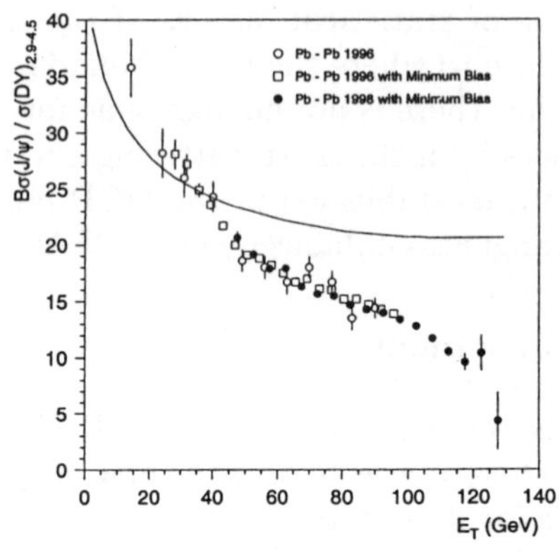

Figure 2. $\sigma_{J/\psi}/\sigma_{DY}$ ratio as a function of E_T (from ref. 13). The curve represent the J/ψ suppression due to ordinary nuclear absorption.

nucleons as a function of wounded nucleons are shown. One can see that, in Pb-Pb collisions, there is an enhanced production of strange particles, and this enhancement is larger for particles of higher strangeness content. It is about 17 for $\Omega + \overline{\Omega}$. Moreover, the enhancements are saturated for $N_{wound} > 100$.

The first preliminary results of the NA57 experiment[10], proposed to study the onset of the enhancements observed in the WA97 experiment, are on the Ξ production in Pb-Pb at 158 A GeV/c. They show (Fig.1), for the Ξ^+, a drop of about 2.5 in the enhancement when N_{wound} varies from 121 to 62.

4 Charmonium suppression

About 15 years ago Matsui and Satz predicted quarkonium bound state suppression as a consequence of color screening in QGP [11]. Also in this case one expects a hierarchical suppression, due to the different binding energies of the $c\bar{c}$ states. In particular, loosely bound ψ' and χ_c should start to be suppressed at a lower temperature with respect to the strongly bound J/ψ state[12].

This signature was studied in detail by the NA38/NA50 experiment[13]. In Fig. 2 the ratio $B_{\mu^+\mu^-}\sigma_{J/\psi}/\sigma_{DY}$ (i.e. the J/ψ per nucleon-nucleon collision) as a function of the transverse energy (E_T) is shown for all the available Pb-Pb data at 158 A GeV/c. The observed two-step suppression pattern, could be simply explained in a deconfinement scenario. The first anomalous step at about $E_T = 40$ GeV is due to the melting of the χ_c mesons, responsible for a fraction of the observed J/ψ yield through their radiative decay (about

30-40% in proton induced collisions). The second drop at about E_T =90 GeV is due to the J/ψ state melting. These transverse energies were associated to an energy density[13] of about 2.3 GeV/fm^3 and 3 GeV/fm^3.

5 Conclusion

In this really short review I have not considered many topics and in particular I have not said anything about the dileptons/photons production studies carried out by the NA45, WA98 and NA50 experiment. You can see for a review ref.[14].

In summary the experimental data collected up to now show that at SPS energies there is charmonium suppression and multistrange hyperon enhancement (signals predicted for QGP evidence). A coherent scenario from partonic to hadronic system up to the final expansion was proposed and at least for Pb-Pb central collisions the formation of a deconfinated state of matter seems very probable[15]. However, a great both experimental and theoretical effort needs to change this "compelling evidence" into "sure evidence". From the 2000 and 2001 data and from (if it will start) new SPS program we hope to have significative contributions to strengthen the actual results and establish the formation (or not) of QGP matter in heavy ion collisions at SPS energies.

References

1. F. Karsch, *Nucl. Phys.* B **83-84**, 14 (2000).
2. T. Alber et al., *Phys. Rev. Lett.* **75**, 3814 (1995).
3. I.G. Bearden et al., *Phys. Rev. Lett.* **78**, 2080 (1997).
4. H. Appelshauser et al., *Eur. Phys. J.* **C2**, 661 (1998).
5. U. Heinz, *Nucl. Phys. A* **661**, 140 (1999).
6. P. Braun-Munzinger, *Nucl. Phys. A* **681**, 119c (2001); F.Becattini et al., *Nucl. Phys.* B **71**, 324 (1999).
7. R. Stock, *Phys. Lett.* B **456**, 277 (1999).
8. P. Koch, B. Muller and J. Rafelski, *Phys. Rep.* **142**, 167 (1986).
9. F. Antinori et al., *Nucl. Phys. A* **681**, 141c (2001) and ref. therein.
10. F. Antinori et al., Quark Matter 2001 Proc.
11. T. Matsui and H.Satz, *Phys. Lett.* B **178**, 416 (1986).
12. F. Karsch, M.T.Mehr and H.Satz, *Z. Phys. C* **37**, 617 (1988).
13. M.C. Abreu et al., *Phys. Lett.* B **477**, 28 (2000); M.C. Abreu et al., Quark Matter 2001 Proc. and ref. therein
14. I. Tserruya, *Nucl. Phys. A* **681**, 133c (2001).
15. U. Heinz and M. Jacob, nucl-th/0002042.

STUDYING EXOTIC NUCLEI THROUGH DIRECT REACTIONS

Yorick BLUMENFELD
Institut de Physique Nucléaire, IN_2P_3/CNRS, 91406 Orsay Cedex, France

Recent improvements in the intensities and optical qualities of radioactive ion beams have made possible the study of direct reactions induced by unstable nuclei. The design and performances of an innovative silicon strip detector array devoted to such studies are described. Elastic and inelastic proton scattering are used to obtain information on density and transition density distributions of exotic nuclei. The structure of the wave functions of halo nuclei can be probed through transfer and knock-out reactions. Achievements in these investigations are illustrated with recent experimental results.

1 INTRODUCTION

The detailed study of the properties of unstable nuclei has been at the forefront of nuclear physics research during recent years. The ultimate goal of such studies is to develop models and interaction potentials that can be applied to nuclei far from stability. Along the way, novel manifestations of nuclear structure should be uncovered, among which can be expected nuclear halos and skins, new regions of deformation, the disappearance of shell closures or the appearance of new magic numbers. With the continuous improvement of intensities and optical qualities of secondary radioactive beams, it has become possible to study nuclei far from stability through reactions such as Coulomb excitation[1] and more recently elastic and inelastic scattering[2], transfer[3] and knock-out[4] reactions.

In this talk I will focus on the the issues that can be addressed through direct nuclear reactions. Elastic proton scattering yields information on the nuclear matter distributions and the effective nucleon-nucleon potentials. Inelastic scattering towards low lying collective states gives access to transition probabilities and nuclear deformation. Comparison of inelastic transition probabilities measured through Coulomb and hadronic excitation should give information on the isoscalar or isovector nature of the states through the ratio of multipole transition matrix elements M_n/M_p. Single nucleon transfer reactions are particularly well suited for probing nuclear shell effects, by selectively populating single particle or single hole states. It has recently been shown that in certain cases, nucleon removal (knock-out) reactions can provide information similar to transfer reactions using much weaker beam intensities.

Direct reactions to study unstable nuclei are performed in inverse kinemat-

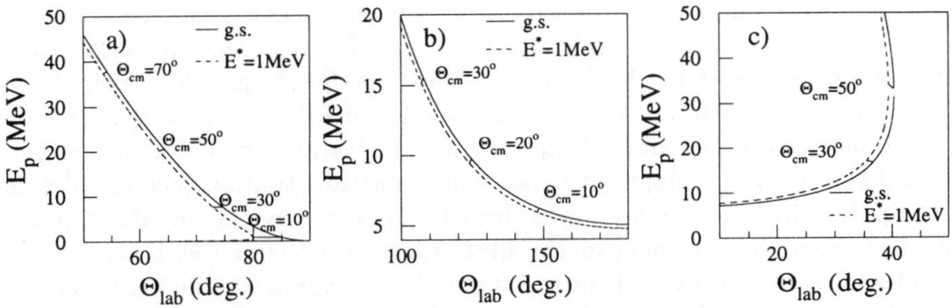

Figure 1: Kinematics for ^{32}Mg(p,p') (a), ^{32}Mg(d,p) (b), and ^{32}Mg(p,d) (c) reactions at 30 MeV/A. The figures show the energy-angle correlation for the recoiling light particle in the case of reactions leading to the ground state (solid line) or a hypothetical excited state at E^*= 1MeV (dashed line). The locations of various center of mass scattering angles are indicated.

ics, where the radioactive beam strikes a target containing the light particles. The kinematics of the reaction are reconstructed from the measurement of the energy and angle of the recoiling particles. The IPN-Orsay, CEA-Bruyères le Châtel and CEA-Saclay have designed and built a novel silicon strip array named "MUST"[5], specifically tailored to such experiments. The characteristics of the array will be described in section 2. The potentialities of elastic and inelastic scattering will be illustrated in section 3 through results obtained at GANIL with ^{20}O and ^{6}He beams. The investigation of the structure of unstable nuclei through transfer and knock-out reactions will be discussed in section 4 in the light of results obtained for the halo nucleus ^{11}Be.

2 EXPERIMENTAL METHODS

In the elastic, inelastic and transfer experiments described in the following, the radioactive beam is produced by fragmentation of a stable primary beam on a thick production target, and analyzed by a fragment separator. The unstable nuclei of interest impinge on a solid CH_2 or CD_2 target. The kinematics of the recoiling light particles are shown on fig. 1 in the case of (p,p'), (p,d) and (d,p) reactions induced by a hypothetical 30 MeV/A ^{32}Mg beam. It can be seen that a precise measurement of the energy and angle of the recoiling light particle will furnish the excitation energy imparted to the nucleus and the scattering angle in the center of mass frame.

The MUST detector[5] has been designed to fulfill the requirements of such experiments. It consists of 8 large area Si-strip, Si(Li), CsI telescopes with associated electronics and data acquisition system. The first stage of the telescopes consists in a 300 μm thick 60 X 60 mm^2 Si-strip detector with a strip pitch of 1 mm in both the horizontal and vertical directions. Each strip detector is backed by a lithium drifted silicon diode (Si(Li)) of approximately 3mm compensated thickness and a 15mm thick CsI crystal. The detector is modular and the set up is adapted to each experimental situation (elastic, pick-up or stripping) in order to cover with maximum efficiency the angular range of interest. Particles stopping in the first stage (protons of less than 6 MeV) are identified by energy and time of flight measurements, while higher energy particles traversing the strip detector are identified with the standard ΔE-E technique.

Because of the large emittance of the secondary beams, it is necessary to perform event by event ray tracing of the incoming nuclei. This is commonly done at GANIL using two low pressure multi-wire proportional counters (CATS)[6] located upstream from the target. These detectors yield a position resolution of approximately 0.3mm in X and Y and also furnish a start signal for time of flight measurements with a resolution of 400 ps. These counters function reliably for counting rates up to 510^5 pps.

In order to select the reaction channel of interest and thus strongly reduce the background in the particle telescopes, the scattered nuclei are detected and identified, either in the focal plane of the SPEG spectrometer[7], through an energy loss measurement in a Bragg chamber and a time of flight measurement between a fast plastic scintillator and one of the tracking detectors, or in a plastic scintillator placed behind the target.

As can be seen from fig. 1, the excitation energy resolution depends both on the energy and the angular resolution of the light particle measurement. A typical value of the excitation energy resolution obtained in such experiments is 500 keV.

3 ELASTIC AND INELASTIC SCATTERING

The study of the oxygen isotopic chain has proven to be particularly exciting. Recently, the neutron drip line was shown to be located at A=24[8]. Therefore, starting from the N=Z nucleus ^{16}O, the neutron drip line is reached by the addition of only eight neutrons, and rapid structural changes are expected. In order to complement electro-magnetic measurements, ^{20}O(p,p') angular distributions were recently measured[2]. A secondary ^{20}O beam at 43 MeV/A with an intensity of 5000 pps was produced at GANIL using a primary ^{40}Ar beam at 77 MeV/A and scattering from a CH_2 target was measured using the MUST array. The scattering of a secondary ^{18}O beam at the same E/A was also measured for comparison. For the ^{20}O case, the elastic and inelastic angular distributions for the 2_1^+ and 3_1^- states known to be located at 1.67 and 5.61 MeV respectively, are displayed on fig. 2. The dotted lines are CCBA calculations performed with the code ECIS[9] using the Bechetti-Greenlees phenomenological optical potential[10]. A very good reproduction of the elastic distribution is obtained. Normalization of the inelastic scattering calculations to the data yield deformation parameters $\beta_2 = 0.55 \pm 0.06$ and $\beta_3 = 0.35 \pm 0.05$.

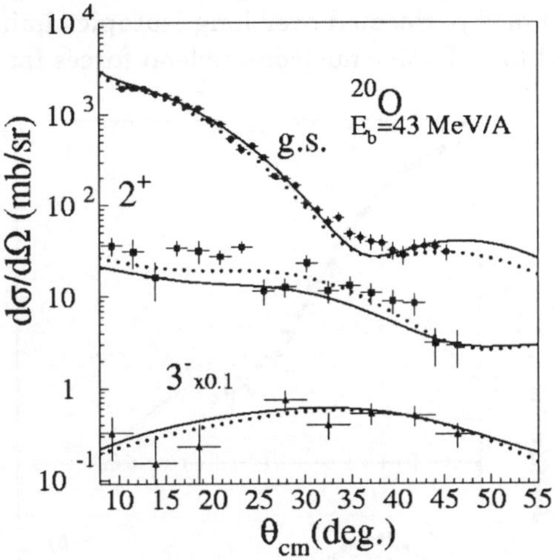

Figure 2: 43 MeV/A ^{20}O(p,p') elastic and inelastic angular distributions for the 2_1^+ and 3_1^- states. The dotted lines correspond to CCBA calculations using the Bechetti-Greenlees optical potential and the solid lines correspond to DWBA calculations using the JLM interaction folded with ground state and transition densities obtained with HF + BCS and QRPA models (see text).

The solid lines represent fully microscopic calculations where neutron and proton densities are obtained from a Hartree-Fock + BCS model and the transition densities are obtained within the framework of the QRPA formalism[11]. The only inputs to the calculation are the SGII Skyrme interaction and the value of the constant pairing gap, Δ=12A$^{-1/3}$MeV. In order to calculate angular distributions complex optical and transition potentials were obtained by injecting the calculated ground state and transition densities into the Jeukenne, Lejeune and Mahaux (JLM) density dependent optical potential[12], which was derived from Brückner-Hartree-Fock nuclear matter calculations. Cross sections were then calculated in a DWBA approach with the TAMURA code[13], and are shown as solid lines on fig. 2. Good agreement with the data is observed, which confirms the validity of the calculated ground state and transition densities. By comparison with electro-magnetic measurements, a value of the ratio of transition matrix elements $M_n/M_p = 3.25 \pm 0.8$ is obtained for the 2_1^+ state. This value, in agreement with the prediction of the QRPA calculation[11], is much larger than $N/Z = 1.5$, demonstrating the strong neutron contribution to the excitation due to the presence of a neutron skin. A similar analysis in the case of the 2_1^+ state of ^{18}O yields $M_n/M_p = 1.10 \pm 0.24$, showing a very rapid structural change when adding 2 neutrons to go from ^{18}O to ^{20}O. Ultimately, such studies performed over long isotopic chains should furnish a solid testing ground for effective nucleon-nucleon forces far from stability.

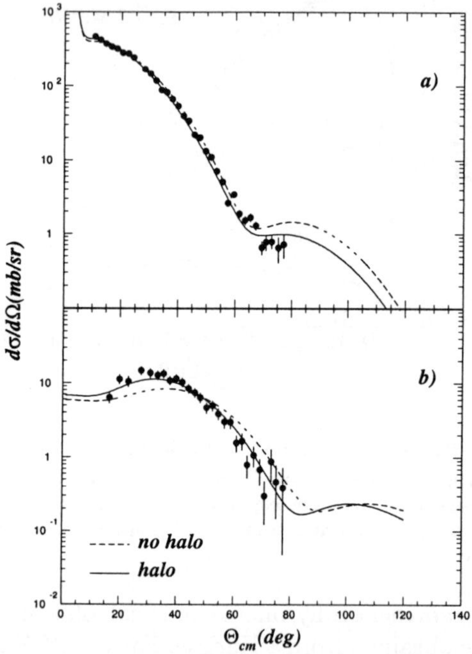

Figure 3: Angular distributions for the elastic (a) and inelastic(b) scattering to the 2^+ state of ^6He located at 1.87 MeV from protons at 40.9 MeV/A. The data are compared to the results of calculations assuming *no halo* (dashed lines) and *halo* (solid lines) densities (see text).

A typical case of a borromean nucleus is ^6He, which has two weakly bound neutrons outside an α-particle core. In order to further probe the halo structure of ^6He, elastic and inelastic scattering on protons was measured at GANIL over a large angular range[14]. A secondary ^6He beam at 40.9 MeV/A was produced using a primary ^{13}C beam at 75 MeV/A and protons recoiling from a CH_2 target were measured using the MUST array. Forward moving ^{4}He and ^{6}He ejectiles were measured in a large area plastic telescope in coincidence with the recoiling protons. The angular distributions for elastic scattering and inelastic scattering to the first 2_1^+ state of ^{6}He located at 1.87 MeV are presented on figs. 3 a) and b) respectively.

The data are compared to theoretical calculations performed using a fully microscopic, complex, non local potential[15] based on large basis shell model calculations of ^{6}He[16]. Two shell model densities are investigated. In the first, labeled *no halo*, complete $6\hbar\omega$ wave functions are used. However, in this case, the binding energy of the last neutron is larger than the experimental separation energy. In the second case, labeled *halo*, the bound state potential is changed so that the 1p-shell binding energy approaches the experimental single neutron separation energy. The neutron density then exhibits an extended halo-like tail.

In the case of elastic scattering, the *halo* (full line) and *non halo* (dashed line) calculations are very similar up to 60° and differ notably at larger angles. The few data beyond this angle are better reproduced by the *halo* description but clearly data at larger momentum transfers are required to probe the halo structure through elastic scattering. The inelastic scattering distribution is remarkably sensitive to the halo structure over the entire angular domain. The inelastic data are better accounted for by the *halo* calculation.

Proton scattering is shown to be an important tool for the study of the structure of unstable nuclei, complementary to electro-magnetic probes as protons interact with both neutrons and protons in the nucleus. Angular distributions are sensitive to the nuclear density and transition density profiles and studies over long isotopic chains should ultimately allow us to assess the validity of microscopic nuclear models at large isospin.

4 TRANSFER AND KNOCK OUT REACTIONS : THE ^{11}Be GROUND STATE

Single nucleon transfer reactions are a powerful tool to obtain level schemes, transfered angular momenta and spectroscopic factors. The first results concerning such reactions induced by unstable nuclei are now becoming available. Here I will report on results from an experiment[3,17] aimed at studying the ground state structure of the well known one neutron halo nucleus ^{11}Be. The large extension of the neutron density distribution indicates that the main component of the $1/2^+$ ground state wave function of ^{11}Be should be an $s_{1/2}$ neutron coupled to the ground state of ^{10}Be. However, the presence of core excited components $[2^+ \otimes d_{5/2}]$ is possible, and their proportion can be measured through the ^{11}Be(p,d)^{10}Be single nucleon transfer reaction.

A ^{11}Be secondary beam with an intensity of 3 10^4 pps was produced by fragmentation of a ^{15}N beam delivered by the GANIL facility and impinged on a 5 mg/cm^2 CH_2 target. The ^{10}Be ejectiles were identified and momentum analyzed in the SPEG spectrometer[7] and the recoiling deuterons were measured in coincidence in the CHARISSA array[18] located between 5 and 35°. In this experiment the kinematic properties of the reaction were deduced from the momentum and angle of the ^{10}Be, the deuteron measurement serving to reduce the background from reactions on the Carbon in the target. Figure 4 displays the focal plane spectrum of ^{10}Be without (top) and with (bottom) coincidences with recoiling deuterons. The three peaks observed in the lower panel, where the background from the Carbon is seen to be fully removed, correspond to the excitation of well known states in the ^{10}Be nucleus. The 2^+ state at 3.37 MeV is well separated from the 0^+ ground state and the unresolved group of four states near 6 MeV.

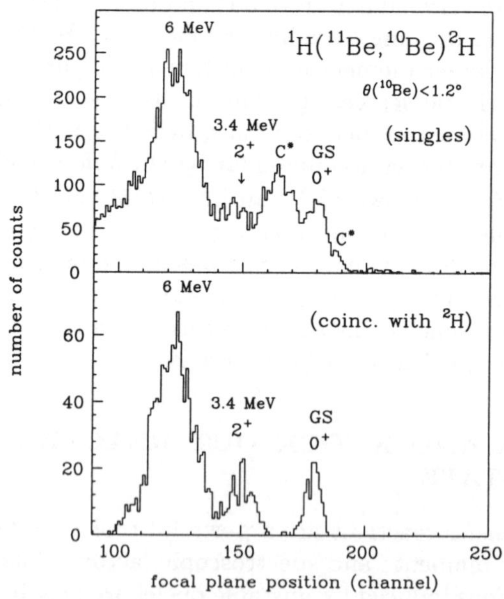

Figure 4: Top: Inclusive momentum spectrum in the SPEG focal plane for ^{10}Be. Bottom: Momentum spectrum in the SPEG focal plane for ^{10}Be in coincidence with deuterons measured in CHARISSA.

Theoretical cross sections were calculated with the zero range code DWUCK4[19], trying various sets of optical potentials. Angular distributions using proton and "adiabatic" deuteron optical parameters generated by the Watson global nucleon potential[20] give the best agreement with the experimental results[3] and are shown on fig. 5, normalized to the data. The resulting ratio of spectroscopic factors $R=S(2^+)/(S(0^+)+S(2^+))$, which measures the core excitation admixture, is 0.32, larger then most theoretical predictions. However, the accuracy of results obtained by standard DWBA for nuclei close to the drip line has to be questioned. It has long been emphasized that correct form factors should be taken as solutions of coupled equations. This is particularly important in the case of ^{11}Be, where the halo neutron is coupled to a ^{10}Be core with a large deformation parameter (β=0.74). A new calculation was performed where the $[0^+\otimes s_{1/2}]$ and $[2^+\otimes d_{5/2}]$ form factors were deduced from coupled channel calculations performed in a vibrational approach using the code CCVIB[21]. The ratio R is then reduced to a value 0.22, in agreement with shell model[22] and core-coupling model[23] predictions.

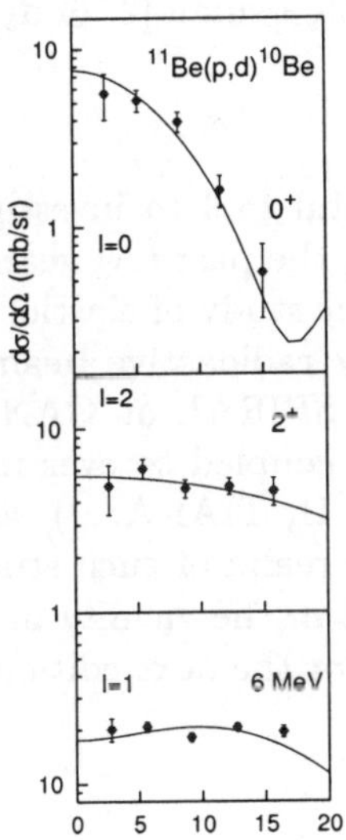

Figure 5: Angular distributions for states of ^{10}Be. The solid lines are DWBA predictions normalized to the data (see text).

Transfer reaction cross sections strongly decrease with increasing incident energy. In addition, the low energies and the inverse kinematics require the use of very thin targets, making their study unacceptably time consuming for beam intensities lower than 10^4 pps. An attractive alternative for the study of the ground state wave function of loosely bound nuclei is the use of one nucleon removal (also called knock out) reactions. Such an experiment was performed recently at the NSCL/MSU to investigate the structure of ^{11}Be[4]. A 60 MeV/A secondary ^{11}Be beam was directed on a 228mg/cm^2 ^{9}Be target surrounded by a position sensitive NaI array. This array detected the γ rays emitted by excited fragments, while the emerging ^{10}Be fragments were identified by the S800 spectrometer. The fraction of ^{10}Be fragments accompanied by a 3.37 MeV γ ray carries the information about the $[2^+ \otimes 1d_{5/2}]$ admixture in the ^{11}Be ground state wave function. The theoretical cross section for a given ^{10}Be core final state and removed neutron j value is assumed to be the product of a spectroscopic factor S and a single particle cross section. The latter is calculated within a spectator-core eikonal three-body model[24]. The experimental result is compatible with a ratio R = 0.20, in very close agreement with that extracted from the transfer reaction study. Both experiments support a picture in which the ^{11}Be wave function is dominated by the $s_{1/2}$ single particle component with a small $[2^+ \otimes d_{5/2}]$ admixture.

5 CONCLUSIONS

Direct reactions are a powerful tool to investigate the detailed structure of weakly bound nuclei. During the past few years, both experimental methods and analysis procedures for the study of elastic, inelastic, transfer and nucleon removal reactions induced by radioactive beams have been perfected. New accelerator facilities such as SPIRAL at GANIL or the Coupled Cyclotron complex at the NSCL/MSU, coupled to ever more efficient detector systems for charged particles (MUST II, TIARA...) and γ rays (EXOGAM, NSCL Ge array...) will extend the realm of such studies to both lower and higher energies and to heavier nuclei in the vicinity of the drip lines. Many exciting new results can be expected for the next edition of the NPBL conference.

Acknowledgments

I wish to thank my colleagues from IPN-Orsay, Florida State University, NSCL/ Michigan State University, PTN-Bruyères le Châtel and SPhN Saclay for their participation in the experiments used to illustrate this paper.

References

1. T. Glasmacher, *Annu. Rev. Nucl. Part. Sci.* **48** 1, 1998.
2. E. Khan *et al. Phys. Lett.* B **490**, 45 (2000).

3. S. Fortier *et al.* *Phys. Lett.* B **461**, 22 (1999).
4. T. Aumann *et al.* *Phys. Rev. Lett.* **84**, 35 (2000).
5. Y. Blumenfeld *et al.* *Nucl. Instrum. Methods* A **421**, 471 (1999).
6. S. Ottini-Hustache *et al.* *Nucl. Instrum. Methods* A **431**, 476 (1999).
7. L. Bianchi *et al.* *Nucl. Instrum. Methods* A **276**, 509 (1989).
8. O. Tarasov *et al.* *Phys. Lett.* B **409**, 64 (1997).
9. J. Raynal, *Phys. Rev.* C **23**, 2571 (1981).
10. F.D. Bechetti Jr. and G.W. Greenlees, *Phys. Rev.* **182**, 1190 (1969).
11. E. Khan and N.V. Giai, *Phys. Lett.* B **472**, 253 (2000).
12. J.P. Jeukenne, A. Lejeune, and C. Mahaux *Phys. Rev.* C **16**, 80 (1977).
13. T. Tamura *et al* *Comp. Phys. Comm.* **2**, 94 (1971).
14. A. Lagoyannis *et al.*, *Phys. Lett. B*, in press.
15. P.J. Dortmans and K. Amos, *Phys. Rev.* C **49**, 1309 (1994).
16. P. Navratil and B.R. Barrett *Phys. Rev.* C **57**, 3119 (1998).
17. J.S. Winfield *et al.* *Nucl. Phys.* A **683**, 48 (2001).
18. R.L. Cowin *et al.* *Nucl. Instrum. Methods* A **423**, 75 (1999).
19. P.D. Kunz, code DWUCK4, U. of Colorado (unpublished).
20. B.A. Watson, P.P. Singh, and R.E. Segel, *Phys. Rev.* **182**, 977 (1969).
21. J. Van de Wiele, code CCVIB, IPN Orsay (unpublished).
22. E.K. Warburton and B.A. Brown *Phys. Rev.* C **46**, 923 (1992).
23. F.M. Nunes, I.J. Thompson, and R.C. Johnson, *Nucl. Phys.* A **596**, 171 (1996).
24. J.A. Tostevin, *J. Phys.* G **25**, 735 (1998).

THE STUDY OF THE REACTION $^{58}Fe + ^{248}Cm \rightarrow ^{306}122$

A.A.BOGATCHEV, M.G. ITKIS, YU.TS. OGANESSIAN, E.M. KOZULIN, I.M. ITKIS, M. JANDEL, J. KLIMAN, G.N. KNIAJEVA, N.A. KONDRATIEV, I.V. KORZYUKOV, L. KRUPA, I.V. POKROVSKI, V.A. PONOMARENKO, E.V. PROKHOROVA, A.YA. RUSANOV AND V.M. VOSKRESENSKI

Flerov Laboratory of Nuclear Reactions, JINR, 141980 Dubna, Russia E -mail: bogachev@nrsun.jinr.ru

F. HANAPPE, B. BENOIT AND T. MATERNA

Universite Libre de Bruxelles, 1050 Bruxelles, Belgium

N. ROWLEY, L. STUTTGE

Institut de Recherches Subatomiques, F-67037 Strasbourg Cedex, France

G. GARDINA

Dipartimento di Fisica dell'Universita di Messina, 98166 Messina, Italy

K.J. MOODY

University of California, Lawrence Livermore National Laboratory, Livermore, California 94551, USA

Interest in the study of the fission process of superheavy nuclei interactions with heavy ions is connected first of all with the possibility of obtaining information, the most important for the problem of synthesis on the production cross section of compound nuclei at excitation energies of about 15 -30 MeV. There is a need for a much more penetrating insight into the fission mechanism of superheavy nuclei. Reaction $^{58}Fe + ^{248}Cm$ leads to the formation of the heaviest compound system ever studied, name ly $^{306}122$ (N=184), i.e. to the formation of the spherical compound nucleus, which agrees well with theoretical predictions [1]. Actually, three ^{58}Fe ion beam induced reactions were taken into account to compare the experimental data: $^{58}Fe+^{232}Th \rightarrow ^{290}116$, $^{58}Fe+^{244}Pu \rightarrow ^{302}120$, $^{58}Fe + ^{248}Cm \rightarrow ^{306}122$.

1 Experimental set-up

The experiments were carried out on the CORSET + DEMON set -up at accelerator U -400, Flerov Laboratory of Nuclear Reactions, JINR (Dubna). The scheme of the set-up is shown in Fig.1.

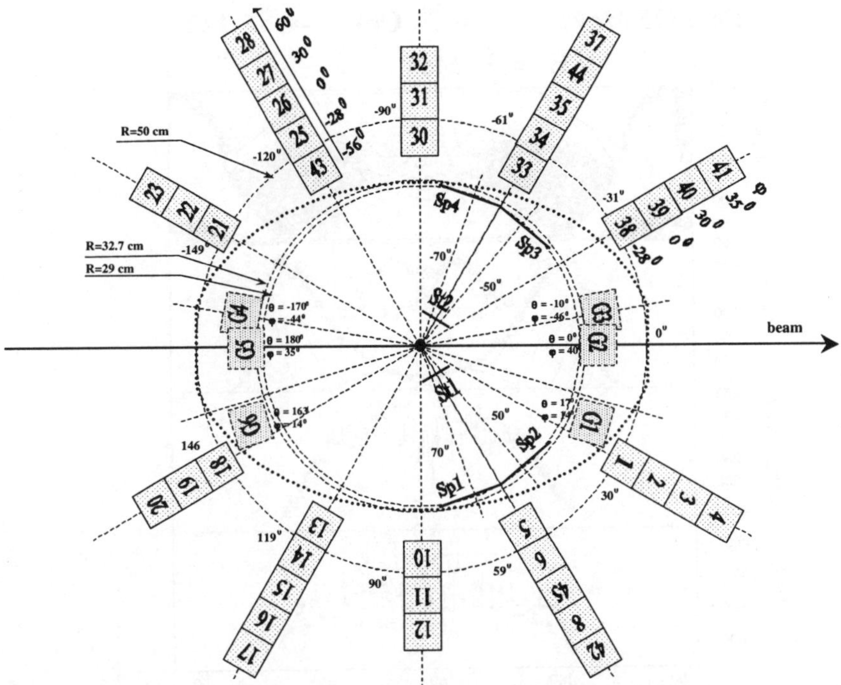

Figure 1. The scheme of the experimental set-up CORSET+DEMON.

In the center of the chamber different targets were placed (^{232}Th, ^{244}Pu and ^{248}Cm). T The energy of the ^{58}Fe ion beam was 325 MeV and 328 MeV. The fission fragments were registered with two time -of-flight arms. One start detector and four stop detectors were set on each arm. Neutron spectrometer DEMON and γ-detectors were mounted around the CORSET chamber [2].

2 Results

The standard experimental data are Total Kinetic Energy – Mass (TKE -M) distributions. The experimental data for reaction ^{58}Fe + ^{248}Cm → ^{306}122 are shown in Fig.2. The beam energy was 325 MeV, that corresponds to the excitation energy $E^* = 33$ MeV.

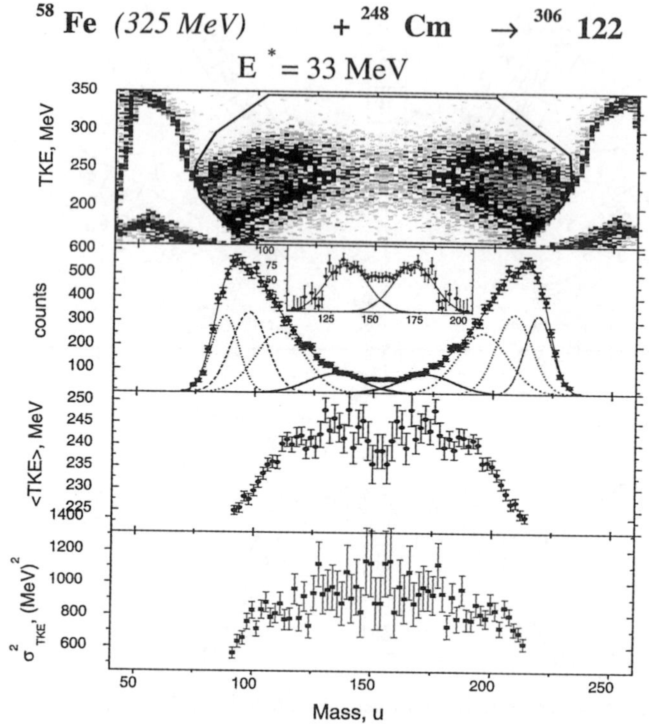

Figure 2. Two-dimensional TKE-M matrix, the mass yield, average TKE and σ^2_{TKE} distributions in dependence on the fragment mass for the reaction $^{58}Fe + ^{248}Cm \rightarrow ^{306}122$.

We observe strong manifestation of the asymmetric mass distribution of $^{306}122$ fission fragments (second and third panels of Fig.2). The light fragment mass of the fusion-fission process (in the central part of the distribution) is situated near the mass number A=132 and the heavy fragment mass is complementary to that of the light fragment.. It can be explained by the effects of the Z=50 and N=82 shells. The heavy fragment mass of the quasi-fission process (second panel of Fig.2) is situated near the mass number A=208. It can be explained by the influence of the shells Z=82 and N=126. The light fra gment mass is complementary to that of the heavy fragment [2]. In Fig.3 the same distributions are shown for the reactions $^{58}Fe+^{244}Pu \rightarrow ^{302}120$ and $^{58}Fe+^{232}Th \rightarrow ^{290}116$.

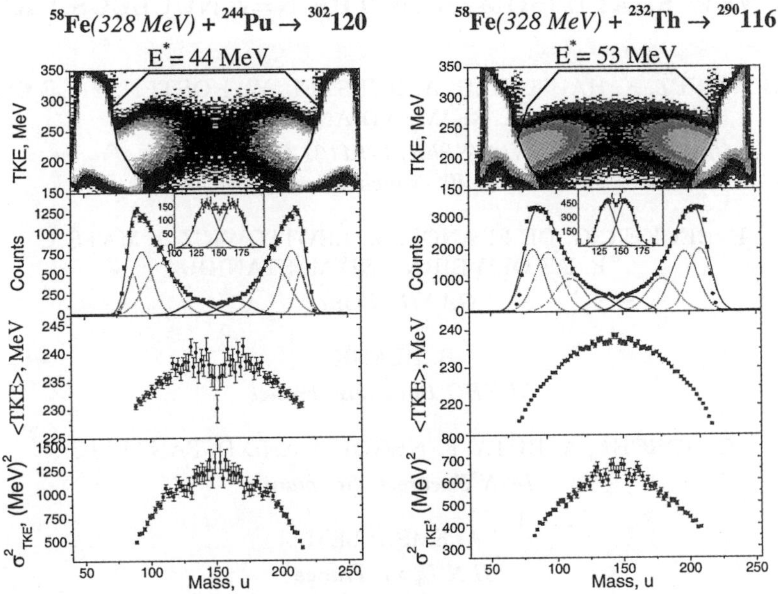

Figure 3. Two-dimensional TKE-M matrix, the mass yield, average TKE and σ^2_{TKE} distributions in dependence on the fragment mass for the reactions $^{58}Fe+^{244}Pu \rightarrow ^{302}120$ and $^{58}Fe+^{232}Th \rightarrow ^{290}116$.

The light fragment mass of the fusion -fission process (in the central part of the distribution) is also situated near the mass number A=132 and the heavy fragment mass is complementary to that of the light fragment. The heavy fragment mass of the quasi-fission process is also situated near the mass number A=208. For the compound nucleus with Z=122 (E^*=33 MeV) the difference peak -valley of TKE (third panels) is about 12 MeV, for the compound nucleus with Z=120 (E^*=44 MeV) this value is about 6-7 MeV, but for the composite system with Z=116 (E^*=53 MeV) this effect is almost suppressed. So, TKE and mass distributions are more symmetric when the excitation energy is increasing.

References

1. Patyk Z., Sobiczewski A., Nucl.Phys. A 533 (1991) 132.
2. Itkis M.G. et al., JINR Preprint E15-2000-295, Dubna, 2000

A NEW SHAPE ISOMER IN THE N=Z NUCLEUS ^{72}KR

E. BOUCHEZ, K. HAUSCHILD, A. HURSTEL, W. KORTEN, Y. LE COZ,
R. LUCAS, M. REJMUND AND CH. THEISEN
CEA Saclay, DAPNIA/SPhN, F-91191 Gif-sur-Yvette, France
E-mail: ebouchez@cea.fr

F. BECKER, G. DE FRANCE, M. LEWITOWICZ, I. MATEA,
F. DE OLIVEIRA AND M. STANIOU
GANIL, France

B. BLANK
CENBG Bordeaux, France

C. BORCEA, A. BUTA, F. NEGOITA AND D. PANTELICA
IFIN Bucarest, Roumania

A. EMSALLEM
IPN Lyon, France

J. GENEVEY AND J. PINSTON
ISN Grenoble, France

F. HANNACHI
CSNSM Orsay, France

P. RAHKILA
Univ. of Jyväskylä, Finland

A new shape isomer in the N=Z ^{72}Kr nucleus has been observed at GANIL with a combined conversion electron and γ-ray spectroscopy. This low-lying 0^+ state is clear evidence for shape coexistence as expected from theoretical calculations. The ground state is interpreted as a pure oblate state.

1 Motivation

Mid-shell nuclei can exhibit very different shapes depending on the proton and the neutron numbers and the single particle structure. For proton number Z=36 and neutron numbers 36<N<42 pronounced shell gaps appear at prolate as well as oblate deformation. They play a major role in the structure and the shape of the light Krypton isotopes. In fact, the different gaps allow the

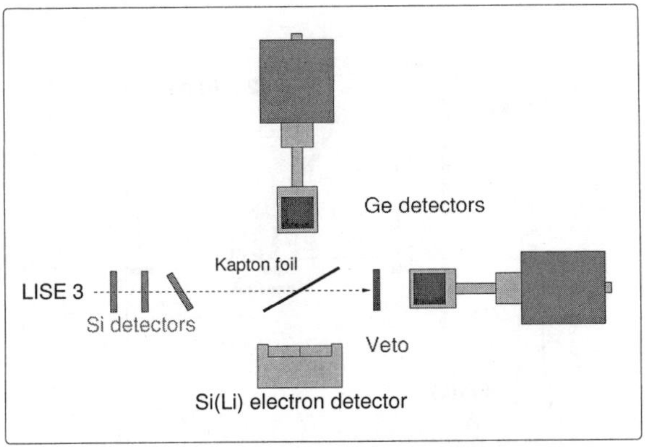

Figure 1. Focal plane detection set-up.

nuclei to minimize their potential energy at oblate or prolate shape. When both very different shapes occur, the phenomenon is called shape coexistence.

Several experiments have been performed to study this phenomenon in even-even Krypton isotopes with N≥38 in ^{74,76,78}Kr[1,2,3,4,5]. Theoretical calculations[6,7] predict that, for these nuclei, the deeper minimum of the potential energy corresponds to the prolate shape. In contrast, in the N=Z nucleus ^{72}Kr, the 0^+ ground state is predicted to be oblate and the excited one prolate. The difference in energy between these two 0^+ states is expected to be around 800 keV[8].

2 Experimental technique

The nucleus ^{72}Kr was produced by the fragmentation of a ^{78}Kr beam at 73 A.MeV on a Be target. The beam was delivered by the GANIL cyclotrons. The fragments were separated by the LISEIII spectrometer[9]. The N=Z nucleus ^{72}Kr represents less than 1% of all the fragments at the focal plane. A stack of three 300μm silicon detectors at the focal plane was used to identify the charge (by energy loss measurement) and the mass (by a TOF measurement) of the fragments. The third silicon detector was tilted to adjust the effective thickness of the matter to precisely implant the nuclei of interest in the last 10μm of a thin Kapton foil. A fourth one was used as a veto. A combined Conversion Electron (CE) and γ ray spectroscopy has been performed at the implantation foil: γ-rays were measured using a set of segmented Clover detectors from the EXOGAM collaboration with 1.3 % efficiency at 1.3 MeV;

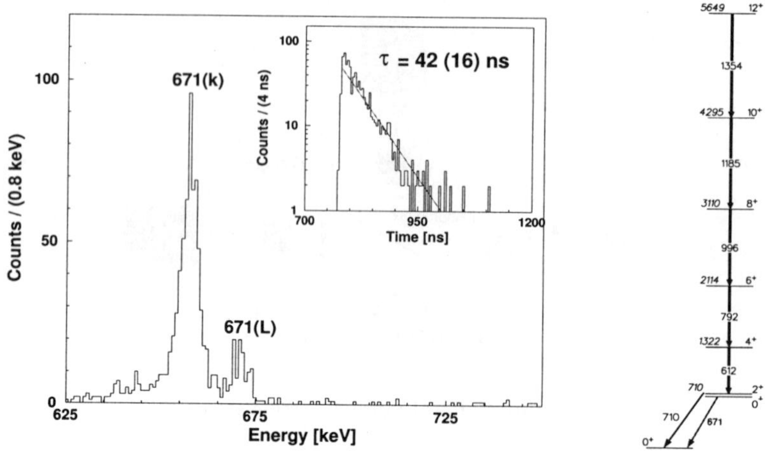

Figure 2. Left: spectrum of conversion electrons detected in coincidence with ^{72}Kr ions; the corresponding time spectrum is shown in inset. Right: partial level scheme[10,11] for ^{72}Kr and the new excited 0^+ state.

to detect the CE electrons, a two fold segmented Si(Li) detector of 4.5 mm thickness with an efficiency of 5.7 % was used (see fig. 1).

With this method isomeric transitions were assigned on an event-by-event basis to a given fragment. Moreover, it was possible to measure strongly converted transitions even with short lifetimes (much shorter than the TOF in the separator). This is because the fragments are fully stripped so that they can not decay by conversion electrons during their flight in the separator until they refill their electronic shells when slowing down in the Si detectors at the focal plane.

3 Results and discussion

The spectrum of conversion electrons in ^{72}Kr is shown in figure 2. The K and L lines of a 671 keV transition with a 42ns lifetime are evident. As no transition has been observed in the γ spectrum and the intensity ratio $\frac{I_K}{I_L}$ is consistent with an E0 transition, the 671 keV transition is interpreted as a fully converted E0 decay of an excited 0^+ state to the ground state.

The partial level scheme[10,11] for ^{72}Kr including the new 0^+ state is also shown in figure 2. The rotational band was interpreted as a prolate band[10] from a comparison with heavier isotopes. This band is regular at high spin.

At low spin, deviations occur due to the shape mixing. The extrapolation of the moment of inertia to the low spin states allows a determination of the energy of the unperturbed 0^+ and 2^+ states. The newly found excited 0^+ state is very close in energy to the expected position of the unperturbed bandhead. The slightly higher energy is due to an interaction with the ground state. This new excited 0^+ state is hence interpreted as the bandhead of the prolate band. A two level mixing calculation has been performed using the energy difference between the experimental and extrapolated 0^+ states. The resulting wave function of the ground state contains a predominant (more than 90%) oblate configuration.

4 Conclusion

We report on a new experimental method to measure conversion electrons from isomeric states produced in fragmentation reactions. A new shape isomer in the N=Z nucleus ^{72}Kr has been observed. With a two-level mixing calculation, we have found that the oblate shape contributes more than 90 % to the ground state and the new excited 0^+ state has been interpreted as the bandhead of the rotational band corresponding to prolate deformation. However, the real shape of a nucleus can only be inferred from quadrupole moment measurements in Coulomb excitation experiments. It will be soon possible to perform Coulomb excitation of ^{74,76}Kr with SPIRAL facilities and of ^{72}Kr at relativistic energy after fragmentation reaction.

References

1. C. Chandler *et al*, Phys. Rev. C **56**, R2924 (1997); Phys. Rev. C **61**, 044309 (2000).
2. F. Becker *et al*, Eur. Phys. J. A **4**, 103 (1999) and Phys. Scripta T **88**, 17 (2000).
3. R. B. Piercey *et al*, Phys. Rev. C **25**, 1941 (1982).
4. A. Giannatiempo *et al*, Phys. Rev. C **52**, 2444 (1995).
5. H. Sun *et al*, Phys. Rev. C **59**, 655 (1999).
6. P. Bonche *et al*, Nucl. Phys. A **443**, 39 (1985).
7. W. Nazarewicz *et al*, Nucl. Phys. A **435**, 397 (1985).
8. M. Girod, priv. comm. (1997).
9. R. Anne *et al*, Nucl. Instrum. Methods Phys. Res. A **257**, 215 (1987).
10. G. de Angelis *et al*, Phys. Lett. B **415**, 217 (1997).
11. S.M. Fischer *et al*, Nucl. Phys. A **682**, 35 (1997).

SPECTROSCOPY OF LIGHT EXOTIC NUCLEI BY (^{7}LI,^{7}BE) CHARGE EXCHANGE REACTION

F.CAPPUZZELLO[1], A.CUNSOLO[1,2], S.FORTIER[3], A.FOTI[2,4], A.LAZZARO[1,2], H.LENSKE[5], A.L.MELITA[1,2], C.NOCIFORO[1,2], L.TASSAN-GOT[3], J.S.WINFIELD[1], S.BONINELLI[1,2], S.ORRIGO[1,2]

[1] *INFN - Laboratori Nazionali del Sud, Via S. Sofia 44, 95123 Catania, Italy*

[2] *Dipartimento di Fisica e Astronomia, Università di Catania, 95129 Catania, Italy*

[3] *I.P.N.-Orsay, Orsay, France*

[4] *INFN - Sezione di Catania, Corso Italia 57, 95129 Catania, Italy*

[5] *Inst. Theo. Physik, Univ. of Geissen, Geissen, Germany*

The (^{7}Li,^{7}Be) charge exchange reaction is performed at low bombarding energy as a probe for spectroscopy of light exotic nuclei. The high resolution experimental technique is based on the interplay of a Tandem Van de Graaf ^{7}Li beam and a magnetic spectrometer. Results are discussed for ^{11}B(^{7}Li,^{7}Be)^{11}Be, ^{15}N(^{7}Li,^{7}Be)^{15}C and ^{14}C (^{7}Li,^{7}Be)^{14}B reactions at 57 MeV.

1 Introduction

Heavy-ion charge exchange reactions (CEX) are powerful tools, which can be used both to investigate the response of near drip-line nuclei to isovector probes and to test models of nuclear forces under the extreme conditions of high charge asymmetry, low nuclear density and low binding energy. One exotic nucleus of great current interest is ^{11}Be [1] due to its one-neutron halo structure. New effects such as the inversion of $2s_{1/2}$ and $1p_{1/2}$ neutron orbitals, the coupling in the ground state of the halo-neutron with the 2^+ excited state of the ^{10}Be core (accounting for about the 20% of the wave function), and the shift to lower excitation energy of the strength associated to the Spin Dipole Resonance (SDR) have been observed via direct reactions. Much less is known about ^{14}B and ^{15}C. The larger separation energy (S_n=0.970 MeV and S_n=1.218 MeV, respectively) compared to the ^{11}Be (S_n=0.504 MeV) one weakens such effects and hinders their clean observation. In fact the question of the existence of one-neutron halo configuration in the ground state of these nuclei is not yet solved. On the other hand, the systematic evolution of these interesting phenomena when going from stable to exotic nuclei, passing through these intermediate cases, must be investigated. It is important, for example, to study the coupling of low-lying states to the particle continuum, which becomes important approaching the one-neutron drip line [2].

In particular we pay attention to the ^{15}C nucleus due to its single particle structure quite similar to the ^{11}Be one. Since, in both cases, the valence neutron is bound to an easily polarizable even-even core, we expect an additional particle-core coupling effect leading to the fragmentation of single particle strength [2]. In [1] it is shown how a many-body approach to the CEX problem on ^{11}Be, based on the QRPA theory, successfully reproduce the observed strength up to around 2 MeV excitation energy. Beyond this limit the observed fragmentation of the strength is not explained in terms of single particle excitation and a more complete approach based on supplementary calculations of Dynamical Core Polarization (DCP) correlations is needed.

Despite the many experimental and theoretical studies, a satisfactory description of the structure of ^{11}Be, ^{14}B and ^{15}C and the reaction mechanisms leading to such nuclei is not yet accomplished. In this paper, we report on investigations based on the (^{7}Li,^{7}Be) reaction to populate low-lying and resonant states of these nuclei.

The (^{7}Li,^{7}Be$_{g.s.}$) transition selects $\Delta T=1$ isospin transfer and, at low incident energies, can proceed via ($\Delta L=0,2$; $\Delta s=0$) transitions mixed with ($\Delta L=0$; $\Delta s=1$) Gamow-Teller and higher ($\Delta L=2$; $\Delta s=1$) total spin transfer components [3]. In contrast, the (^{7}Li,^{7}Be$_{1/2-}$) transition is mainly dominated by $\Delta s=1$ transfer. Interesting information can be extracted from ^{7}Be spectra under the conditions of: i) weak linear momentum transfer, ii) dominance of the one-step CEX mechanism and iii) negligible contribution of the tensor interaction. For the ^{11}B(^{7}Li,^{7}Be)^{11}Be reaction such conditions are fulfilled at forward angles even at low bombarding energy [4]. Moreover, according to refs. [3], the ratio between the cross sections at zero degrees of the two ^{7}Be CEX transitions is directly connected to the probability of nucleonic spin transfer by the factor G, vanishing for non nucleonic and being near 0.46 for nucleonic spin transfer.

2 Experimental set-up and results

Two experiments were performed at the IPN-Orsay Tandem laboratory using a 57 MeV ^{7}Li^{+++} beam. In the first a 95% enriched ^{11}B self-supporting target 130 μg/cm^2 thick was used and the ^{7}Be ejectiles have been detected by the Split-Pole magnetic spectrometer. A preliminary measurement done with a ΔE-E telescope of silicon detectors has been performed in the second experiment [5] to investigate the ^{15}N(^{7}Li,^{7}Be)^{15}C and ^{14}C(^{7}Li,^{7}Be)^{14}B reactions. The thickness of the two targets was 85 μg/cm^2. Supplementary ^{12}C and ^{14}N background runs were done.

In Fig. 1 the obtained ^{7}Be spectra are shown for the different cases. Eight states of ^{11}Be, as well as the ground state, are clearly observed at the energies of 0.32, 1.77, 2.67, 3.41, 3.89, 3.96, 6.05 and 9.5 MeV. Detailed analysis of these states is presented in reference [1]. The high energy resolution of the ^{7}Be (~ 50 keV) spectra allows to separate the (^{7}Li,^{7}Be$_{g.s.}$) and (^{7}Li,^{7}Be$_{1/2-}$) transitions. In Tab. 1 the values of the G factor measured around 0 degrees for different states of ^{11}Be and ^{12}B are compared. The results for ^{11}Be indicate the prevalence of nucleonic spin transfer in the CEX mechanism, while the results for ^{12}B indicates an enhanced contribution from the sequential two-steps mechanisms. That can be explained by the larger overlap between the potential and the interacting nuclei wave functions for stable nuclei [6].

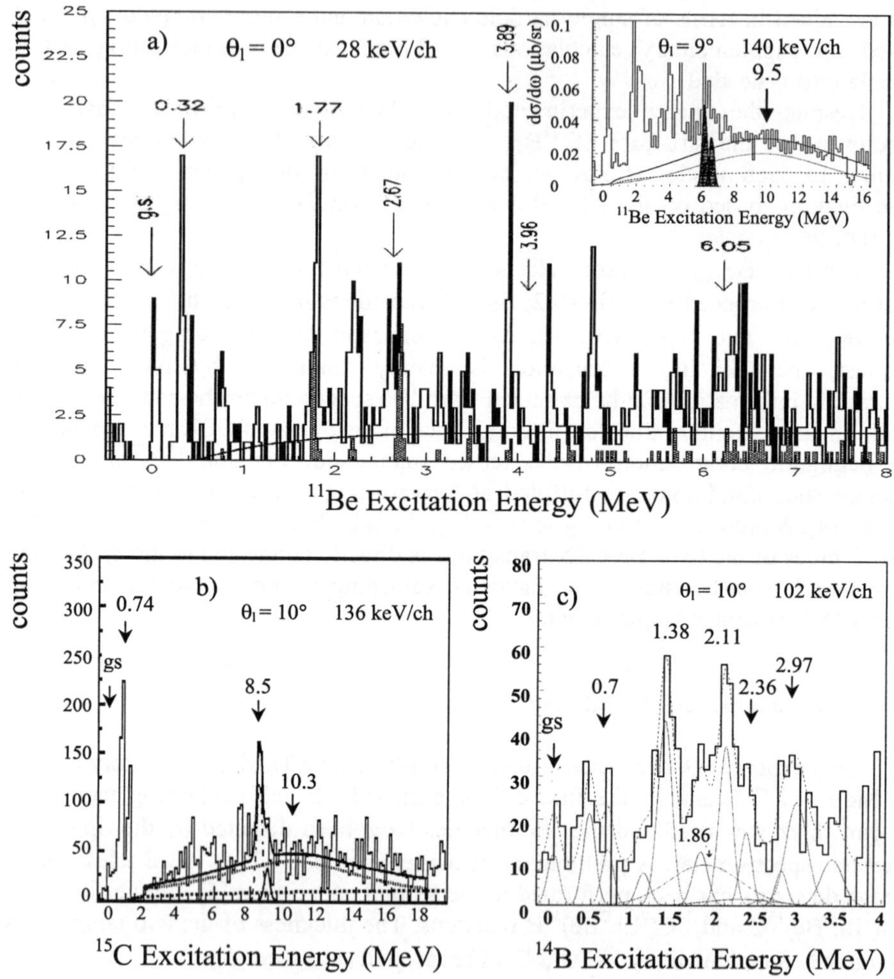

Figure 1. Experimental excitation energy spectra for; a) ^{11}Be states observed at 0°. There the ^{11}B(^{7}Li,^{7}Be)^{10}Be+n 3-body phase space curve has been drawn, while the shadowed histograms represents the contribution from ^{12}C impurities. In the panel, the fit of the peak at 6.05 and the correspondent one associated to the excitation of the ^{7}Be is shown; the continuous line is obtained by adding the 3-body phase space distribution to the gaussian fit of the broad bump at 9.5 MeV; b) ^{15}C states observed at 10°; the solid line is obtained by adding the ^{15}N(^{7}Li,^{7}Be)^{14}C+n 3-body phase space distribution to the lorentzian fit of the broad bump at 10.3 MeV; c) Low lying ^{14}B states observed at 10°.

Table 1. Values of G factors measured at very forward angles.

	^{11}Be							^{12}B		
Energy (MeV)	gs	0.32	1.77	2.67	3.89	3.96	6.05	gs	0.95	1.67
G (θ≅0°)	0.4	0.35	0.51	0.35	0.47	0.4	0.56	0.15	0.08	0.4

Regarding the ^{15}C and the ^{14}B states a few conclusions can be drawn due to the presence of strong background associated to the target impurities. In the ^{15}C case the ground and the first excited state are clearly seen together with a structure at 8.5 MeV (Γ~0.3 MeV). Preliminary DCP calculations for ^{15}C [2] show the presence of a considerable excited-core component at around 8 MeV. Note the similarity between the 140 keV/ch spectra shown in Fig.1a and 1b, as expected. Finally, as shown in Fig.1c, the experimental spectrum is well fitted taking into account the ^{14}B known states.

Conclusions

The selectivity and the high energy resolution achieved with the (^{7}Li,^{7}Be) reaction at Tandem energies, provides useful information about the structure of light neutron-rich nuclei. Its application to the ^{11}Be case allowed the observation of both single particle and structures related to the core polarization and Spin Dipole Resonance. The preliminary results on the ^{15}C indicates a similar behaviour for this nucleus, for which a narrow state far from the neutron emission threshold and a broad bump in the region where the SDR mode could be expected are observed. For ^{14}B all the presently known states have been observed. We believe that a deeper knowledge may be achieved extending to these nuclei the experimental technique used for the ^{11}Be investigation.

References

1. Cappuzzello, F. et al., *Phys. Lett.* **B**, in press, and reference therein.
2. Lenske, H., contribution to this conference.
3. Jänecke, J. et al., *Phys. Rev* .**C54** (1996) 1070.
4. Cappuzzello, F. et al., 9th Intl. conf. on nuclear reaction mechanisms, Varenna, 2000, Ed. E. Gadioli, Publications of the Univ. of Milan, Sup. N^o. **115**, p. 631.
5. Boninelli, S. tesi di laurea (2001) and Orrigo, S. tesi di laurea (2001).
6. Lenske, H. et Schieder, G., *Eur. Phys. Jour.* **2** (1998) 41.

ISOSPIN EFFECTS ON INSTABILITIES AND FRAGMENTATION MECHANISMS

M. COLONNA, V. BARAN, M. DI TORO, V. GRECO
Laboratorio Nazionale del Sud, Via S. Sofia 44, I-95123 Catania, Italy and Università degli Studi di Catania
E-mail: colonna@lns.infn.it

We discuss some properties of instabilities in asymmetric nuclear matter, underlying in particular the link between the nature of unstable fluctuations and the properties of the interaction. We show that, conversely to what happens in symmetric nuclear matter, chemical instabilities may appear as isoscalar-like unstable oscillations, similarly to mechanical instabilities, leading to the formation of a more symmetric liquid phase, surrounded by a neutron rich gas phase. Then we present numerical simulations of multifragmentation processes in asymmetric nuclear matter and in central heavy ion collisions at intermediate energies. In the same context, but in the case of semi-peripheral reactions, the dynamics of the so-called neck region is discussed.

1 Introduction

Nuclear matter (NM) is a binary system formed by neutrons and protons. This feature has stimulated a quite exciting research field: nuclei far from β-stability, dynamical effects coming from large charge asymmetry values, properties of neutron stars. In particular, using exotic beams, NM with high isospin asymmetry can be created transiently, offering the possibility to study also chemical instabilities. Indeed, recent experiments on nuclear reactions involving a large variety of neutron to proton number ratios have shown that products originating from nuclear multifragmentation depend significantly on the isospin asymmetry of the colliding nuclei [1].

In the first part of this article we will focus on the behaviour of Asymmetric NM (ANM) at sub-normal densities where we can foresee scenarios for a dynamical formation of fragments with particular isotopic contents, due to instabilities. Then we will present numerical simulations of fragmentation mechanisms occurring in central and semi-peripheral collisions of neutron-rich nuclei, discussing in particular the properties of Intermediate Mass Fragments (IMF). The sensitivity of the results to the Equation of State (EOS) used in the calculations will be discussed.

2 Chemical and mechanical instabilities

2.1 The EOS symmetry term

The behaviour of the symmetry term of the nuclear matter EOS is poorly known in regions far from normal density. In Fig.1 we report the density dependence of the symmetry contribution to the mean-field potential for neutrons (top curves) and protons (bottom curves) , for three different Skyrme effective interactions, SIII (asy-soft), Skyrme-Bonn (obtained in order to reproduce the EOS of a microscopic nuclear matter calculation with the Bonn A potential [2], asy-stiff) and the stiffer interaction introduced in [3]. While

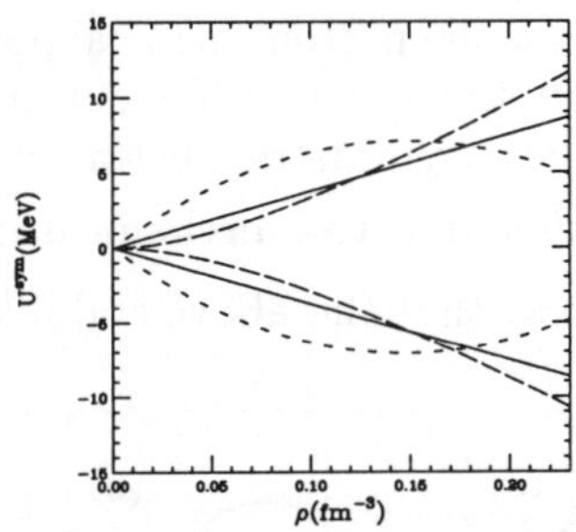

Figure 1. Symmetry contribution to the mean-field potential of neutrons (top curves) and protons (bottom curves), according to SIII (short-dashed), Skyrme-Bonn (full line) and the interaction introduced in ref.[3] (long-dashes).

all interactions give close values at normal density, quite large differences are present in low density and particularly in high density regions. One of the aims of this paper is to show that relevant information on these density ranges, of large importance in astrophysics, can be obtained from terrestrial experiments with radioactive beams.

In the next section we will discuss isospin effects on the low density region of the nuclear matter phase diagram.

2.2 Analytical analysis

In the context of multifragmentation the instability of NM (the spinodal region, where the system becomes unstable against long wave-length but small amplitude fluctuations) plays a crucial role. These aspects were discussed repeatedly in the past, [4,5,6,7,8] but the binary character will induce new features, as discussed in the following. In the framework of the Landau theory for two component Fermi liquids the spinodal border was determined by studying the stability of the collective modes obtained solving two coupled Landau-Vlasov equations for protons (p) and neutrons (n). In terms of the appropriate Landau parameters F_0, the stability condition can be expressed as (see ref.[9]),

$$(1 + F_0^{nn})(1 + F_0^{pp}) - F_0^{np} F_0^{pn} > 0 \; . \qquad (1)$$

It is possible to show that this condition is equivalent to the following thermodynamical condition [10]:

$$\left(\frac{\partial P}{\partial \rho}\right)_{T,y} \left(\frac{\partial \mu_p}{\partial y}\right)_{T,P} > 0 \; . \qquad (2)$$

discussed in [11,12], where μ_p is the proton chemical potential and y is the proton fraction. In Fig. 2 we show the spinodal line obtained from eq. (1) (continuous line with dots) which for asymmetric nuclear matter ((b) and (c)) is seen to contain the line corresponding to "mechanical instability", $\left(\frac{\partial P}{\partial \rho}\right)_{T,y} < 0$ (crosses). We want to stress that the above stability condition (eq.(2)) does

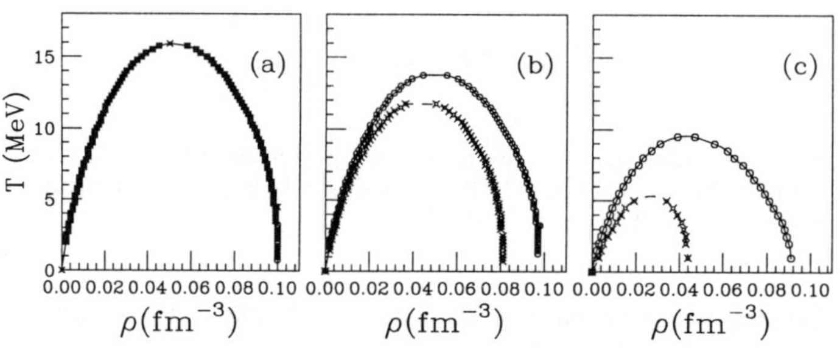

Figure 2. Spinodal lines corresponding to chemical (circles) and mechanical (crosses) instability for three values of proton fraction y. Panel (a) corresponds to symmetric nuclear matter.

not tell the nature of the fluctuations against which a binary system becomes chemically unstable. Indeed, the thermodynamical condition $\left(\frac{\partial \mu_p}{\partial y}\right)_{T,P} < 0$ cannot distinguish between two very different situations that can be encountered in nature: an attractive interaction between the two components of the mixture ($F_0^{np}, F_0^{pn} < 0$), as in the case of nuclear matter at low density, or a repulsive interaction between the two species.

Then it is possible to prove [10], based on a thermodynamical approach of asymmetric Fermi liquid mixtures, that chemical instabilities are triggered by isoscalar-like fluctuations (neutrons and protons move in phase) in the first, i.e. attractive, situation and by isovector fluctuations (neutrons and protons move out of phase) in the second one. In asymmetric nuclear matter the phase transition is thus due to isoscalar-like fluctuations that induce chemical instabilities, while the system is never unstable against isovector fluctuations. The chemical character of the instability rests on the fact that, in asymmetric NM, isoscalar-like fluctuations lead to a more symmetric high density phase, since neutrons and protons move in phase but with different amplitude. However, it should be noticed that this isospin fractionation effect happens also for asymmetric unstable systems prepared inside the mechanical instability region. Actually we observe a smooth transition from chemical to mechanical instabilities, as we can see from Fig.3, where we plot the mixing angle β ($tg\beta = \nu_n/\nu_p$, where ν_n, ν_p are neutron and proton density distribution oscillations) and the function χ related to the isospin variation δI of the liquid phase, respectively ($\delta I = \nu_p[\chi - 1]$), as a function of density. We note that χ always keeps values smaller than 1, leading to $\delta I < 0$ and hence to a more symmetric liquid phase.

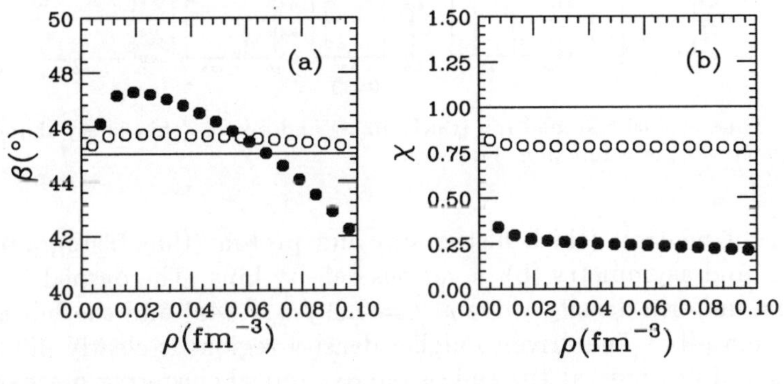

Figure 3. Density dependence of the mixing angle β, (a), and of the function χ (b) for three proton fractions, $y = 0.5$ (solid), $y = 0.4$ (circles), and $y = 0.1$ (full circles) at T=1MeV.

2.3 Numerical results: heated nuclear matter in a box

Numerical simulations have been performed in order to follow all stages of the fluctuation growth in the fragment formation process. In the numerical approach we consider nuclear matter in a box of size $L = 24\ fm$, imposing periodic boundary conditions.

We follow a phase space test particle method to solve the Landau-Vlasov dynamics (using Gaussian wave packets) [9,13,14]. An initial temperature is introduced by distributing the test particle momenta according to a Fermi distribution. We have followed the space-time evolution of test particles initialized at asymmetry I=0.5, density $\rho^{(0)} = 0.09 fm^{-3}$ and T=5 MeV. In such a way we start with a system prepared inside the chemical instability region (see Fig.2). The initial density perturbation was created automatically due to the random choice of the test particle positions. We report in Fig.4 the time

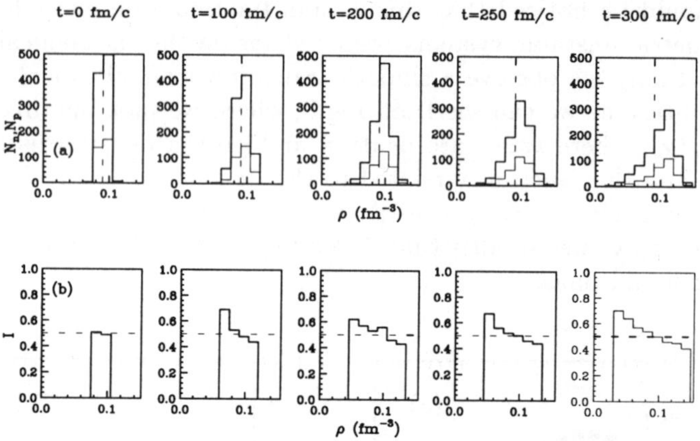

Figure 4. Time evolution of neutron (thicks lines) and proton (thin lines) abundance (a) and asymmetry (b) as a function of density.

evolution of neutron (thick histogram) and proton (thin histogram) abundance (a) and asymmetry (b) in various density bins. The dashed line shows the initial uniform density value $\rho^{(0)} = 0.09 fm^{-3}$ (4a) and the initial asymmetry I=0.5 (4b). The drive to higher density regions is clearly different for neutrons and protons: at the end of the dynamical clustering mechanism we have different values of the asymmetry in the liquid and in the gas phase (see the panel at 300 fm/c). This result is the same of ref.[9], where the dynamics of mechanical instabilities was studied, demonstrating that the nature of fluctuations associated with chemical and mechanical instabilities is essentially the same.

3 Central heavy ion collisions

We have studied the reaction $^{124}Sn +^{124} Sn$ at 50 MeV/A, where new data are under analysis at $NSCL - MSU$ [15]. The study of the dynamical evolution of the reaction has been performed considering a stochastic extension of the Landau-Vlasov dynamics. So we will consider stochastic mean-field calculations, where isospin and fluctuation effects are suitably accounted for [8,14,16]. We will comment first around 500 events generated in semi-central collisions ($b = 2fm$). In this case the reaction mechanism corresponds to a bulk fragmentation. We can identify three main stages, each characterized by very particular features of the isospin dynamics, since the system explores different density regions. After the compression phase (until 60-70fm/c) follows a fast expansion (until 110-120fm/c) and then during the last stage the systems breaks up into pieces (see the time evolution of the IMF multiplicity in Fig.5). The values of density and temperature calculated at the beginning of the third stage correspond to situations inside the unstable region of the NM phase diagram. Volume instabilities have time to develop and we expect a kinetic mechanism of formation of the liquid phase of the spinodal decomposition type. At $260fm/c$, we observe a saturation in the time dependence of the average number of produced fragments, hence this time would correspond to the freeze-out configuration.

The results obtained with the *asy-stiff* symmetry term are shown in Fig.5, while the *asy-soft* results are presented in Fig.6. We performed for some quantities a separate investigation in a central region having a linear dimension of 20 fm centered around the CM of the system. Since it corresponds to the active volume in which fragmentation takes place, in this way we can obtain more detailed information about this process. We define as belonging to the liquid phase the regions with density higher than 0.03 fm^{-3}; regions with lower density belong to the gas phase.

Each figure is organized in this way:

Left column, from top to bottom: Time evolution of: *Mass* in the liquid (up) and gas (down) phase ; *Asymmetry* $I = (N-Z)/(N+Z)$ in the "central" gas (squares joined by solid line), and total gas (squares joined by dashes), "central" liquid (circles joined by solid line) and IMF (clusters with $3 < Z < 23$, stars). The horizontal line shows the initial average asymmetry; *Mean Fragment Multiplicity* $Z \geq 3$. The saturation of this curve defines the freeze-out configuration, as discussed also before.

Right column, properties of the "primary" fragments at the *Freeze-Out Configuration*, from top to bottom: *Charge Distribution*, *Asymmetry Distribution* and *Fragment Multiplicity Distribution* (normalized to 1).

We observe a quite different behaviour, concerning expecially the IMF isotopic content, depending on the EOS one uses. More asymmetric fragments are formed in the stiff case. This can be nicely understood by looking at the behaviour of the symmetry potential (Fig.1). At low density larger

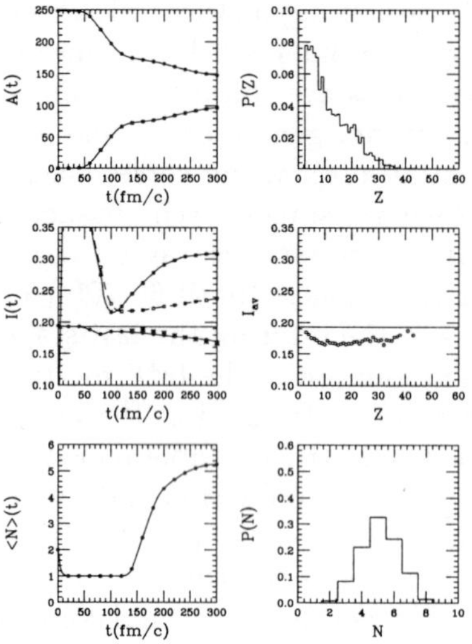

Figure 5. $^{124}Sn +^{124} Sn$ $b = 2fm$ collision: time evolution and freeze-out properties. See text. *ASY-STIFF* EOS

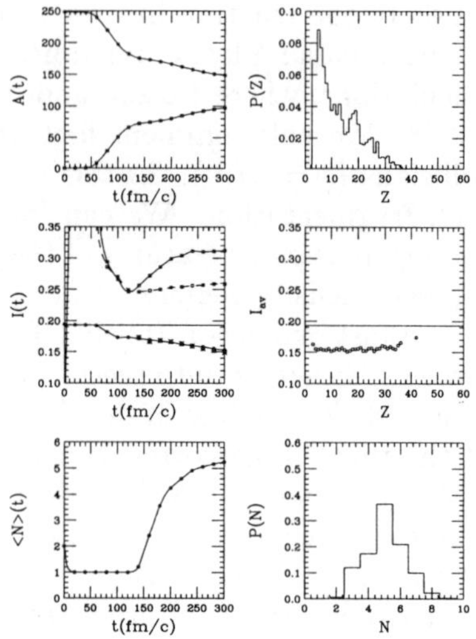

Figure 6. $^{124}Sn +^{124} Sn$ $b = 2fm$ collision: time evolution and freeze-out properties. See text. *ASY-SOFT* EOS.

isospin effects are expected in the soft case since the symmetry energy and its derivative are larger than in the stiff case. Hence the isospin fractionation mechanism is more effective in the soft case leading to a more symmetric liquid phase. This kind of information can be used to disentangle between the different trends proposed for the symmetry energy; In fact, the isotopic content of IMF's, once the secondary de-excitation process has been taken into account, can be directly compared to data.

3.1 Semi-peripheral collisions

To illustrate the dynamics of semi-peripheral collisions we have considered the reaction $^{124}Sn +^{64} Ni$ at 35 MeV/A, b= 6 fm. This experiment has been recently performed at LNS by the REVERSE collaboration [17]. For $b = 6fm$ (semiperipheral collision) we observe a quite different fragmentation mechanism with respect to the central collisions. Now in the overlap region a neck structure is developing. During the interaction time it heats and expands but still remains in contact with denser and colder regions of projectile like (PLF) and/or target like (TLF) type and we are dealing essentially with surface instabilities. In some events small fragments may originate from this structure, while in other cases, this region can be re-absorbed by the PLF and/or the TLF.

In Fig.7-8 we present, for semi-peripheral reactions, the same kind of analysis as illustrated above (Fig.5-6). Now we observe that small IMF's are quite more neutron-rich than the larger ones. In fact they mostly originate from the neutron rich neck region. However, there is a clear difference between the asy-stiff and asy-soft predictions. In the soft case the effect is quite reduced. This can be again nicely related to the behaviour of the symmetry energy (Fig.1), but around normal density, since we are facing now surface rather than volume instabilities. Now the larger value of the symmetry energy in the stiff case is responsible for a larger migration of neutrons towards the less dense neck region, that will become more neutron rich.

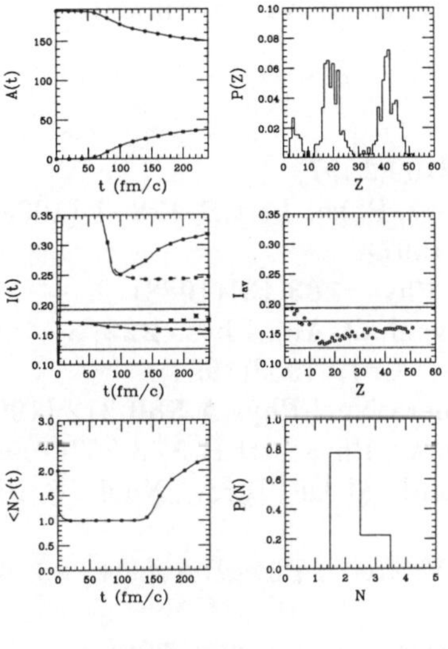

Figure 7. $^{124}Sn +^{64} Ni$ $b = 6fm$ collision: time evolution and freeze-out properties. See text. *ASY-STIFF* EOS

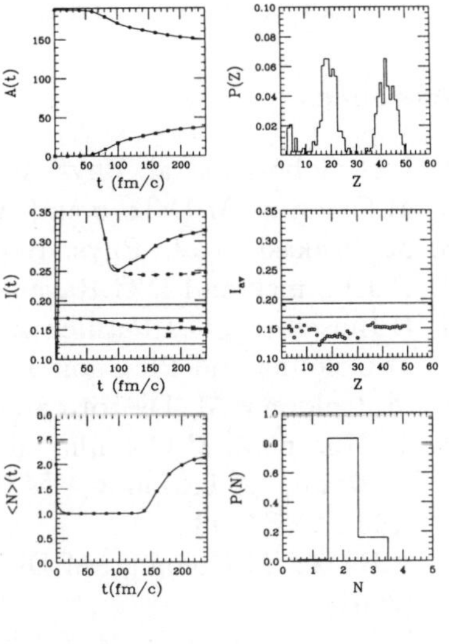

Figure 8. $^{124}Sn +^{64} Ni$ $b = 6fm$ collision: time evolution and freeze-out properties. See text. *ASY-SOFT* EOS.

4 Conclusions

We have discussed new features of fragmentation mechanisms occurring in asymmetric nuclear matter. A new chemical component in the spinodal region is expected, with observable signatures on the isotopic content of the dynamically produced fragments. This effect will be enhanced in the overlapping zone of semi-peripheral collisions (the neck region) where larger shall be the NM asymmetry.

Starting from simulations of reaction dynamics performed with a new transport code where isospin and fluctuation effects are suitably accounted for, we have shown that dissipative heavy ion collisions at medium energies are rather sensitive to the density dependence of the symmetry contribution to the nuclear EOS.

We would like to stress that these effects can be clearly seen also with "exotic" but not radioactive beams.

A possibility is emerging of obtaining in terrestrial accelerator laboratories important information on the symmetry term of large astrophysical interest. It appears essential to have good charge asymmetric beams available at intermediate energies and to perform more exclusive experiments.

References

1. L.G. Sobotka et al.,Phys. Rev. C **55**,2109(1997).
2. M.Colonna, M.DiToro and A.B.Larionov, Phys. Lett.B **428**, 1 (1998).
3. M. Prakash et al., Phys. Rep. **280**,1 (1997).
4. C.J.Pethick and D.G.Ravenhall, Ann. Phys. **183**,131(1988).
5. H.Heiselberg, C.J.Pethick and D.G.Ravenhall, Ann.Phys. **223**, 37(1993).
6. M.Colonna and Ph.Chomaz, Phys. Rev. C **49**,1908(1994).
7. M. Colonna, M. Di Toro and A. Guarnera,Nucl.Phys.A **580**,312 (1994).
8. A. Guarnera, M.Colonna and Ph.Chomaz, Phys.Lett.B **373**,267(1996).
9. V.Baran, A.Larionov, M.Colonna and M.Di Toro, Nucl. Phys. **A632**,287(1998).
10. V.Baran, M.Colonna, M.Di Toro and V.Greco, Phys.Rev.Lett. 86, 4492 (2001).
11. L.D. Landau and E.M. Lifshitz, *Statistical Physics*, pag. 208.
12. H.Mueller and B.D.Serot, Phys. Rev.C **52**,2072(1995).
13. TWINGO code: A.Guarnera Ph.D.Thesis, Univ.Caen , July 1996
14. M.Colonna, M.DiToro, G.Fabbri and S.Maccarone, Phys.Rev. **C57**, 1410(1998).
15. H.S.Xu et al., Phys.Rev.Lett.85, 816(2000).
16. M.Colonna et al., Nucl.Phys. **A642**, 449(1998).
17. A.Pagano et al., Nucl. Phys. A 681, 331c(2001.

GROUND-STATE PROPERTIES OF NEUTRON-RICH NUCLEI IN THE SD-SHELL

D. CORTINA-GIL[1,2],J. FERNANDEZ-VAZQUEZ[1], K. MARKENROTH[3], F. ATTALLAH[2], T. BAUMANN[4], J. BENLLIURE[1], M.J.G. BORGE[5], L. CHULKOV[2,6], C. FORSSÉN[3], L.M. FRAILE[5], H. GEISSEL[2], J. GERL[2], K. ITHASHI [7], R. JANIK[8], B. JONSON[3], S. KARLSSON[3], S. MANDAL[2], M. MEISTER[3], X. MOCKO[8], T. OHTSUBO[2], A. OZAWA[9], V. PRIBORA[6], K. RIISAGER[10], G. SCHNEIDER[11], H. SCHEIT[12], G. SCHRIEDER[13], B. SITAR[8], A. STOLZ[11], P. STRMEN[8], K. SÜMMERER[2], X. SZARKA[8], S. WAN[2], H. WEICK[2]

[1] *Universidad de Santiago de Compostela, Spain,* [2] *GSI Darmstadt, Germany, Chalmers Tekniska Högskola, Göteborg, Sweden,* [4] *NSCL, MSU, USA,* [5] *IEM CSIC, Madrid, Spain,* [6] *Kurchatov Institute, Moscow, Russia,* [7] *Tokyo Institute of Technology, Japan,* [8] *Comenius University, Bratislava, Slovakia,* [9] *RIKEN, Japan,* [10] *Aarhus Universitet, Denmark,* [11] *TU Muenchen, Germany,* [12] *Max-Planck-Institut, Heidelberg, Germany,* [13] *TU Darmstadt, Germany*

The fragment separator FRS(GSI) was used to measure longitudinal momentum distributions and one-nucleon removal cross-sections of relativistic fragments after breakup for different secondary beams in coincidence with gamma rays from the de-excitation of those fragments. The combination of in-beam gamma spectroscopy at relativistic energies with secondary nuclear reactions has added new information that will help us to better understand the ground state properties (definite assignment of spins and parities of the ground state) of the exotic nuclei . The proton-rich isotope ^{8}B is presented as a test case to show the power of the experimental technique. New experimental results on momentum distributions and one-nucleon removal cross-sections in coincidence with gamma ray detection for N, O and F isotopes will also be discussed.

Drip-line studies have proved the existence of nuclear halo states. However, the existence of proton halos is a more controversial topic. The presence of Coulomb potential for the halo proton confine the wave function to the nuclear interior. In this paper, the case of ^{8}B, yet the only confirmed proton halo nucleus, will be examined with new experimental data from knock-out reactions, which include coincidences between γ rays and ^{7}Be fragments.

We have recently extend these studies to heavier neutron-rich isotopes in the sd-shell region. The preliminary results obtained until now could be useful to understand the evolution of nuclear structure when we approach the drip-line and to extract detailed structural information for those until now not very well known nuclei.

The secondary beams were produced by nuclear fragmentation of relativistic stable beams (^{12}C and ^{40}Ar) at 1 GeV/nucleon from the heavy ion synchrotron (SIS) at GSI. A thick Be target was placed at the entrance of the FRS [1]. The first half of the FRS was set to transport the secondary beams to the intermediate focal plane F2, where the breakup reaction on a carbon target, was studied. The second half of the FRS was set to the magnetic rigidity of the fragments arising from one-nucleon knock-out reactions. Both the secondary nuclei arriving at the intermediate focal plane(F2) and the fragments arising from one-nucleon knock-out reaction at the final focus(F4) were fully identified and the emitted γ detected in coincidendence using 32 NaI detectors placed behind the breakup target at the intermediate focus(F2).

We have measured in this experiment the longitudinal momentum distribution and the corresponding one-proton removal cross-section of ^{7}Be fragments after ^{8}B knock-out. These results will not be discussed here, we refer the interested reader to [2,3,4]. We will directly present the coincidence technique and the procedure we follow to deduce the ground state configuration of ^{8}B.

In order to disentangle between the different ^{7}Be contributions (g.s or excited state) to the ground state of ^{8}B, we set a gate in the gamma spectrum around the observed peak at 429 keV (^{7}Be($1/2^-$) $\rightarrow$ ^{7}Be($3/2^-$)). Under this condition we obtain the coincidence between the core longitudinal momentum distribution and the γ-s in the peak.

This contribution could be then corrected for the neutron background and for the estimated efficiency of our detector (3%) and the contribution to the total ^{8}B wave function from ^{7}Be core excite contribution was recovered. The last step was to subtract this core excited contribution from the total longitudinal momentum distribution and in this way obtain the ground state ^{7}Be contribution to the ^{8}B wave function. The total longitudinal momentum and the one-proton removal cross-section agree with our previous measurements and confirm the proton halo character of ^{8}B. The new "coincidence" results show the success of this technique at relativistic energies at the same time that the relative ratio between the two components of the ^{7}Be to the ^{8}B$_{g.s}$ wave function could be evaluated. This relative ratio allowed to normalized each contribution to the measured one-proton removal cross-section.

The final distributions are shown in Fig. 1 for the breakup in a C target. The errors assigned to the FWHM's take into account statistical errors, intrinsic resolution of the spectrometer and location straggling in the target. The relative weight of the ^{7}Be($1/2^-$) has been measured for the first time at relativistic energies and shows an excellent agreement with theoretical

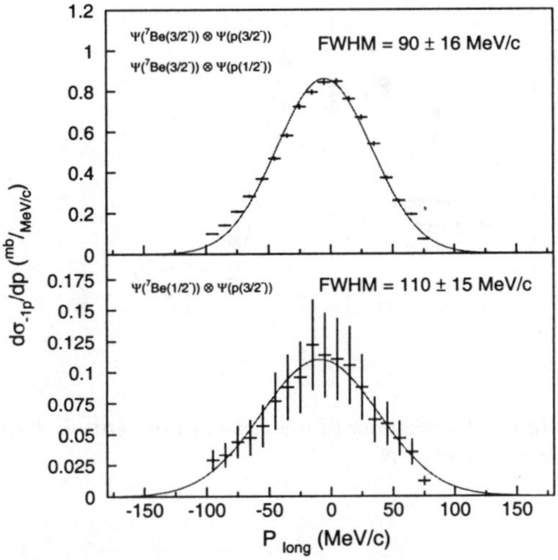

Figure 1. *Longitudinal momentum distributions of* ^7Be *fragments after one-proton removal of* ^8B *on a carbon target. The top picture corresponds to* $^7Be_{g.s.}$ *and the bottom one to* $^7Be_{exc.}$ *contribution to the* ^8B *g.s wave function.*

calculation [5]

We have extended these studies to heavier n-rich isotopes in the sd shell. The first step has been to obtain the total longitudinal momentum distributions for a large number of n-rich isotopes. The main results are shown in Fig 2 that represents the FWHM of the measured fragment longitudinal momentum distribution as a function of the neutron number of the preojectile (N).

We clearly observed that the FWHM for the lighter isotopes is larger than the one obtained for the heavier ones. This trend is easy to understand, when the neutron number N reaches the value 14 the system has enough neutrons to start the filling of the 2s1/2 shell. This will mean that the radial wave function exhibited by those states(l=0) will extend far away compared with others occupying a 1d5/2 orbit(l=2), and in consequence the longitudinal momentum distribution will be narrower. However, it is surprising and still not yet understood the fact that the d-behavior is not recovered when the system reaches N=17 (^{26}F). It will be interesting to extend these studies to

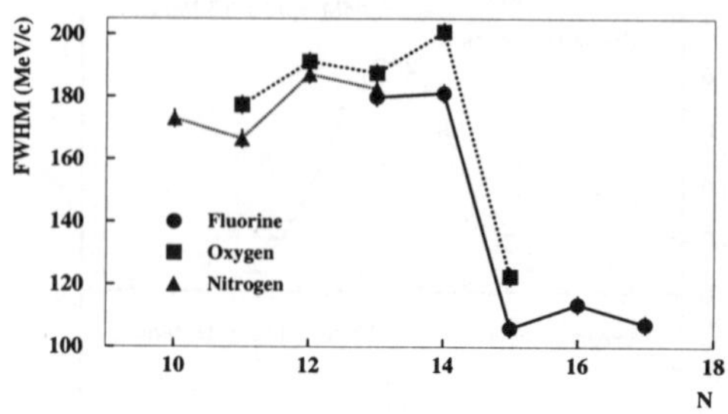

Figure 2. *FWHM evolution of the fragment longitudinal momentum distributions measured as a function of the neutron number N.*

the F isotopic chain to confirm this trend and to study the evolution of magic numbers when we approach the drip-line beyond N=20.

The next step would be to exploit the coincidence technique in order to get quantitative information. We will only comment on the preliminary results obtained for ^{23}O. It is still an open question if this nucleus exhibit a neutron halo structure in its ground state. From the narrow width of the total ^{22}O longitudinal momentum distribution one could deduce a strong mixture of s-states. The γ-spectra obtained in coincidence with the ^{22}O allow us to disentangle the coincidence between the contribution of ^{22}O in different excited states to the ground state one. The longitudinal momentum distributions obtained for all the excited states exhibit a large width (larger than the total) that has been associated to d-wave orbits. In conclusion the component of ^{23}O built on the $^{22}O_{g.s}$ will be associated to an s-vawe but the data analysis is still in progress and the complete spectroscopic information is not yet available.

References

1. H. Geissel et al., Nucl. Instr. and Methods B70 (1992) 286
2. M.H. Smedberg et al., Phys. Lett. B452 (1999) 1
3. D. Cortina-Gil Eur. Phys. Jour A (2001).
4. D. Cortina-Gil In preparation.
5. Y. Paferniova and M. Zuhkov, private comunication.

CLUSTER STATES IN BE ISOTOPES

P. DESCOUVEMONT
Physique Nucléaire Théorique et Physique Mathématique,
Université Libre de Bruxelles, CP229, B-1050 Brussels, Belgium
E-mail: pdesc@ulb.ac.be

The ^{9,10,11}Be nuclei are described in a microscopic multicluster model involving $\alpha+\alpha+n$, $\alpha+\alpha+n+n$ and $\alpha+\alpha+n+n+n$ configurations, respectively. The effect of α clustering is shown to decrease from ^{9}Be to ^{11}Be. A discussion of the band structure is presented. In ^{11}Be, the existence of a $3/2^-$ band, suggested recently, is not supported by the model.

1 Introduction

Clustering is a well known effect in many light nuclei [1]. In particular, the role of the α particle is well established and is at the origin of most cluster models [2]. Recent work using the antisymmetrized molecular dynamics (AMD) concludes [3,4,5] that α clustering shows up in light systems without assuming this cluster structure a priori. Many models have been developed to investigate cluster states in light nuclei. The idea of the AMD is essentially based on the molecular model developed by Okabe and Abe [6] and by Seya *et al.* [7] to analyze systems composed by two α particles and external neutrons or protons. More recently, microscopic multicluster models have been applied to the spectroscopy of p shell nuclei [8,9].

The cluster structure of Be isotopes has attracted much attention in recent years. An analysis of available experimental data by von Oertzen [10] suggests different rotational bands in Be isotopes where α clustering should play a dominant role. This result received further experimental support from Bohlen *et al.* [11] who analyzed high excited states of ^{11}Be and find evidence for a $3/2^-$ rotational band with spins up to $J = 19/2^-$.

2 The microscopic model

In a microscopic model, all nucleon coordinates (space, spin and isospin) are included in the hamiltonian, which essentially depends on an effective nucleon-nucleon interaction. We use the Minnesota [12] (hereafter referred to as MN) and Volkov $V2$ [13] potentials as central parts. The Coulomb force and a zero-range spin-orbit interaction are also introduced. We refer to Ref. [14] for details.

To solve the Schrödinger equation associated with the A-body Hamiltonian, we use the multicluster Generator Coordinate Method [8,15]. The Be wave functions are defined by two α particles and external neutrons (see Fig. 1). The distances between the centres are the generator coordinates and define the basis states. As shown in Fig.1, the number of free parameters is 3, 6 or 9 for ^{9}Be, ^{10}Be and ^{11}Be, respectively.

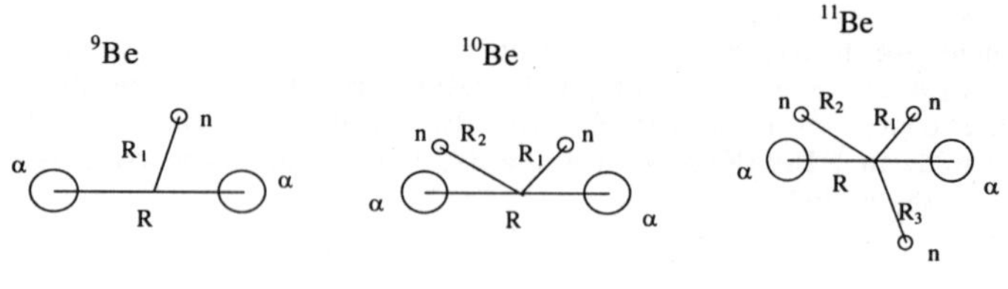

Figure 1. Cluster structure of Be isotopes.

3 Spectroscopy of Be isotopes

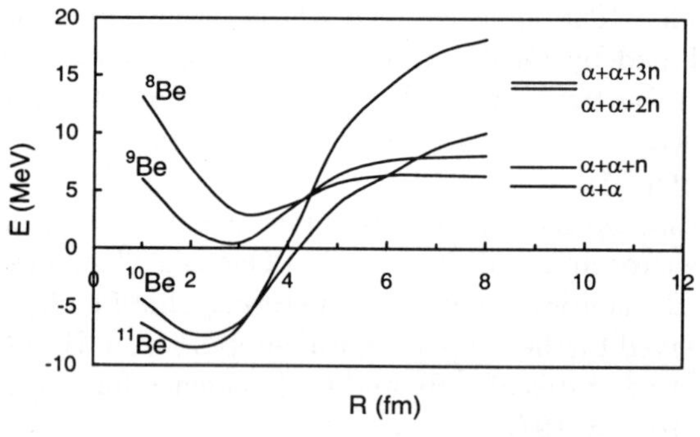

Figure 2. Energy curves of the ^{8,9,10,11}Be ground states. The right side of the figure shows the breakup channels (including the $\hbar\omega/4 = 5.61$ MeV c.m. kinetic energy).

We define the energy curves as the energy of the system for a fixed value of the $\alpha + \alpha$ distance (generator coordinate R). Diagonalization is performed with respect to the other generator coordinates. Energy curves are quite useful in two-cluster calculations, where they bring information on the structure of the system, such as clustering effects or level ordering. In the present multicluster model, they give information regarding α clustering in Be isotopes.

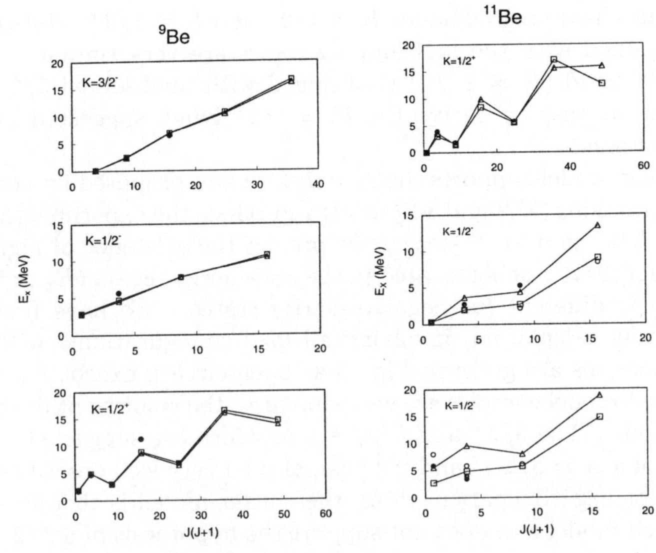

Figure 3. Band structure of ^{9}Be and ^{11}Be. The MN and V2 results are given as squares and triangles, respectively. For ^{11}Be, open circles represent a shell-model calculation.

In Fig. 2, we gather the energy curves of the ground states, and we also consider the ^{8}Be energy curve. On the right side of Fig. 2, we present the different breakup thresholds where the residual kinetic energy has been added. It is clear that the $\alpha + \alpha$ clustering decreases from ^{8}Be to ^{11}Be and that, simultaneously, the breakup threshold energy increases. Both effects are somewhat correlated.

For ^{9}Be and ^{10}Be, the r.m.s. radius is 2.49 (2.36) fm and 2.40 (2.26) fm with the V2 and MN potentials, respectively; the difference between theory and experiment (2.45 $\pm$ 0.01 fm and 2.30 $\pm$ 0.02) is weak. However the ^{11}Be radius [2.39 (2.29) fm] is strongly underestimated by the model (experiment gives 2.73 $\pm$ 0.05 fm). It is well known that the large experimental value is explained by the weak binding energy of ^{11}Be with respect to the ^{10}Be+n channel [16]. This property provides a large amplitude of the wave function at large distances, and therefore enhances the r.m.s. radius. Our aim here is essentially to investigate the band structure and the existence of high-spin states. A more precise description of ^{11}Be low-lying states would require an explicit treatment of the ^{10}Be+n channel [16].

The ^{9}Be and ^{11}Be spectra are presented in Fig. 3. We refer to Ref. [14] for a discussion of the ^{10}Be spectrum. For ^{9}Be, in addition to the ground-state band $K = 3/2^-$, we also have excited bands $K = 1/2^-$ and $K = 1/2^+$. Notice that the energies obtained with the MN and V2 forces are very similar. A second negative-parity band ($K = 1/2^-$) is obtained with unobserved $5/2^-$ and $7/2^-$ resonances. In positive parity, the $K = 1/2^+$ band, suggested by von Oertzen [10] is confirmed.

In ^{11}Be, the present model supports the $K = 1/2^+$ band proposed by von Oertzen [10]. Both interactions (MN and V2) nicely reproduce the experimental energies of the $1/2^+$, $3/2^+$ and $5/2^+$ states. We predict the existence of high spin states, one of them ($9/2^+$) being located in the low-energy region ($E_x \simeq 5$ MeV). The situation is different for negative-parity states. We have first performed a shell-model calculation, involving all $0\hbar\omega$ configurations, with the MN potential. Energies are given in Fig. 3 as open circles; except for a second $1/2^-$ state, shell-model energies are very similar to the results obtained in the multicluster model. The $3/2_1^-$ and $5/2_1^-$ states which are suggested to be the first members of a $K = 3/2^-$ band [10,11] can also be very well described by the shell-model. For negative-parity states, the cluster model is therefore very similar to the shell-model and does not support the hypothesis of a $3/2^-$ band [11].

References

1. H. Furutani *et al.*, Prog. Theor. Phys. Suppl. **68**, 193 (1980).
2. D. Brink, Proc. Int. School "Enrico Fermi" 36, Varenna 1965, Academic Press, New-York (1966) 247.
3. Y. Kanada-En'yo, H. Horiuchi and A. Ono, Phys. Rev. **C52** 628, (1995).
4. Y. Kanada-En'yo, H. Horiuchi and A. Doté, J. Phys. **G24**, 1499 (1998).
5. Y. Kanada-En'yo, H. Horiuchi, A. Doté, Nucl. Phys. **A687**, 146c (2001).
6. S. Okabe and Y. Abe, Prog. Theor. Phys. **61**, 1049 (1971).
7. M. Seya, M. Kohno and S. Nagata, Prog. Theor. Phys. **65**, 204 (1981).
8. M. Dufour and P. Descouvemont, Nucl. Phys. **A605**, 160 (1996).
9. Y. Ogawa *et al.*, Nucl. Phys. **A673**, 122 (2000).
10. W. von Oertzen, Z. Phys. **A357**, 355 (1997).
11. H.-G. Bohlen *et al.*, Prog. Part. Nucl. Phys. **42**, 17 (1999).
12. D.R. Thompson *et al.*, Nucl. Phys. **A286**, 53 (1977).
13. A.B. Volkov, Nucl. Phys. **74**, 33 (1965).
14. P. Descouvemont, to be published.
15. K. Arai *et al.*, Phys. Rev. **C54**, 132 (1996).
16. P. Descouvemont, Nucl. Phys. **A615**, 261 (1997).

CHANGES IN NEUTRON MAGIC NUMBERS OF NEUTRON-RICH NUCLEI

Z. Dlouhý, D. Baiborodin, J. Mrázek, G. Thiamová
Nuclear Physics Institute ASCR, CZ-250 68 Řež, Czech Republic
E-mail: dlouhy@ujf.cas.cz
For the GANIL-Orsay-Dubna-Řez-Bucharest collaboration

We present a survey of experimental results obtained at GANIL (Caen, France) resulting in an appearance of a new magic number, N=16, in very neutron-rich nuclei. Two neutron separation energies derived from recent mass measurements of neutron-rich nuclei together with the measurements of instability of the doubly magic nucleus ^{28}O give a very clear evidence for the existence of the new shell closure N=16. This shell appears between $2s_{1/2}$ and $1d_{3/2}$ neutron orbitals for neutron-rich nuclei from carbon to neon.

Deformations, shape coexistence or variations in the spin orbit strength as a function of N/Z ratio could result in the modification of magic numbers in very neutron-rich nuclei [1]. A breaking of magicity has already been observed at the N=20 shell closure where an "island of inversion" in shell ordering has been shown to exist [2]. Such behaviour has very wide consequences that have resulted in the instability of the doubly magic nuclei ^{10}He [3,4] and ^{28}O [5]. The anomalous behaviour of the binding energy near the shell closure N=20 close to the neutron-drip line is also closely connected to this question. More recently, the determination of the lifetime and of the deformation of ^{44}S has indicated the existence of a similar effect at N=28.

Recently, in an experiment on the LISE3 spectrometer at GANIL, we have used the fragmentation of the neutron-rich projectile ^{36}S to produce and study very neutron-rich nuclei in vicinity of a doubly magic nucleus ^{28}O. However, no events were observed corresponding to even ^{26}O and ^{28}O, odd oxygen isotopes ^{25,27}O and also ^{24,25}N. Till now, the heaviest experimentally found oxygen isotope remains ^{24}O. Our finding that ^{28}O is particle unstable [5] fairly supports the idea that the onset of the deformation found in the Ne-Al region causes the breaking of magicity of the N=20 shell closure in ^{28}O. The instability of ^{25,26,27,28}O and ^{24,25}N was also confirmed by the experiment performed by Sakurai et al. [6] at the fragment separator RIPS at RIKEN in which, however, a new isotope, ^{31}F, was observed for the first time. The calculated and observed yields of isotopes are in a good agreement and provide a strong evidence for the particle instability of ^{24,25}N, ^{25,26,27,28}O and ^{30}F.

We can summarize that the heaviest experimentally found isotopes of carbon, nitrogen and oxygen [5,6] are ^{22}C, ^{23}N and ^{24}O, respectively, with the same

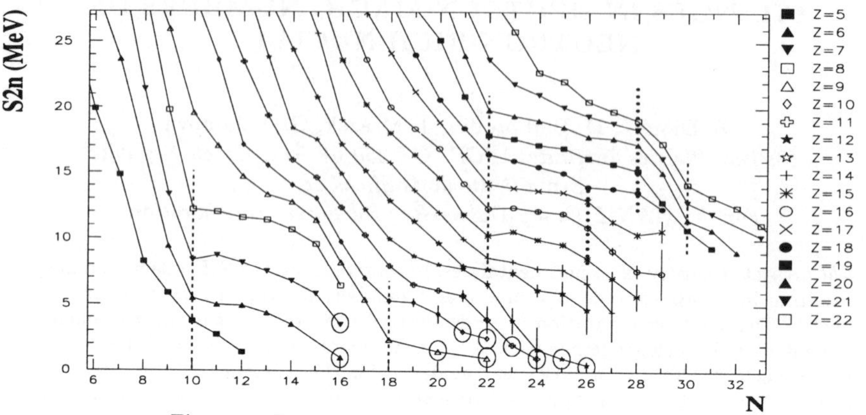

Figure 1: Two-neutron separation energy S_{2n} versus N

neutron number, N=16, while the heaviest isotope of fluorine was found to be ^{31}F with N=22. It should be noted that it is a rather interesting behaviour among the light nuclei. Usually, in the region further away from the shell closure the neutron numbers of heaviest isotopes of neighbour elements are gradually increasing with Z. Therefore, the sudden step in the largest neutron number from N=16 for carbon, nitrogen and oxygen to N=22 for fluorine may correspond to a substantial change in shell structure.

The particle stability of nuclei is directly related to the masses and nuclear binding energies, which are very sensitive to the existence of shells and may provide clear signatures of shell closures. An experiment on mass measurement of neutron-rich nuclei using a direct time of flight technique was undertaken by Sarazin et al. [7] in order to investigate the N=20 and N=28 shell closures for nuclei from Ne (Z=10) to Ar (Z=18) and thus to bring some clarifications concerning the behaviour of magic numbers far from stability. The nuclei of interest were produced by the fragmentation of a 60 AMeV ^{48}Ca beam on a Ta target. The two-neutron separation energies S_{2n} derived from the measured masses are displayed in Fig.1. The new data [7] are presented with error bars while the others, except the encircled data, are taken from Audi and Wapstra[8]. The Ca, K and Ar isotopes show a behaviour typical for the filling of shells, with the two shell closures at N=20 and N=28 being evident at the corresponding sharp decrease of S_{2n} for the next two isotopes and a moderate decrease of S_{2n} for subsequent points as the filling of the next shell starts to influence S_{2n}. The sharp drop at N=22, shown by the dashed vertical line and corresponding to the shell N_{sh}=20 is clearly visible through all the Si-Ca region, while going

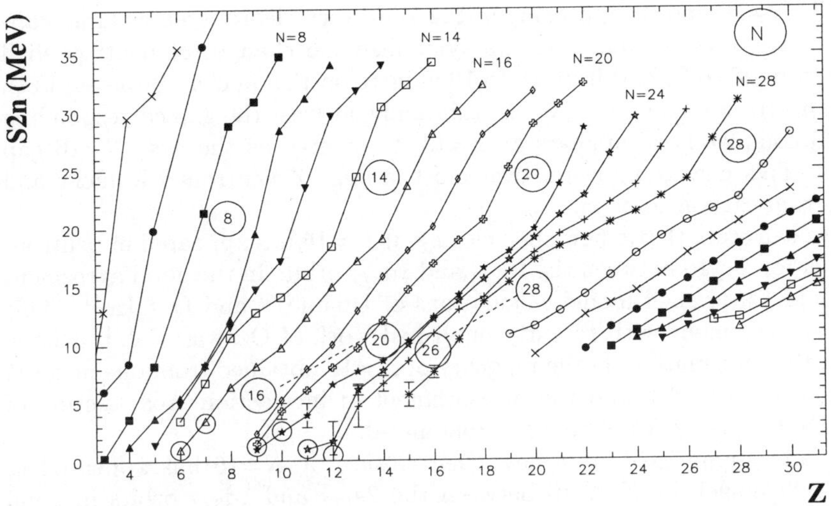

Figure 2: Experimental S_{2n} values versus proton number Z. Dashed lines symbolize the changes of neutron shell closures.

to lower Z to the Al-Na region this drop seems to move towards the lower N.

This was the reason why we made an attempt to clarify the situation of two-neutron separation energies in this region. We used the fact that several particle stable nuclei [5,6,9] were found to exist in this region, however, their masses are not known yet. Nevertheless, their S_{2n} values must be positive and therefore, we included the expected S_{2n} values of the heaviest particle stable isotopes ^{23}N, ^{22}C and ^{29,31}F as well as some others to the graph, they are marked by circles. The inclusion of the S_{2n} values for ^{29}F and ^{31}F was most important, because this allowed us to observe the sharp drop of ^{27}F value followed by a moderate decrease of S_{2n} values for ^{29}F and ^{31}F giving a very clear evidence for the existence of the new shell closure at N=16 for fluorine. A similar behaviour confirming the N=16 shell closure can be seen at the neon isotopes that exhibit a moderate decrease of S_{2n} values for ^{29}Ne and ^{30}Ne. We have already mentioned that in the Al-Na region the sharp drop in S_{2n} values seems to move towards larger N with increasing Z. Now, we can make the firm conclusion that it stabilizes for F and Ne at the vertical line at N=18 (new shell N_{sh}=16).

It should be noted that the evidence for a new magic number N=16 follows also from Fig.2 where the S_{2n} values are plotted versus atomic number Z. The position of various possible shells or pseudo-shells are also shown in the figure.

The shells N=20 and 28 appearing in Fig.1 are very clearly seen as large gaps in Fig.2. The dashed lines in Fig.2 symbolize the changes of neutron shell closures from 28 to 26 and from 20 to 16 in neutron-rich nuclei. However, both gaps are narrowing going to lower Z, till finally, at least the gap corresponding to N=20 completely disappears at Z=13, to emerge as the new N=16 gap at Z=10. This new gap governs over most of light Z neutron-rich nuclei and extends from carbon to neon.

So we can state that a new shell closure at N=16 has appeared in neutron-rich nuclei for $Z \leq 10$ between the $2s_{1/2}$ and $1d_{3/2}$ orbits in the good agreement with Monte Carlo shell model calculations of Utsuno [10] and Otsuka [11]. This is in good agreement with recently published work of Ozawa et al. [12] where, however, the information on the magicity of N=16 obtained from one neutron separation energies S_{1n} and the measurement of interaction cross section of radioactive beams σ_I is not so well pronounced.

We can summarize that a new shell closure at N=16 has appeared in neutron-rich nuclei for $Z \leq 10$ between the $2s_{1/2}$ and $1d_{3/2}$ orbits in good agreement with Monte Carlo shell model calculations. This fact is strongly supported by the instability of C, N and O isotopes with $N > 16$, and confirms the magic character of N=16 for the neutron-rich nuclei in the region $6 \leq Z \leq 10$, while the shell closure at N=20 tends to disappear for $Z < 14$. The existence of the new shell closure at N=16 in neutron-rich nuclei is supported also by the behaviour of the calculated shell correction in the O-Al region.

This work was done in the GANIL-Orsay-Dubna-Řez-Bucharest collaboration and was supported by GA of the Czech Academy of Sciences N$^{\underline{o}}$ 1048102.

References

1. Z. Ren et al., Phys. Lett. **B380**, 241 (1996).
2. C. Thibault et al., Phys. Rev. **C12**, 644 (1975).
3. A. Korsheninnikov et al., Phys. Lett. **B326**, 31 (1994).
4. A.N.Ostrowski et al., Phys. Lett. **B338**, 13 (1994).
5. O. Tarasov et al., Phys. Lett. **B409**, 64 (1997).
6. H. Sakurai et al., Phys. Lett. **B448**, 180 (1999).
7. F. Sarazin et al., Phys. Rev. Lett. **84**, 5062 (2000).
8. G. Audi and A.H. Wapstra, Nucl. Phys. **A624**, 1 (1997).
9. H. Sakurai et al., Phys. Rev. **C54**, R2802 (1996).
10. Y. Otsuno et al., Phys. Rev. **C60**, 054315-1 (1999).
11. T. Otsuka private communication
12. A.Ozawa et al., Contr. to RNB2000, 3-8 April 2000, Divonne (France); Phys. Rev. Lett. **84**, 5493 (2000).

DYNAMICAL MODEL OF FISSION FRAGMENT ANGULAR DISTRIBUTION

V.A.DROZDOV, D.O. EREMENKO, O.V. FOTINA, S.YU. PLATONOV, O.A.YUMINOV

Institute of Nuclear Physics, Moscow State University, Moscow, Russia
E-mail: eremenko@p6-lnr.npi.msu.su

G.GIARDINA, A.TACCONE

Istituto Nazionale di Fisica Nucleare, Sezione di Catania and Dipartamento di Fisica dell' Universita. Messina, Italy
E-mail: giardina@nucleo.unime.it

A dynamical model of fission fragment angular distributions is suggested. The model allows one to calculate fission fragment angular distributions, prescission light particle multyplicities, evaporation residue cross sections etc. for the cases of decay of hot and rotating heavy nuclei. The experimental data on angular anisotropies of fission fragments and prescission neutron multiplicities are analyzed for the ^{16}O + ^{208}Pb, ^{232}Th, ^{248}Cm and ^{238}U reactions at the energies of the incident ^{16}O ions ranging from 90 to 160 MeV. This analysis allows us to extract both the nuclear friction coefficient value and the relaxation time for the tilting mode. It is also demonstrated that the angular distributions are sensitive to the deformation dependence of the nuclear friction.

In the present contribution it is suggested a dynamical model of fission fragment angular distributions which takes into account the stochastic aspects of nuclear fission. The model may be useful tools in analysis of the experimental data on fission fragment angular distributions, prescission light particle multiplicities etc. for the cases of decay of hot and rotating heavy compound nuclei when a validity of the traditional transition state model is questionable.

The dynamics of induced nuclear fission is considered in the stochastic approach [1] by the one dimensional Langeven equations. The potential energy of fissioning nucleus is calculated within the liquid drop model by means of the fast method suggested in [2]. The initial spin distribution was calculated using the parameterization based on the surface friction model [1]. Light particle emission was simulated by means of the Monte-Carlo method usually used in the Langeven simulations of induced fission [1]. The calculations of the fission fragment angular distributions were performed assuming that for every Langeven sample completed by fission there an effective point and the nuclear properties (temperature and deformation) at this point determines the angular distribution of fission fragments. However, the stochastic nature of fission results in an ensemble of such points by which the avcraging has to make. The position of every transition point was determined by the relaxation time of the tilting mode τ_K which was treated as a free parameter. With this aim the fission time of every Langeven sample we subdivided by intervals which are equal to τ_K. The average deformation for every interval was

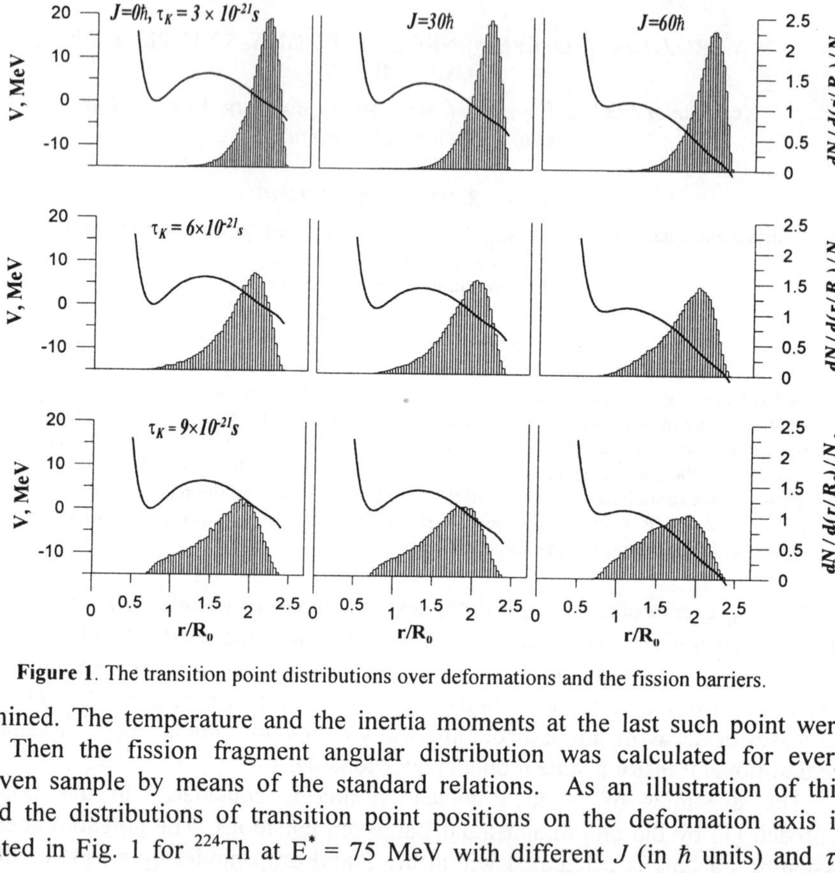

Figure 1. The transition point distributions over deformations and the fission barriers.

determined. The temperature and the inertia moments at the last such point were fixed. Then the fission fragment angular distribution was calculated for every Langeven sample by means of the standard relations. As an illustration of this method the distributions of transition point positions on the deformation axis is presented in Fig. 1 for ^{224}Th at $E^* = 75$ MeV with different J (in $\hbar$ units) and τ_K values. As one can see these distributions have essential variances and the average deformation for the transition point is between the saddle and scission deformations. The distribution broadening is observed with increase of both J and τ_K.

In the present work we analyzed the experimental data on the angular anisotropy of fission fragments and the prescission neutron multiplicity for the ^{16}O + ^{208}Pb, ^{16}O+^{232}Th, ^{16}O + ^{248}Cm and ^{16}O + ^{238}U reactions first assuming deformation independences of β and m. For the inertia coefficient the reduced mass corresponding to symmetric fission ($m=A/4$) was used. Since, neutron multiplicity does not depend on the τ_K value we obtained the reduced friction value $\beta = 4\times10^{21}s^{-1}$ fitting the prescission neutron multiplicity and than varied the τ_K value to describe the experimental angular anisotropy as accurate as possible. In Fig. 2 the calculation results are presented in comparison with the experimental data [3]. The best description of the experimental data is achieved with $\tau_K = 6\times10^{-21}s$. It is seen that for the ^{16}O + ^{208}Pb reaction the experimental data are between two calculations within the model of the transition states in the saddle point. It means that the transition state model in principle can describe the experimental angular anisotropy for this reaction

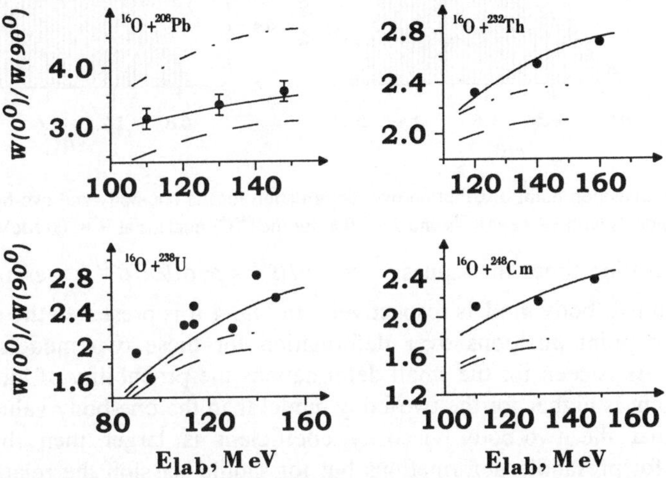

Figure 2. The angular anisotropy of fission fragments. The points are the experimental data, the solid lines are the calculation result within the present model, the dashed-dotted and dashed lines are the calculation within the model of transition states in the saddle point with and without consideration of the presaddle neutron emission, respectively.

in the energy range under consideration. However it seems, that our calculation reproduces the slope of the experimental energy dependence of the angular anisotropy better than the transition state model. On the other hand the ^{16}O+^{232}Th, ^{16}O + ^{248}Cm and ^{16}O + ^{238}U reactions lead to compound nuclei which are heavier than one in the case of the ^{16}O+^{208}Pb reaction. As is seen from Fig. 3 for these cases the angular distributions of fission fragments can not be reproduced in the frames of the standard transition state model. It is connected with the formation of compound nuclei having angular momentum dependent fission barriers which are smaller than temperature values for the considerable part of the initial distributions over angular momentum. It is seen that for the ^{16}O+^{232}Th, ^{248}Cm and ^{238}U reactions our model allows a good description of the experimental data.

In the frames of the model angular distribution of fission fragments is connected with the distribution of the transition points over deformation. It is obviously that the probability of a realization of the transition point is determined by the time scales for different stages of fission. The last means that angular distributions of fission fragments have to be sensitive to the deformation dependence of nuclear viscosity. In the present work we used the one-body (only wall contribution) and two-body models of nuclear dissipation to illustrate this sensitivity. In these calculations for the deformation dependent inertia coefficient Werner-Wheeler method is used.

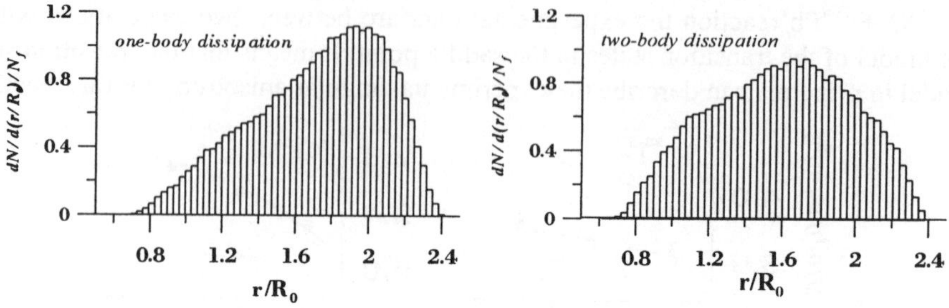

Figure 3. The transition point distribution over deformation for the one-body and two-body dissipation models with models with $\tau_K = 8\times10^{-21}$s and $J = 40\,\hbar$ for the ^{248}Cf nucleus at E^*=100 MeV

For the relaxation times the values $\tau_K = 7.\times10^{-21}$ s and 9×10^{-21} s were obtained for the one- and two-body models respectively. In Fig. 4 it is presented the distributions of transition point positions over deformation for these two models of nuclear dissipation. As is seen for the small deformations the probability of locating of the effective point is higher for the two-body model than the one-body values. This is a result of that the two-body viscosity coefficient is larger then the one-body coefficient for presaddle deformations but for saddle-scission the relation between them relation is opposite [1]. All of the preceding allows us to conclude that the angular distributions of fission fragments are sensitive to the deformation dependence of the nuclear friction coefficient.

We conclude that the dynamical model may be useful in analysis of fission fragment angular distributions for the cases of decay of hot and rotating heavy compound nuclei especially when a validity of the traditional transition state models is questionable. Analisys of the experimental data for a set of complete fusion-fission reactions allowed one to obtaine the relaxation time for tilting mode. It was demonstrated that the angular distributions are sensitive to deformation dependence of the nuclear friction coefficient.

References

1. Fröbrich P., Gontchar I.I. Langevin description of fusion, deep–inelastic collisions and heavy–ion–induced fission. *Phys. Rep.* ***292*** *(1998) pp.131–237.*
2. Lestone J.P. Fast method for obtaining finite range corrected potential energy surfaces. *Phys. Rev.* ***C51*** *(1995) pp.580–585.*
3. Back B.B., Betts R.R., Gindler J.E., Wilkins B.D., Saini S., Tsang M.B., Gelbke C.K., Lynch W.G., McMahan M.A., Baisden P.A. Angular distribution in heavy–ion induced fission. *Phys. Rev.* ***C32*** *(1985) pp.195–213.*

THE REACTION ^{6}He+^{64}Zn AROUND THE COULOMB BARRIER

P.FIGUERA, A.DI PIETRO, A.MUSUMARRA, F.AMORINI, G.CARDELLA, J.LU, M.PAPA, M.G.PELLEGRITI, F.RIZZO, S.TUDISCO

INFN LNS and Sezione di Catania, Universita' di Catania, Via S.Sofia 44 I-95123 Catania Italy, E-mail:figuera@lns.infn.it

T.DAVINSON,H.MAHMUD, C.RUIZ, A.C.SHOTTER

Department of Physics and Astronomy, University of Edinburgh, UK

C.ANGULO, S.CHERUBINI, A.NINANE

Institut de Physique Nucleaire and CRC Louvain la Neuve, Belgium

M.MILIN, N. SOIC

Rudjer Bošković Institute, Zagreb, Croatia

R. RAABE, L. WEISSMAN

Institut voor Kern-en-Stralingsfysica, University of Leuven, Belgium

The reaction ^{6}He +^{64}Zn was studied in order to investigate the effects of the halo structure on the fusion cross-section at low bombarding energy. The evaporation residues were identified by measuring off-line the X-ray emission which follows their Electron Capture decay. The reaction ^{4}He+^{64}Zn was also measured and the results are presented.

1 Introduction

Controversial effects on the fusion cross-section have been predicted by different theoretical models in reactions induced by halo nuclei at low bombarding energies (i.e.[1]). These models agree that the larger spatial extent of halo nuclei, and the coupling with possible low lying resonant states, would increase fusion cross-section. However different models disagree about the role played by the break-up of the loosely bound halo nucleons on the fusion cross-section. Break-up in fact can be considered either as a loss of flux for fusion decreasing the cross-section, or as an additional open channel which enhances the cross-section [e.g. 2-4].

The experimental investigation is very difficult owing to the low intensity of radioactive beams, the small fusion cross-section and the low evaporation residue (E.R.) recoil energies. Moreover indirect E.R. detection, via on-line or off-line γ-ray spectroscopy is very difficult due to the high background compared with the very low γ yields. The systems studied so far are ^{6}He+^{209}Bi [2-4], ^{6}He+^{238}U [5],

^{11}Be+^{209}Bi [6]. In these experiments fission cross-section or E.R. α radioactivity were measured and apparently controversial results were obtained. In the reactions induced by ^{6}He beams a strong enhancement of the fusion cross-section below the Coulomb barrier was observed but this was not seen in the ^{11}Be induced reaction.
We recently studied at the CRC laboratory of Louvain la Neuve the reaction ^{6}He+^{64}Zn at two beam energies around the Coulomb barrier. The choice of the ^{64}Zn as target was made, with the help of statistical model calculations, in order to produce evaporation residues (E.R.) which decay by Electron Capture (E.C.) therefore emitting low energy atomic X-rays. Different points in the fusion excitation function were measured, at fixed beam energy, activating a stack of targets and detecting off-line the X rays following the E.R. decay. As a comparison, but also to check the normalisation procedure, the reaction ^{4}He+^{64}Zn was investigated at the same centre of mass energy.

2 Experimental technique

The experiments were performed at Louvain la Neuve with an average ^{6}He beam current of $3*10^{6}$ pps using the experimental set-up sketched in figure 1.

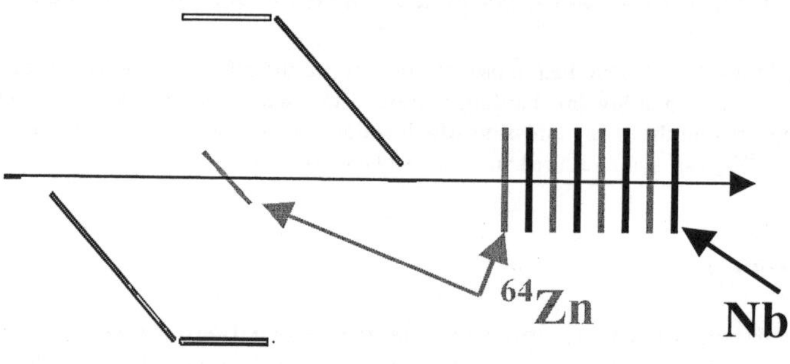

Figure1. Experimental set-up.

A thin ^{64}Zn target, forming an angle of 45^{0} respect to the incident beam, was surrounded by an array of silicon strip detectors devoted to the detection of light charged particles. The setup covered a solid angle of about 2π and was highly segmented into 224 independent strips. The detected light particles were discriminated by using the ToF technique.
A stack of four ^{64}Zn targets followed by ^{93}Nb catchers was placed downstream the thin ^{64}Zn target and used for the activation technique. Different stacks were irradiated with ^{4}He and ^{6}He beams. X-rays coming from the E.C. decay of the E.R. were measured off-line using a Si(Li) detector surrounded with a lead shield. The advantages of detecting X-rays instead of γ-rays are: 100% intrinsic detection

efficiency in the X energy region of interest (few KeV) and the possibility of a very effective background suppression. We were able to measure with a counting rate of less than 1 count per hour.

3 Results and discussion

The elastic scattering angular distributions for the two reactions $^{6,4}He+^{64}Zn$ were measured. Optical model calculations were performed using the code PTOLEMY. The results are shown in fig.2. The total reaction cross-sections extracted from the elastic scattering data are: σ_r=360±70 mb and σ_r=1420±280 mb for the reaction $^{6}He+^{64}Zn$ at E_{cm}=9.1 and 12.4 MeV respectively and σ_r=720±140 mb for $^{4}He+^{64}Zn$ at E_{cm}=12.4 MeV.
Protons and α angular distributions for the $^{6}He+^{64}Zn$ reaction were also extracted showing that *p* are emitted from fusion evaporation processes whereas different reaction mechanisms contribute to the αs emission.

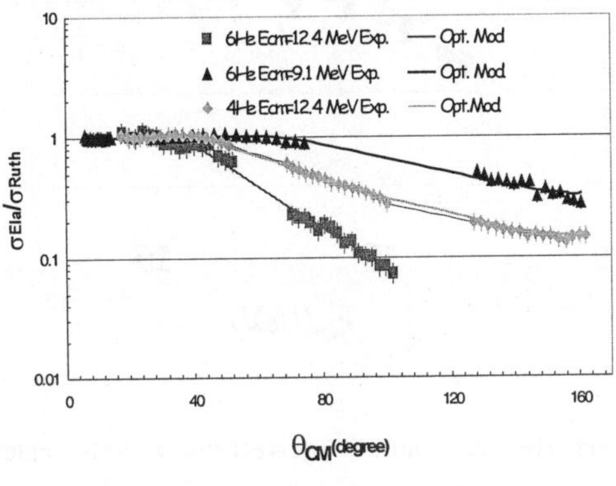

Figure2. Elastic scattering angular distributions.

Residues produced after $^{4,6}He+^{64}Zn$ fusion evaporation process are radioactive and decay by Electron Capture with different half-lives ranging from 18 minutes to 270 days. After E.C. decay, X-rays are emitted. By detecting these X-rays we extracted the cross-section for the production of the different E.R.. X-ray energies characterise different elements but not different isotopes. In our case it was possible to discriminate isotopes following the activity curve of each element as function of time, and unfolding the contributions of the different isotopes having different half lives. We started by analysing the targets irradiated with ^{4}He beam. A previous measurement existed [7] and therefore we could compare our results with that of [7]. We identified two radioactive E.R. the same as in [7] ^{67}Ge and ^{67}Ga. In the ^{6}He

activation experiment we identified 5 E.R.: ^{65}Zn,$^{66,67,68}Ga$ and ^{68}Ge. The fusion cross sections were obtained by summing up the contributions of all E.R. identified. Preliminary results are shown in figure 3. Here is shown a comparison between the fusion excitation function for $^{4}He+^{64}Zn$ measured in [7] (squares), $^{4}He+^{64}Zn$ measured in this work (open triangles) and $^{6}He+^{64}Zn$ measured in this work (diamonds). In this analysis the efficiency was extracted by normalising the $^{4}He+^{64}Zn$ data to the ones of [7]. From this preliminary results it seems that at low energy (the calculated Coulomb barrier being around 11 MeV) the data sets for the two reactions $^{6,4}He+^{64}Zn$ are very similar. They differ at the higher energy point. At this energy however, there could be the opening of an additional E.R. channel which is stable. Analysis of the evaporated light charged particle data is in progress to verify this possibility.

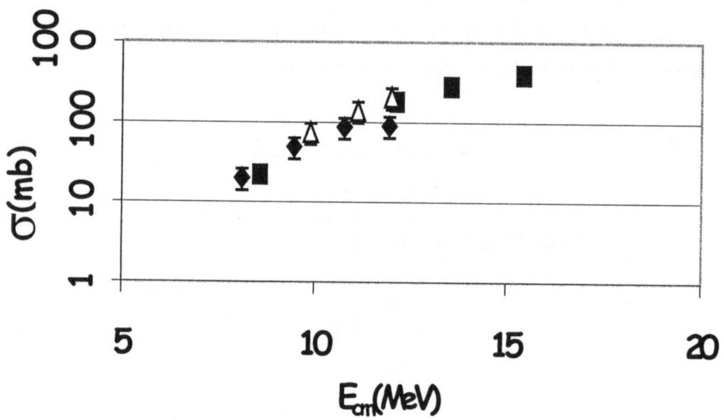

Figure 3. Excitation function $^{4}He+^{64}Zn$ from [7] (squares) this work (triangles), $^{6}He+^{64}Zn$ this work (diamonds).

References

[1] Signorini C. et al., Nucl. Phys. **A616** (1997) pp. 262c-277c

[2] Aguilera E.F. et al, Phys.Rev.Lett. **84** (2000) pp. 5058-5061

[3] DeYoung P. A. et al., Phys. Rev. **C58** (1998) pp. 3342-3444

[4] Kolata J.J. et al., Phys.Rev.Lett. **81** (1998) pp. 4580-4583

[5] Trotta M. et al., Phys.Rev.Lett. **84** (2000) pp. 2342-2345

[6] Signorini C. et al., Eur. Phys. Jour. **A2** (1998) pp. 227-231

[7] Ruddy F. H. et al., Nucl. Phys. **A127** (1969) pp. 305-322

NEW OPPORTUNITIES WITH BEAMS OF RARE ISOTOPES IN THE U.S.

C. KONRAD GELBKE

National Superconducting Cyclotron Laboratory,
Michigan State University, East Lansing, MI 48824-1321, USA
Email: gelbke@nscl.msu.edu

Major new opportunities for research with beams of rare isotopes in the United States will be discussed. For the near term future, the new Coupled Cyclotron Facility (CCF) at the National Superconducting Cyclotron Laboratory (NSCL) will provide fast beams of rare isotopes with intensities higher by about 3 orders of magnitude than previously available in the U.S. The scope of the CCF construction project, the planned facility layout, and key research questions that can be addressed with fast beams of rare isotopes will be presented. On a longer time scale, large additional gains, especially for medium-mass and heavy elements, will be possible with the Rare Isotope Accelerator (RIA) now under discussion. The writing group for the new (2001) long-range plan for nuclear science in the United States has recommended construction of RIA as its highest priority for new construction. Main features of the RIA concept will be highlighted.

1 Introduction

Some distance from the valley of stability lies a frontier beyond which the properties of nuclei are unknown. Approximately 300 isotopes are stable, but thousands more are continually created in the cosmos and play a fleeting, though important role in stellar processes. Technical advances now make it possible to produce many of these short-lived nuclei in the laboratory and to study their often-unusual properties. These are of considerable intrinsic scientific interest and are needed to provide a solid understanding of the nuclear processes that underlie the evolution of stars and the origin of the elements in the cosmos, see, e.g., the 1999 White Paper *Opportunities in Nuclear Astrophysics -- Origin of the Elements* [1].

The importance of the physics and the nature of the technology necessary to advance this scientific frontier were examined in detail by the DOE/NSF Nuclear Science Advisory Committee (NSAC) in preparation of its 1996 Long Range Plan (LRP) [2] and by the National Research Council (NRC) Committee on Nuclear Physics in its assessment of the field, *Nuclear Physics, The Core of Matter, The Fuel of Stars* (1999) [3]. At the international level, the nuclear physics working group of the Organization for Economic Cooperation and Development (OECD) Megascience Forum addressed this issue in its 1999 report [4]. The conclusions of these groups strongly support exploration of nuclei far from the valley of stability, and the Megascience Forum report recommends that "a new generation of high-intensity RNB (Radioactive Nuclear Beam) facilities of each of the two basic types,

ISOL (Isotope Separator On-Line) and In-Flight, should be built on a regional basis."

The 1996 LRP emphasized that: "The scientific opportunities made available by world-class radioactive beams are extremely compelling and merit very high priority." As the highest priority for new initiatives, the LRP strongly recommended the "immediate upgrade of the MSU facility to provide intense beams of radioactive nuclei via fragmentation" and, in addition, the "development of a cost-effective plan for a next generation ISOL-type facility and its construction when RHIC is essentially complete." The LRP noted further that "the fragmentation and Isotope Separator On-Line (ISOL) techniques are complementary in the species and energies of the beams produced. Thus they drive different aspects of the science."

Construction of the NSCL Coupled Cyclotron Facility (CCF) was expeditiously approved by the National Science Foundation in 1996 and has now been completed (ahead of schedule). First beams of rare isotopes from the CCF have been delivered for research in June 2001.

The scientific case for a next-generation ISOL facility was further elaborated in the 1997 Report, *Scientific Opportunities with an Advanced ISOL Facility* [5]. In 1998, NSAC was charged to establish a task force that should provide a technical analysis of the various options for a new facility dedicated for the research outlined in this 1997 report. Initially, the task force focused on evaluating specific plans for constructing ISOL facilities at Oak Ridge and Argonne National Laboratories. After months of deliberations, however, MSU members of the task force suggested an alternative concept, utilizing projectile fragmentation/fission plus separation in-flight as the prime production mechanism for rare isotopes. By stopping the separated fragments in an appropriate medium (e.g., a gas cell), the rare isotopes would be available for experiments at rest or for re-acceleration. The Task Force charged the MSU group to explore this concept in more detail. This work led to the RIA concept that was ultimately recommended by the Task Force. Most notably, the ISOL Task Force recommended:

- "The design and construction of a Rare-Isotope Accelerator (RIA) facility that provides unprecedented beams of a diverse assortment of nuclei. The scientific potential of the RIA facility will be maximized by integrating multiple techniques for producing and separating, then accelerating and utilizing these rare isotopes. RIA will be based on a highly flexible superconducting linac driver capable of providing 100 kW, 400 MeV/nucleon beams of any stable isotope from hydrogen to uranium. The broad assortment of short-lived secondary beams needed for the experimental program will be produced by a combination of techniques: projectile fragmentation, target fragmentation, fission, and spallation."
- "That an additional important opportunity be provided: fast in-flight separated beams of rare isotopes. This will extend the scientific reach of the RIA facility. We recommend that the RIA design accommodate this capability."

The schematic layout of the RIA facility envisioned by the ISOL Task Force is shown in Figure 1. Rare isotopes at rest in the laboratory will be produced by conventional ISOL target fragmentation, spallation, or fission techniques and, in addition, by projectile fragmentation/fission and stopping in a gas cell. Upon extraction, these stopped isotopes can be used at rest for experiments in Area 3, or they can be accelerated either to energies below the Coulomb barrier and used in Area 2 or above the Coulomb barrier and used in Area 1. Fast beams of rare isotopes, produced by projectile fragmentation or fission, can also be used directly after separation in a high-resolution fragment separator (area 4).

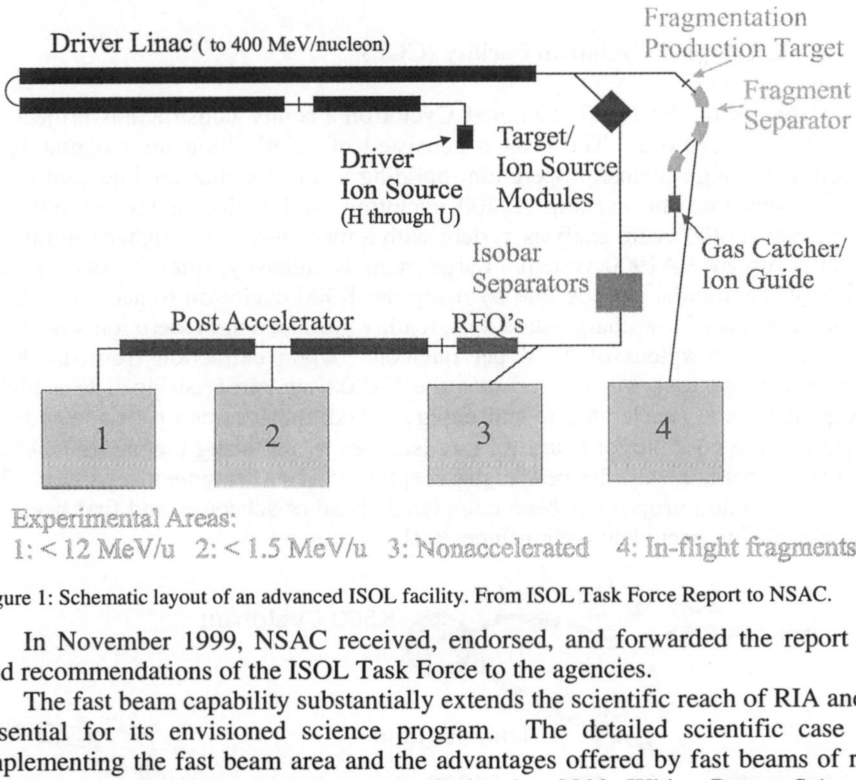

Figure 1: Schematic layout of an advanced ISOL facility. From ISOL Task Force Report to NSAC.

In November 1999, NSAC received, endorsed, and forwarded the report [6] and recommendations of the ISOL Task Force to the agencies.

The fast beam capability substantially extends the scientific reach of RIA and is essential for its envisioned science program. The detailed scientific case for implementing the fast beam area and the advantages offered by fast beams of rare isotopes were elaborated in great detail in the 2000 White Paper *Scientific Opportunities with Fast Fragmentation Beams from RIA* [7]. The most compelling advantages offered by fast beams can be summarized in rather simple terms [7]:

- Economic production of medium-energy ($E/A \geq 50$ MeV) beams of rare isotopes, without re-acceleration;
- Increased luminosity from the use of thick secondary targets (by up to a factor of 10,000);
- Reduced background from in-flight tracking and identification of individual isotopes in the beam on a particle-by-particle basis;

- Efficient particle detection from strong forward focusing;
- Short beam development times and low losses due to fast (sub-microsecond) and chemistry-independent separation and transport to the experiment.

In summary, RIA combines the advantages of both the conventional thick-target ISOL technique and the transmission-target projectile fragmentation/fission technique – a paradigm change as compared to the original recommendation of the 1996 LRP. In the following, I will highlight the key features of the NSCL CCF and its expected reach, summarize key science questions to be addressed with beams of rare isotopes, and provide an update on the current status of RIA.

2 NSCL Coupled Cyclotron Facility (CCF)

Key elements of the NSCL Coupled Cyclotron Facility construction project are illustrated in Figure 2. The project consisted of refurbishing the existing K500 cyclotron for high intensity operation, building a transfer line for injection of the K500 beams into the existing K1200 cyclotron, and replacing the present low-acceptance A1200 beam analysis system with a more powerful, higher-rigidity and higher-acceptance A1900 system. Large gains in intensity, often by two or three orders of magnitude, are possible by using the K500 cyclotron to accelerate high-intensity beams of low charge-state ions, readily produced in modern ion sources, to energies of a few tens of MeV per nucleon. Upon extraction from the K500 cyclotron, these ions are injected into the K1200 cyclotron, stripped to a higher charge state, and accelerated to full energy. Additional gains up to one order of magnitude are possible for beams of rare isotopes by replacing the previous A1200 fragment separator with the new high-acceptance A1900 fragment-separator. The CCF construction project has been completed ahead of schedule, and first beam for experiments has been delivered in June 2001.

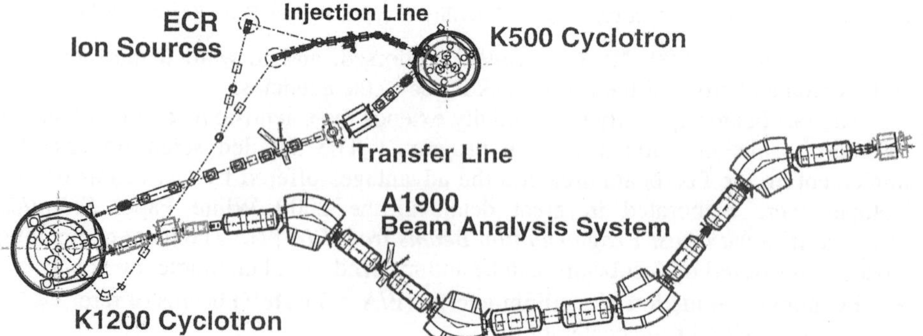

Figure 2: Key elements of the NSCL Coupled Cyclotron project.

The floor plan of the high bay experimental area is shown in Figure 3. The CCF has the capability of accelerating ions of all elements to energies from approximately 10 to 200 MeV/nucleon (the exact energy range depends on the A/Z ratio of the isotope). These energetic primary beams can be used directly, or they can be converted into rare isotopes that are separated in-flight by the A1900 fragment separator and then delivered to one of six experimental vaults.

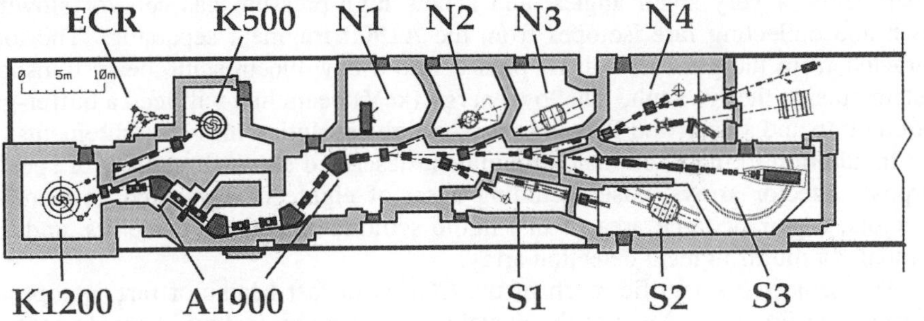

Figure 3: Floor plan of the NSCL high bay experimental area showing the two superconducting cyclotrons (K500 and K1200 on the left), the superconducting A1900 beam analysis system and subsequent beamlines, the various experimental vaults, and the newly constructed areas.

The S1 vault contains a new multi-purpose beamline and, on another line, the reaction product mass separator (RPMS) that is used for low-background studies of nuclei far from stability. The Wien filter of the RPMS provides additional purification of radioactive beams, particularly for reaction studies with proton-rich nuclei. An experimental station for measuring nuclear magnetic moments is situated in S1 just upstream from the RPMS.

The "Superball" in the S2 vault is a high-efficiency neutron multiplicity meter that contains approximately 17 m^3 of Gd-doped scintillator for the detection of neutrons in 4π geometry. The large internal scattering chamber of the Superball can house the Miniball and Miniwall 4π charged-particle arrays, thus allowing simultaneous measurements of neutron and charged particle multiplicities on an event-by-event basis in 4π geometry.

The S3 vault contains the S800, a superconducting high-resolution magnetic spectrograph with specifications well matched to experiments with radioactive ions (energy resolution $E/\delta E = 10^4$; maximum rigidity $B\rho_{max} = 4$ Tm; momentum acceptance $\Delta p/p = 5\%$; and solid angle $\Delta\Omega = 20$ msr). The S800 beamline can be used to dispersion-match the beam, or it can be operated as a fragment separator. Both the S800 spectrograph and its beamline deflect the ions vertically.

The N2 vault contains the 4π Array, a low-threshold "logarithmic" 4π charged particle detector consisting of thirty-two position sensitive parallel plate multiwire detectors, backed by segmented Bragg ionization chambers, backed in turn by an array of 170 phoswich detectors, each consisting of a fast-slow plastic scintillator combination.

The N3 vault is the major general-purpose vault in the NSCL facility as it contains the 92" scattering chamber. The large cylindrical multi-purpose scattering chamber (diameter: 231 cm, length: 271 cm) is used for a variety of reaction studies and radioactive beam experiments. The chamber shell can be lifted from the vault to provide a large free space.

The newly configured N4 vault will contain the "sweeper magnet", a high-acceptance magnetic spectrometer for charged-particle neutron coincidence experiments at very small angles, and a new high-pressure gas cell for slowing down and collecting rare isotopes from the A1900 fragment separator. The ions extracted from the gas cell will be passed into a new electrostatic beam transport system outside the N4 vault. The low energy (keV) beam line will feed a buffer-gas cooling trap and a precision Penning trap for high-resolution mass measurements.

In addition to the fixed major equipment described above, a number of special purpose detector arrays exist, including a set of eighteen segmented germanium detectors, a pair of large area (4 m^2) liquid-scintillator neutron-detectors, and the Miniball 4π multifragment detection array.

The projected scientific reach of the CCF with fast beams of rare isotopes is illustrated in Figure 4. Most of the nuclei in the astrophysical rapid proton capture process (rp-process) as well as the lighter nuclei in the rapid neutron capture process (r-process) will become accessible at the CCF. Studies of the heavier r-process nuclei will only be possible with a more powerful facility, such as RIA.

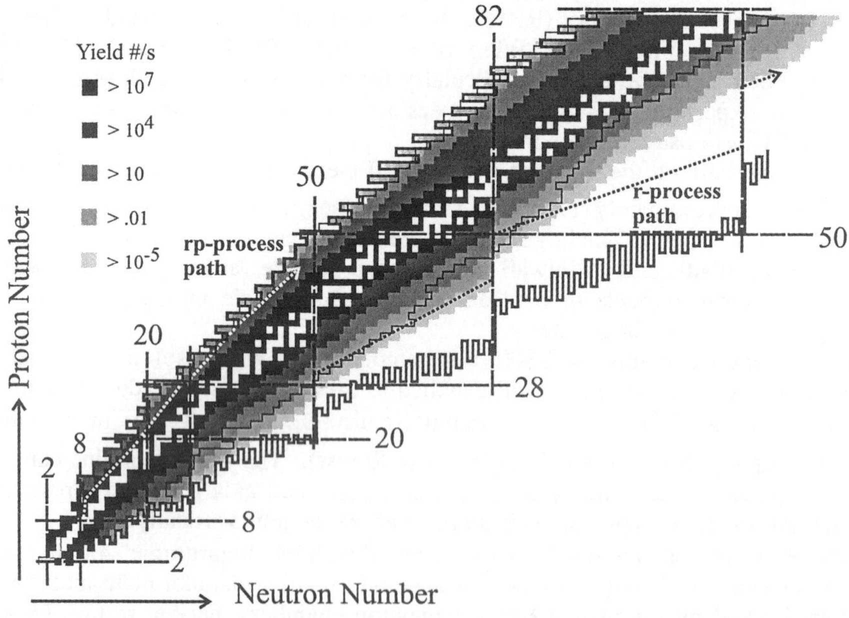

Figure 4: The scientific reach of the NSCL Coupled Cyclotron Facility is illustrated in terms of the projected intensities for fast beams of rare isotopes. Note of caution: Far from stability, the predicted intensities become increasingly uncertain due to a current lack of experimental data. (A color version of this plot can be found at the NSCL website: http://www.nscl.msu.edu/research/ribeams/home.html.)

3 Key Science Questions

Research at the NSCL and at the future RIA facility will focus on the study of nuclei far from the line of stability and address the following important questions:

- *What is the origin of the elements of the cosmos – and what are the nuclear processes that determine the fate of stars and that lead to the synthesis of the elements of which we are made?* Nuclear processes in stars similar to our Sun create some elements, but many others must be created by a very different set of processes which require much hotter, explosive environments, such as those found in nova or supernova explosions. Experiments at the CCF will provide important data needed for improved modeling of the nuclear processes that take place in nova and supernova environments. Together with modern astronomical observations, this research will help to unravel the mystery of the origin of the elements.
- *Which combinations of protons and neutrons can form a nucleus?* The answer to this basic question is known only for elements lighter than oxygen ($Z \leq 8$). It is a major challenge to predict the limits of stability; it requires very accurate solutions of the mesoscopic quantum problem for strongly interacting particles. With beams from the CCF, it may be possible to determine the limits of stability up to sulfur ($Z=16$); precision mass measurements will help develop a better theory of nuclear binding.
- *What are the properties of nuclei with extreme proton-to-neutron ratios?* In recent years, it has been discovered that some very light rare isotopes (e.g., the rare isotope ^{11}Li) have an extended region of nearly pure neutron matter in their outer periphery – a condition that does not exist in any nucleus close to the line of stability. This halo state and related neutron skins are expected in heavier nuclei and will be explored at the NSCL. These nuclei provide an extreme test for nuclear models and at present are not well characterized. Elsewhere, such neutron matter exists only inside neutron stars.
- *What are the properties of bulk neutron-rich matter under extreme conditions of temperature and density?* Large objects composed of neutron-rich nuclear matter, neutron stars, are created in the cosmos during supernova explosions triggered by the gravitational collapse of the cores of massive stars that have run out of nuclear fuel. While the existence of neutron stars is firmly established by astronomical observations, neither their properties nor the sequence of events which leads to their formation are well understood, in part due to our incomplete knowledge of the equation of state (EOS) of neutron-rich nuclear matter. Experiments at the NSCL can provide important information about the symmetry-energy term of the EOS at supra-normal densities.
- *How must existing theoretical models be changed to describe the properties of very neutron or proton-rich isotopes?* Present models have been fine-tuned to reproduce the properties of nuclei close to the line of stability, but their predictions become increasingly uncertain with increasing neutron excess. They cannot reliably predict the properties of the very neutron-rich nuclei that

play a central role in the formation of heavy elements. It seems likely that long-held concepts, such as the shell structure of nuclei, will undergo major revisions as one moves away from the line of stability.

- *How are the properties of nuclei related to the basic nucleon-nucleon interaction?* This question is closely related to one of the most challenging problems of modern science – how can one derive the properties of composite objects from those of their building blocks? Solutions of such mesoscopic problems require advanced computational techniques of ever increasing sophistication, the development of which will have many cross-disciplinary applications. Nuclei, and especially those with unusual properties, provide an ideal test bed for problems of mesoscopic quantum physics and a continuing stimulus for the development of new theoretical approaches.

For the coming decade, the NSCL coupled cyclotron facility will provide extraordinary opportunities for research with fast (and stopped) beams of rare isotopes and for developing and utilizing the experimental techniques that will be needed for research at RIA. The development of the RIA facility concept has already strongly benefited from the experience gained at the NSCL. The future research program at RIA is the logical extension of the ongoing research at the NSCL, especially into the region of heavier nuclei where RIA will offer intensity gains by several orders of magnitude. Consistent with its educational mission, the NSCL will play a major role in training the next generation of scientists who will be needed for a successful research program at RIA.

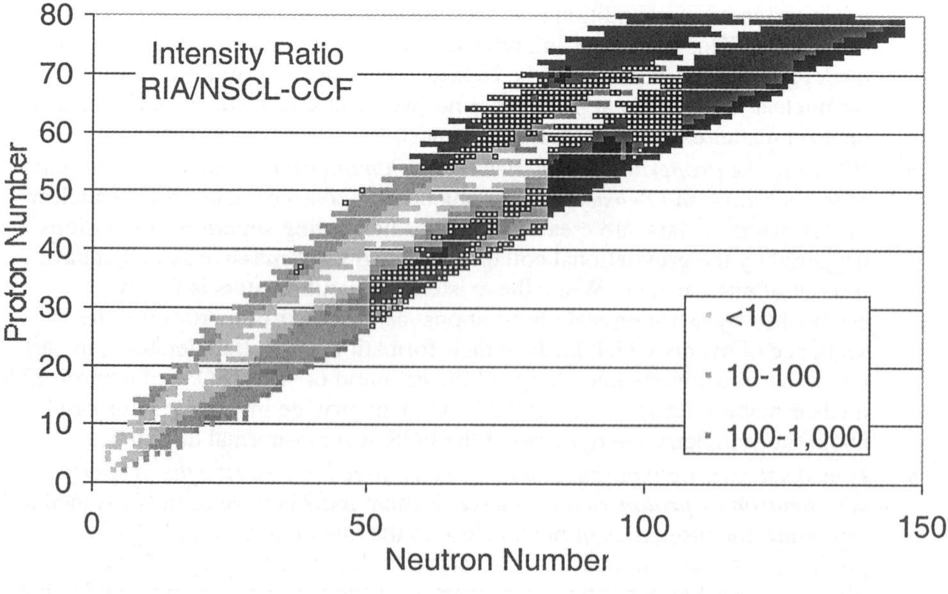

Figure 5: Intensity gains for fast beams of rare isotopes provided by RIA relative to the intensities form the CCF.

4 The Rare Isotope Accelerator (RIA)

The additional gains in intensity provided by RIA beyond those expected for the CCF are illustrated in Figure 5. For lighter (A < 100) nuclei, the predicted intensity gains from RIA are often less than one or two orders of magnitude. With ten years lead-time, much work in this region of nuclei can be done at the NSCL or at other facilities abroad with comparable capability. Even after the construction of RIA, the NSCL will remain an attractive and cost-effective option for conducting research with fast and stopped beams of rare isotopes in the lighter mass region. However, for heavier (A > 100) nuclei, RIA exceeds by far the intensities possible at the CFF, often by many (3 – 5) orders of magnitude. In this mass domain, RIA will provide access to nuclei, including many important r-process nuclei that cannot be realistically studied at the CCF.

Two institutions have expressed their interest to host the new RIA facility, Michigan State University and Argonne National Laboratory. Figures 6 and 7 show the preliminary facility layouts at MSU and ANL -- with the caveat that both institutions are still optimizing the specifics of their facility plans and that significant modifications are likely. MSU would provide sufficient land on its campus to allow an optimized layout with sufficient space for a stretched driver linac and future facility upgrades or additions. At present, ANL plans to make use of the existing Atlas facility and build a folded driver linac.

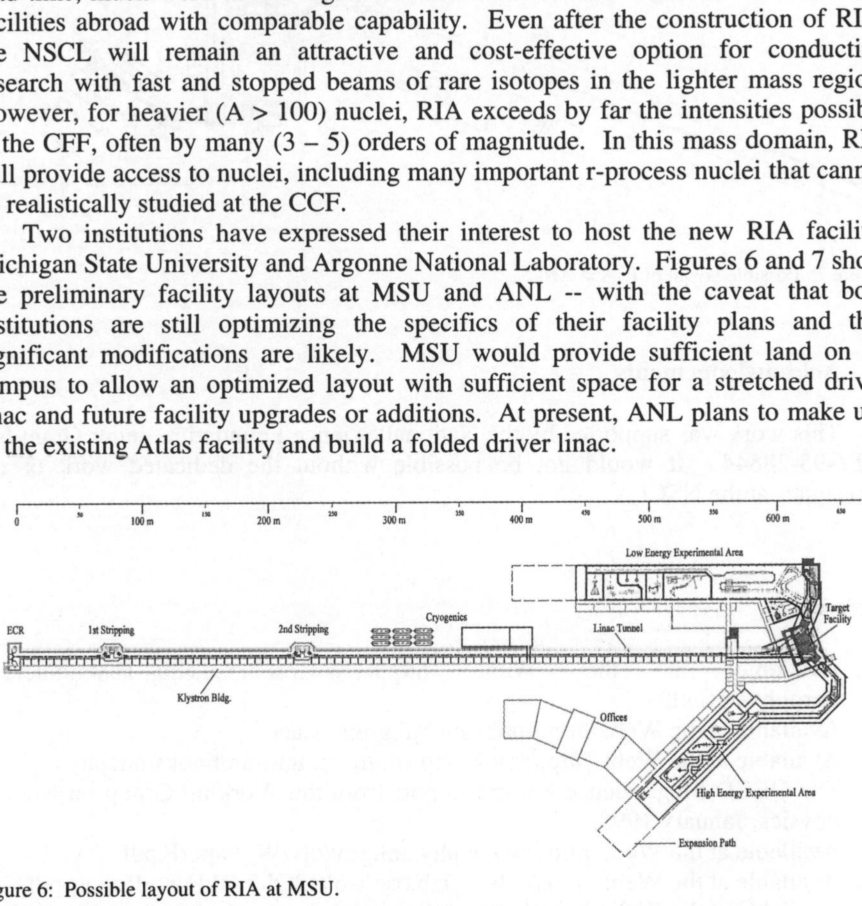

Figure 6: Possible layout of RIA at MSU.

In September 2000, NSAC was charged to recommend a new long-range plan for nuclear science. A new cost estimate for RIA construction, pre-operations, and operations was prepared in January 2001 by a joint team from MSU and ANL. After a series of town meetings, the Long-Range Plan Writing Group met in March

2001 and formulated its priority recommendations. The highest priority recommendation of the U.S. nuclear science community is the exploitation of the extraordinary opportunities for scientific discoveries made possible by recent investments; RIA is recommended as the highest priority for major new construction. However, NSAC emphasizes that RIA will require significant funding above the nuclear physics base budget. While RIA has an excellent chance of becoming a reality, it will not be trivial to secure its funding.

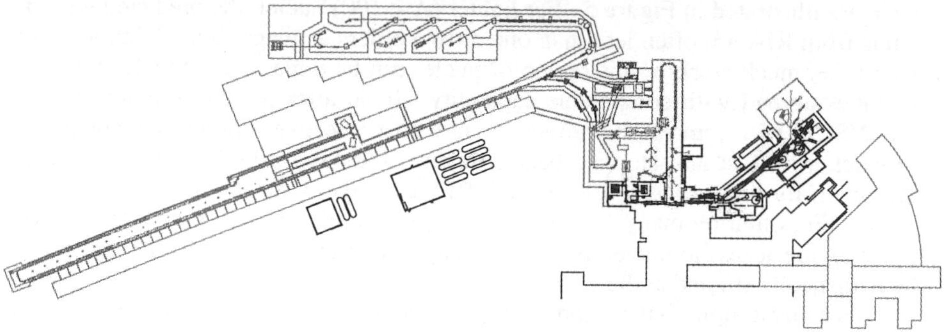

Figure 7: Possible layout of RIA at ANL.

5 Acknowledgements

This work was supported by the National Science Foundation under Grant No. PHY-95-28844. It would not be possible without the dedicated work of my colleagues at the NSCL.\

References

1. Available at the Web: http://www.nscl.msu.edu/~austin/nuclear-astrophysics.pdf
2. Available at the Web: http://pubweb.bnl.gov/~nsac/
3. Available at the Web: http://www.nap.edu/readingroom/books/nucphys
4. The OECD Megascience Forum: Report from the Working Group on Nuclear Physics, January 1999.
5. Available at the Web: http://www.phy.anl.gov/div/W_PaperF.pdf
6. Available at the Web: http://srfsrv.jlab.org/isol/ISOLTaskForceReport.pdf
7. Available at the Web: http://www.nscl.msu.edu/research/ria/whitepaper.pdf

QUATERNARY FISSION

F. GÖNNENWEIN AND P. JESINGER
Physikalisches Institut, Auf der Morgenstelle 14, D-72076 Tübingen, Germany
E-mail: goennenwein@uni-tuebingen.de

M. MUTTERER
Institut für Kernphysik, TU Darmstadt, D-64280 Darmstadt, Germany

A.M. GAGARSKI AND G.A. PETROV
PNPI of RAS, Gatchina, Leningrad district 188350, Russia

W.H. TRZASKA
University of Jyväskylä, FI-40500 Jyväskylä, Finland

V. NESVIZHEVSKI AND O. ZIMMER
ILL, BP 156, F-38042 Grenoble Cedex, France

In quaternary fission the two fission fragments are accompanied by two light charged particles (LCP). The process has been studied for $^{233}U(n,f)$ and $^{235}U(n,f)$ induced by cold neutrons. For (α,α) and (α,τ) as LCPs true and pseudo quaternary fission are disentangled. In true quaternary fission the LCPs are born right at scission, while the pseudo process is mediated by the particle-unstable ternary isotopes ^{8}Be and $^{7}Li^{*}$, respectively.

In quaternary fission the two main fission fragments are accompanied by the emission of two further charged particles. In a sense quaternary fission is just a generalisation of ternary fission. At least in low energy fission under discussion here, the fragments from quaternary decays have similar masses and charges as in binary fission, while the additional particles are very light. In low energy quaternary fission no particles heavier than α-particles have been observed so far. Therefore, the additional particles will be addressed as light charged particles (LCP) in the following. The quaternary decay being very rare, the process has barely been studied in the past.

In the study to be described here quaternary fission induced by cold neutrons in the fissile isotopes ^{233}U and ^{235}U was investigated. The experiments were performed at the High Flux Reactor of the Institut Laue-Langevin in Grenoble. The layout of the experiment is given in Fig. 1. Target and detectors are placed in a plane at right angles to the neutron beam. Fission fragments are intercepted by multiwire proportional chambers (MWPC) facing opposite sides of the target being thin and transparent to fragments. From

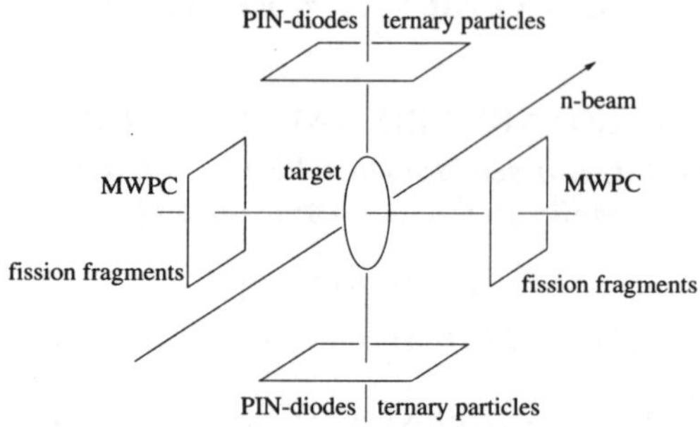

Figure 1. Schematic view of the experimental setup of target and detectors

the MWPC signals the time-of-flight difference of coincident fragments is deduced which allows to assess whether, in Fig. 1, the light or the heavy fragment is flying to the left or right. The LCPs are detected by PIN diodes. A total of 32 diodes was arranged in two arrays of 16 diodes each. One array was viewing the target from top, the other from bottom. Each diode was 3 × 3 cm^2 in size. The majority of diodes was equipped electronically to measure along with the energy deposited also the drift times of charge carriers in the diodes. For given energy the time signal carries information on the particle type. It was thus possible to identify unambiguously α-particles and hydrogen isotopes. Within the hydrogen isotopes by far the most abundant LCPs are tritons (τ). To avoid radiation damage by fission fragments the diodes had to be shielded by thin Al foils. This introduces, on the one hand, a low energy cutoff in the detection of light LCPs like αs and τs, and, on the other hand, prevents heavier LCPs to be observed. The experimental setup thus allowed to measure in coincidence with fission fragments the appearance of one or two (or more) αs and τs and to analyse their distribution in space relative to the fission axis.

For targets with some 5 mg of fissile material and a cold neutron flux of 3×10^8 neutrons/cm^2 about 5×10^3 and 10^4 quaternary events from fission of 234U* and 236U*, respectively, were registered in 6 weeks of measuring time. The results obtained for the two U-isotopes are very similar. Though it was not feasible to identify double hits within time intervals of some ns in one single diode, the granularity of the two arrays enabled to observe two LCPs impinging on the same array or on opposite arrays within a short time window. Taking advantage of particle identification coincidence patterns could be found separately for (α, α) and (α, τ) coincidences but not for (τ, τ) coincidences where count rates were too low. While the rates of chance coincidences are very much the same for double hits in the same array or in opposite arrays, the number of true coincidences is definitely larger for events in one and the same diode array. This is explained in the following by noting that "true" quaternary fission is in competition with "pseudo" quaternary fission. In the first case two LCPs are emitted right at scission, while in the second case it is a ternary yet particle unstable LCP which is ejected at scission and only on its flight to the detectors is decaying into lighter particles. Examples for such particle unstable ternary LCPs which can mimic quaternary fission are ^{8}Be decaying into two α-particles or ^{7}Li* which, from an excited state, will decay into an α and a τ. The task then is to disentangle true and pseudo quaternary fission. This is done under the basic assumption that all LCPs, either from ternary or quaternary fission, emanate from the neck between the two main fragments which is formed when the fragments start to recede from each other. For the so-called equatorial ternary LCPs, which are emitted roughly at right angles relative to the fission axis and which contribute the lions share to the ternary yield, this view is nowadays generally accepted. The angular distribution of LCPs from quaternary fission as observed in the present experiments follows the same pattern and, therefore, it appears justified to assume that also the majority of quaternary LCPs are ejected from the neck region. In the next step it is conjectured that the two LCPs showing up in true quaternary fission are emitted independently. Therefore the two LCPs should be distributed with equal probability in azimuthal angle around the fission axis. In the layout of the present experiment (s. Fig. 1) this means that the probability for the LCPs to impinge on the same or on opposite diode arrays should be the same. By contrast, for quaternary fission mediated by originally a single ternary LCP, the double hits in the same array should have a higher probability than for opposite arrays. This is best seen in the somewhat idealised case that the intermediate ternary LCP has a lifetime long enough to be fully accelerated in the Coulomb field of the two main fragments, but short enough to decay before arriving at the detectors. If, in addition, the decay energy is small in comparison with the kinetic energy gained in the Coulomb field, then the opening angle between the decay products will be small and the two LCPs are observed in the same array. In reality the situation may become more involved when the lifetime for particle decay is so short that the intermediate LCP is not fully accelerated before decay.

The evaluation of data along the above lines takes it for granted that quaternary events seen in opposite arrays of the LCP diodes have to be attributed to true quaternary fission. Pseudo quaternary events, by contrast, are to be evaluated as coincident events in one single array, of course after due subtraction of the contribution from true quaternary fission assumed to

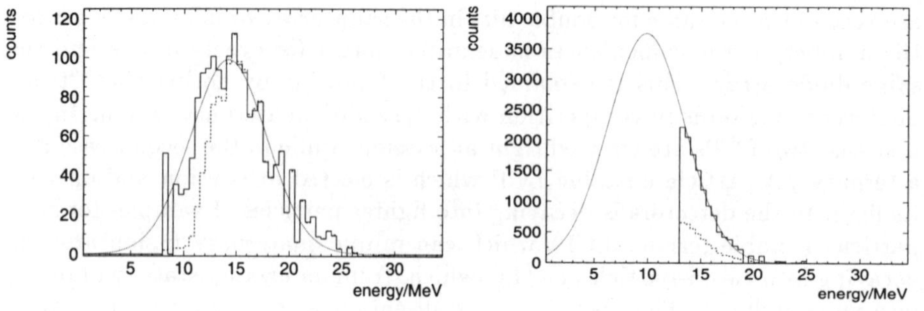

Figure 2. Energy distribution of αs from true (left) and pseudo (right) quaternary fission

be isotropic in azimuthal angle with respect to the fission axis. This ansatz for separating the two components of quaternary fission was checked by a Monte Carlo simulation. The polar angular distributions of true quaternary LCPs were taken to be identical to the distribution of ternary LCPs which are well known from experiment. For pseudo quaternary fission the intermediate ternary LCPs ^{8}Be (both in its ground and its first excited state decaying into 2αs) and ^{7}Li* (decaying from its 2nd excited state into $\alpha + \tau$) were taken into account. Except perhaps for the excited state in ^{8}Be* it is rather well justified to consider the decay to occur for virtually fully accelerated LCPs. The angular distribution of the resulting secondary LCPs, and in particular the opening angle between the two LCPs, is then found by simulating the ternary kinetic energy distribution and the decay of the primary LCP, the latter being taken as isotropic in the CM-frame of the LCP. It turned out that the observed angular correlation between the two LCPs from quaternary fission can be satisfactorily reproduced giving thus credit to the ansatz.

As a first example of results obtained after disentangling true and pseudo quaternary fission, the energy distributions of α-particles from (α, α)-coincidences are shown in Fig. 2. To the left the energy distribution of αs from true quaternary decay in the reaction ^{233}U(n,f) is given. The dotted histogram represents experimental raw data. The histogram drawn as a full line is the energy distribution after having applied corrections for energy losses in the target, backing and the Al protecting foil in front of the diodes. The continuous line is a Gaussian fitted to the experimental distribution bringing to evidence the cut-off for α-energies below 8 MeV. The average energy is found from the fit to be $\langle E \rangle = (14.2 \pm 0.8)$ MeV and the FWHM is (8.5 ± 0.8) MeV.

The average α-energy in the ternary decay was determined in the present experiment to be (15.7 ± 0.3) MeV and it is thus seen that the average α-energies from the two decay types differ by less than 2 MeV.

The right hand side of Fig. 2 shows the energy distribution of α-particles from the decay of an intermediate ^{8}Be for the reaction ^{235}U(n,f). The dotted histogram depicts raw data while the solid line histogram represents data having been corrected for energy losses and, in addition, for "hidden" events. Hidden events are $\alpha - \alpha$ coincidences showing up in the Monte Carlo simulation with narrow opening angles between the two LCP and impinging on the same diode. Those events will not be registered in the acquisition of data, i.e. they are lost in experiment. The correction becomes the more important the smaller the α-energies are and, therefore, the procedure was stopped for energies below 13 MeV when the correction factors become large and no longer reliable. Again a Gaussian has been fitted to the available data. From the fit average α-energies and FWHMs of (9.9 ± 1.0) MeV and (8.3 ± 1.0) MeV, respectively, are deduced for the ^{235}U(n,f) reaction under study. Especially the average α-energy is surprisingly low. For the reaction ^{233}U(n,f) the average energy measured is (9.0 ± 1.0) MeV, hence confirming the former result. Compared to ternary αs the average energy is down by about 6 MeV. These low α-energies point to kinetic energies of the decaying ^{8}Be nucleus of about 20 MeV, a figure which fits well into the systematics of ternary LCP energies as a function of LCP masses and charges. The differences in average energy are a further distinct feature differentiating true from pseudo quaternary fission.

Finally, results on yields obtained for the two reactions investigated are summarised in the table below. Ternary α-, true and pseudo (α, α)- and (α, τ) yields are given relative to binary fission. It is evident from the table that quaternary fission is indeed a very rare process with a by far predominant yield of pseudo quaternary fission mediated by ^{8}Be. Nevertheless, the existence of true quaternary fission with (α, α)- and (α, τ)-pairs in coincidence with fission fragments is clearly demonstrated by the present experiments for the first time.

	α	^{8}Be	(α, α)	^{7}Li*	(α, τ)
^{234}U*	2.1×10^{-3} $\pm 4\%$	1.9×10^{-7} $\pm 20\%$	5.0×10^{-8} $\pm 20\%$	7.0×10^{-9} $\pm 30\%$	6.4×10^{-9} $\pm 30\%$
^{236}U*	1.7×10^{-3} $\pm 4\%$	1.4×10^{-7} $\pm 20\%$	3.9×10^{-8} $\pm 20\%$	6.3×10^{-9} $\pm 30\%$	4.2×10^{-9} $\pm 30\%$

The support of the present work by grants from INTAS/Brussels and RAS/Moscow is gratefully achnowledged.

PRELIMINARY RESULTS AND FUTURE ACTIVITIES AT THE GARFIELD APPARATUS

F. GRAMEGNA, P.F.MASTINU, L.VANNUCCI
Laboratori Nazionali di Legnaro, Via Romea 4,35020 Legnaro(Pd)
Italy
E-mail: gramegna@lnl.infn.it

M. BRUNO, M.D'AGOSTINO, L. FIANDRI, A.LANCHAIS, G.VANNINI
I.N.F.N.Sezione di Bologna e Dip. Fisica dell'Università di Bologna, Via Irnerio, Bologna
Italy
E-mail: bruno@bo.infn.it

G.CASINI, M.CHIARI,A.NANNINI
I.N.F.N.Sezione di Firenze e Dip. Fisica dell'Università di Firenze, L.go E. Fermi, Firenze
Italy
E-mail: casini@fi.infn.it

A.BONASERA,S.CAVALLARO,A.COSMANO
Laboratori Nazionali del Sud e Dip. Fisica dell'Università di Catania, Catania,
Italy
E-mail: Bonasera@lns.infn.it

A. MORONI
I.N.F.N.Sezione di Milano, Via Caloria 16, Milano,
Italy
E-mail: arialdo.moroni@mi.infn.it

A.ORDINE
I.N.F.N.Sezione di Napoli, Via Cinthia,Napoli,
Italy
E-mail: ordine@na.infn.it

U.ABBONDANNO, P.M. MILAZZO, G.M. MARGAGLIOTTI
I.N.F.N.Sezione di Trieste e Dip. Fisica dell'Università di Trieste, Via Valerio2, Trieste
Italy
E-mail: abbondanno@ts.infn.it

A new apparatus has been designed and built to study reaction mechanisms in the energy regime of the ALPI linear accelerator of the Laboratori Nazionali di Legnaro (E/A = 5 – 20 MeV). In this paper the importance of studying these mechanisms will be underlined, no more as a problem limited to a narrow energy range or a single process, but as a continuous trend from low to high energies and from the physics of stable nuclei to that one regarding instabilities. With this remarks in mind, a first experiment has been performed studying the reaction $^{32}S+^{58}Ni$ at 11AMeV. Preliminary results show that important information can be derived on multi-body emission, which can contribute to renew the interest in this energy regime.

1 Introduction

GARFIELD is a new 4π apparatus, which has been built and installed at the Laboratori Nazionali di Legnaro [1]. The main goal of the physics program to be performed with GARFIELD is to investigate the reaction mechanisms in the energy range accessible with the ALPI linear accelerator of L.N.L. and to revisit some processes. This energy range has been studied since a lot of time and in fact:

- All those phenomena which could be directly reached and characterized with few detectors and almost inclusive measurements has been already faced and solved;
- Nevertheless several open questions are still there, even if it may result very difficult to overcome the experimental difficulties needed to extract new, interesting and complete information;
- It is no more a world of fashion and often people consider worthwhile in this regime only to study γ-ray spectroscopy, mixing up the importance of studying the nuclear structure with an experimental technique.

The GARFIELD collaboration intends to stress that there are still things to be done, which, being deeply linked to processes found at higher energies, can be studied through new analysis methods and more complex and complete apparatuses.

The main topics of the GARFIELD programs are mainly devoted to:

- Dissipative mechanisms: they are still alive at higher energies (up to 100 MeV/n), where other kinds of phenomena are present (multi-fragmentation, pre-equilibrium emission, neck emission etc.). A better understanding at lower energy, where they predominate, can be helpful even for a comprehension of their role in the Fermi region, where they compete with other processes.
- Multi-body processes: their presence and their role are not yet well defined at low energy. Surely there are sequential processes strongly contributing to their cross-section, which have to be studied in a more detailed way. Moreover, a threshold of the many-body channel has in any case to be experimentally determined, especially for central collisions. Besides that, a new interest has grown up lately on the bases of recent statistical model calculations, which predict a possible low energy phase transition.

2 Experimental Set-Up

The GARFIELD apparatus, which is lodged in a large scattering chamber in the III Hall of the Laboratori Nazionali di Legnaro, is mainly composed by three parts:

1. An annular three stage telescope, covering from θ=3° to θ=18°. It is made by 8 Axial Ionization Chambers, an annular strip silicon detector (8 strips x 8 sectors), 16 CsI(Tl) crystals, with photodiode readout.
2. Three position sensitive parallel plate detectors (20x20 cm^2), some of which coupled to Si(Li) detectors, placed respectively at 150 cm from the target in the forward direction for Projectile-Like Fragment

detection and at almost 50 cm, at correlated angles for Target-Like Fragment detection.

3. Two drift chambers, which are made by ≈ 200 double stage telescopes, where the residual energy signal is given by CsI(Tl) crystals, while the ΔE signal is provided by metallic micro-strips edged on glass (Micro-strip gas chambers) [2], which collect and pre-amplify the primary ionization produced by the reaction products in the gas. Each sector is composed by four CsI(Tl) plus four micro-strips signals, so that the angular resolution is 7.5° in ϕ and 1° in θ (this last obtained from drift time measurements).

Figure 1. One of the GARFIELD drift chambers.

3 Experimental Motivation

The aim of the first experiment performed was to characterize, at low center of mass energy, the decay of systems formed in central collisions by the incomplete fusion of the two reaction partners, looking for the onset of multi-fragment production.
From a theoretical point of view, multi-fragmentation models [3] predict anomalies of thermo-statistical observable, due to the opening of the phase space to the phase transition at about 1.5-2.5 AMeV excitation energy. Theory predicts also, in correspondence to this anomaly a huge increase in the variances of the experimental observable so far connected. When dealing with nuclei, which are finite and isolated systems, due to the conservation of charge, mass and energy, the signal of a first

order transition is given by an "S" shape in the caloric equation of state T (ε^*). At a first order transition the system cools down with rising energy and this is explained by the opening up of new fragmentation (and so de-excitation) channels, which need more energy to open up.
Up to now, at lower energies, the experimental apparatuses were not complete enough to perform precise measurements.
By looking at the predictions of models working under different assumptions (prompt fragmentation or sequential emission), it is clear that the many IMF decay channel is a good probe, together with the study of the caloric curve through a precise excitation function, to investigate the population of the accessible phase space.
Indeed, let us consider a system of mass A=90 and Z=45 in the energy range above considered: multi-fragmentation models, for which all the fragments are supposed to be present at the same time, predict a non-negligible probability of multi-fragment emission at about the same energy where they predict an S-shape in the caloric curve. It is predicted a ratio P3/P2≈4%, where P3 is the probability of events with an intermediate mass fragment ($Z \geq 3$) multiplicity M_{imf}=3, while P2 is the probability of binary emission (M_{imf}=2).
On the contrary, if we consider binary sequential models [4], at each step only two-body channels are permitted and the evaporation of light particles is the most probable, unless the decaying system is rotating (the system is there deformed and the probability of emitting fragments increases). For the GEMINI code at L=0 no three body events are then predicted and the charge distribution is U-shaped. Only for high values of angular momentum (L=48hbar, for the system considered, that is a value very close to Lmax=55hbar) the elemental distribution N(Z), the probability distribution and the charged particle multiplicity result on the average quite similar to the prediction of Statistical Multi-fragmentation Model (SMM).
Nevertheless, the fact that the distributions are similar on average does not imply that these distributions will be similar even event by event.
Indeed, when looking at the higher momentum of the charge distribution connected with the fluctuations of the partitions or when studying the correlations among the charges of the three largest fragments in each event (Dalitz plots), it is clear that the different origin of the fragment production becomes distinguishable.
The same and complementary information can be given by correlation functions as a function of the relative velocity and/or angle between the fragments.

4 Preliminary results

Motivated by the above theoretical background, overall by the stimulating model predictions and by the lack of coincidence data at low energy, a first system leading to a compound system of A≈ 100 has been studied to start exploring the phase space of decaying systems at ALPI.

In particular the ^{32}S + ^{58}Ni and ^{32}S + ^{64}Ni systems at 11 AMeV have been investigated. The data reduction has started by identifying the charge of the detected particles and by calibrating the energy of the micro-strip detectors.

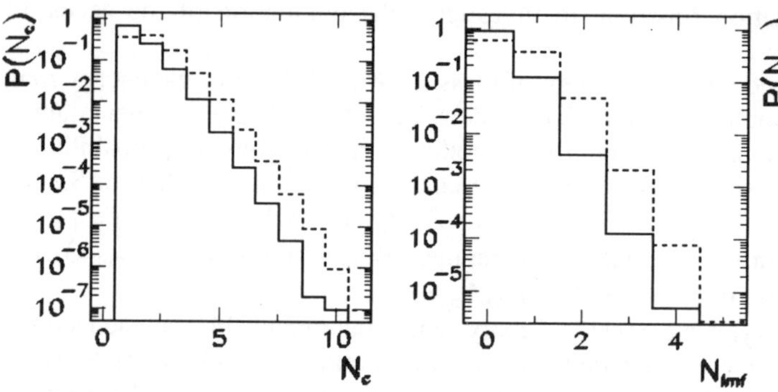

Figure 2: Charged particles (left panel – part a) and Intermediate Mass Fragment (Z>2) (right panel – part b) probability distributions. Dashed lines are the distributions detected by the whole apparatus; solid lines represent the distribution detected by the GARFIELD drift chamber.

Fig. 2a shows the probability distribution of charged particles Nc. The dashed line means the whole apparatus (annular detector covering θcm=7°- 40° and drift chamber, covering θcm=50°-110°), while the continuous line is the GARFIELD drift chamber only. The distributions are almost similar to what is expected from the theory.

Looking at the probability distribution of the number of intermediate fragments emitted (Nimf), as shown in Fig. 2b, we still have a good agreement with the theory (SMM or GEMINI with L=Lmax), if we look only to the drift chamber (continuous line), while an overproduction is found when including also the annular detector (dashed line). This can be explained under the hypothesis that the drift chamber selects somehow the central collisions, while contributes of semi-peripheral collisions are present when the annular is included. Indeed this can be also evidenced by looking at the elemental distribution under the constrain Nimf≥3.

The peak of the quasi-projectile is present in the distribution of the whole apparatus, while a smoother distribution is observed for the GARFIELD drift chamber only, as shown in Fig. 3.

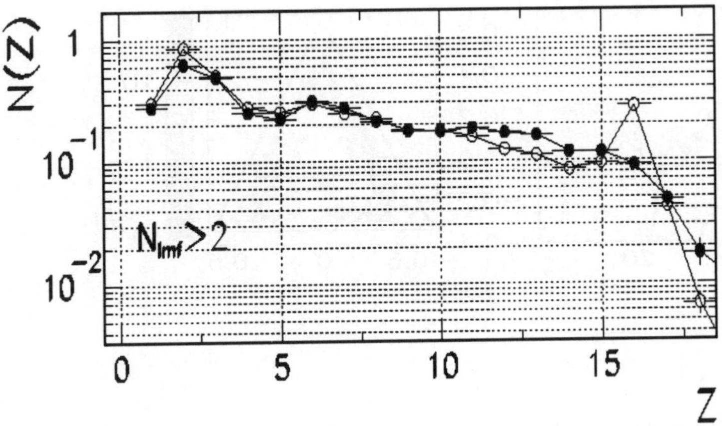

Figure 3. Charge distribution of light charged particles and fragments detected in the whole apparatus (open symbols) and in GARFIELD (full symbols), for events with at least three fragments. Both distributions are normalized to the total number of events.

If a further constraint is included on the total charge detected (Ztot≥32), the distributions of the three largest fragments in each event become more and more similar one to the other. This can be seen in a powerful way by looking at the Dalitz plot, which results filled up in the central part.
The comparison with the theory show a quite impressive similarity with the SMM predictions, while rules out the GEMINI sequential emission, which, for high values of angular momentum, can only predicts very asymmetric events, essentially made by two Lithium plus a large fragment. The three distributions are compared in Fig 4, fig. 5 and Fig. 6 respectively.

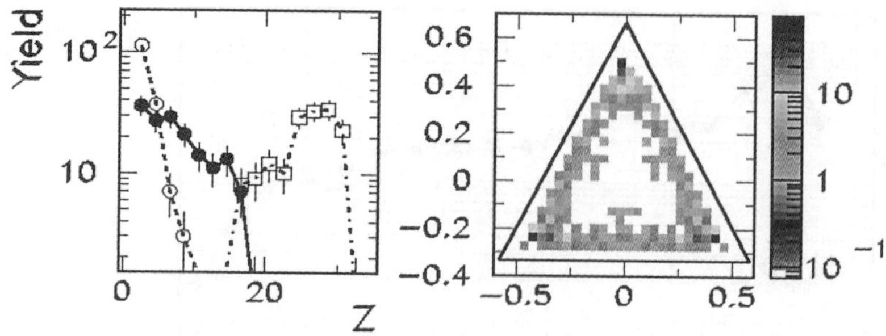

Figure 4. GEMINI predictions: Charge distribution of the three largest fragments in each events (left panel) and corresponding Dalitz plot (right panel).

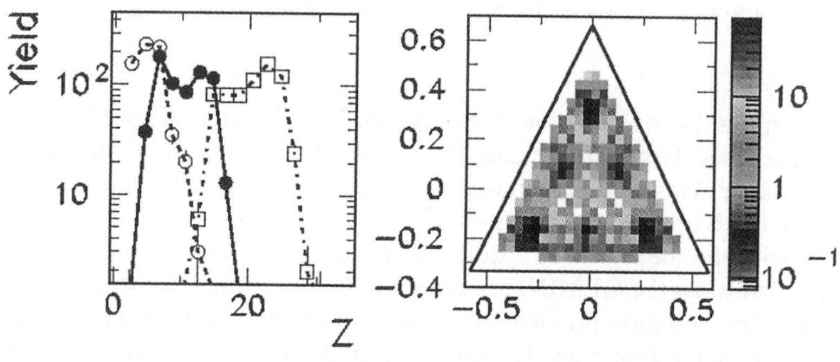

Figure 5. SMM predictions: Charge distribution of the three largest fragments in each events (left panel) and corresponding Dalitz plot (right panel).

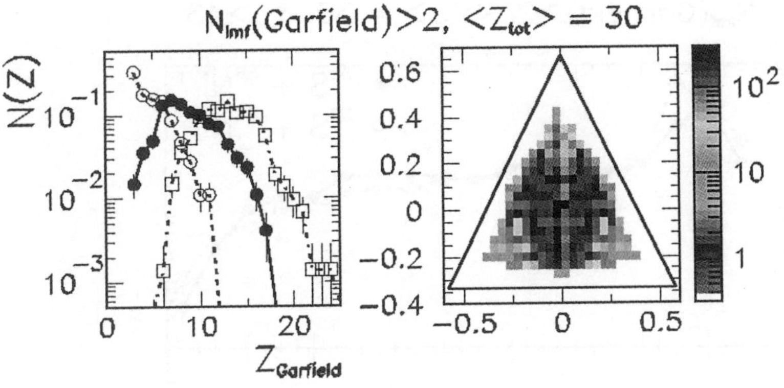

Figure 4. Experimental data: Charge distribution of the three largest fragments in each events (left panel) and corresponding Dalitz plot (right panel).

Multi-fragment events need, however, further investigation to be better characterized, in order to define both their emission source and their production mechanisms. Velocity and angular correlation are therefore necessary and will be performed in the next future. This part of analysis is still under development: it will give us more insight in the understanding of the nature of multi-fragment production at low energy.

Indeed other interesting findings come from the analysis of the charge distribution of fragments detected at large C.M. angles, by changing the neutron content of the system as shown in Fig. 7. As stated at the beginning, a small statistics was collected on the $^{32}S+^{64}Ni$ at the same incoming beam energy (11 AMeV) of the previous system. The two reactions measured $^{32}S+^{58}Ni$ and $^{32}S+^{64}Ni$, characterized respectively by a value of N/Z=1.05 and N/Z=1.18, showed a quite different behavior. In the first reaction, in fact, oscillations greater than the statistical errors around Z=6 and Z=12 are evident, while in the second case the oscillations are smoothed and the production of Z=3 is larger by a factor 2.

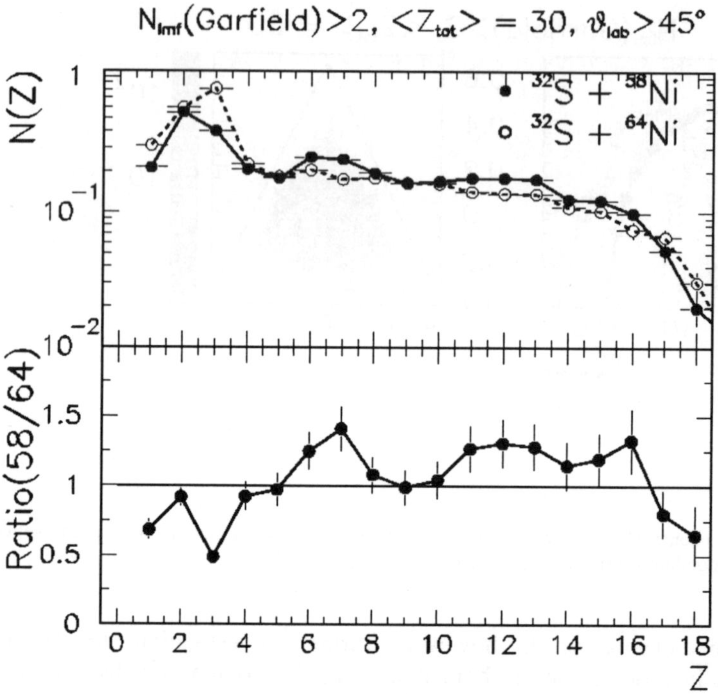

Figure 7. Charge distribution (upper panel) in the reaction 32S+58Ni and 32S+64Ni, for events with at least three fragments detected in GARFIELD. The lower panel shows the ratio of the two-charge distribution.

Effects of suppression/enhancement in the charge distribution could be ascribed to the opening/closure of decay channels, due to the energy conservation constraint on the phase space and have to be carefully investigated by means of models.

5 Conclusions

A new detector has been built and installed at LNL, Padua. Measurements performed at the ALPI accelerator with an 11 AMeV ^{32}S beam on Nickel targets have shown that interesting features exists at these energies, which can give more enthusiasm in the analysis and study of this energy range. Multi-fragment production has been evidenced and the characterization of the reaction mechanisms will be performed through a complete and exclusive analysis, by means of multi-correlation methods. Structure effects can also be responsible for the difference in the opening of decaying channels at low excitation energies, which can be evidenced in playing with the entrance channel N/Z ratio.

6 Acknowledgements

We thank A. Cortesi, R. Cavaletti, A. Boiano, R. Baccomi, P. Del Carmine and G. Tobia for their skillful assistance in the preparation of the apparatus, both from the mechanical point of view and for the electronics. Without their help, it would have been impossible to join our results. We want also to thank the staff of the LINAC Accelerator and the people of the User Service of LNL who helped us and sustained our work along all these years.

References

1. F. Gramegna et al., GARFIELD: a General ARray for Fragment Identification and for Emitted Light particles in Dissipative collisions – to be published; F. Gramegna et al. from LNL, Annual Reports 1994 to LNL, Annual Report 2000
2. F. Gramegna et al. Nucl. Instr. And Methods **A389**(1997)474
3. D.H.E. Gross et al. Zeit. Fur Phys. **D39**(1997)75; Rep.Prog.Phys. **53**, (1990)605; Phys. Rep **279**(1997)119.
   J.P. Bondorf et al. Phys. Rep. **257**(1995)133
   P. Chomaz and F. Gulminelli, **A647**(1999)153; Phys. Rev. Lett. **85**(2000)3587
   A.Bonasera, M.Bruno, C.O.Dorso and P.F. Mastinu, Rivista Nuovo Cimento Vol.23 N.2.
4. R.J. Charity et al., Nucl. Phys. **A483**(1988)371

MEAN-FIELD CALCULATIONS OF SUPER-HEAVY ELEMENTS

P.H. HEENEN
Service de Physique Nucléaire Théorique, U.L.B.-C.P.229, B-1050 Brussels, Belgium

Several methods based on effective interactions or Lagrangians are available today. Although different in many respects (use of zero range or finite range interactions, relativistic or non relativistic framework, different treatments of pairing correlations), their applications to super heavy nuclei have shown converging results which still have to be incorporated in macroscopic approaches. Many efforts are also actually devoted to the improvements of the effective interactions, especially of the pairing force. I shall review the main results obtained thanks these methods in the last years.

1 Introduction

Thanks to the developments of new techniques and to the increase of computational power, mean-field methods based on effective 2-body interactions can now be systematically applied to the study of super-heavy nuclei. In this talk, I will shortly introduce the main mean-field methods whose only phenomenological ingredient is an effective nucleon-nucleon interaction or an effective Lagrangian (in the case of the relativistic mean-field approach). I will present calculations of fission barriers which demonstrate that self consistency between all the nucleons has to be taken into account. I will compare some predictions with experimental data and show the predictive power of self consistent methods. I will finally discuss the deformation properties of super-heavy nuclei.

2 The methods

The microscopic mean-field methods can be classified according to the way the interactions between the nucleons is treated. The general framework can be either non relativistic or relativistic. In the first case, the nucleon-nucleon interaction can be either zero range or finite range.

The most widely used zero-range interaction is the Skyrme force, for which several parametrizations have been introduced. Some of the most recent parametrizations have been adjusted by Chabanat et al.[1]. They are identified by the letters SLy followed by a number and are usually considered as the most reliable forces to study unknown regions of the nuclear chart. They differ from previous adjustments by a very careful study of nuclear and neutron matters, with the hope to develop an interaction valid also for applications in astrophysics. The main constraints that have been used to built the SLy forces are:

- a good reproduction of the saturation point of symmetric infinite nuclear matter (E /A around -16MeV, r_0 around 0.16fm^{-3})
- a compression modulus of nuclear matter around 230MeV
- a symmetry energy around 32MeV
- a constraint on the reproduction of a fundamental equation of state of pure neutron matter
- the binding energies and rms radii of doubly magic nuclei (^{16}O, ^{40,48}Ca, ^{56}Ni, ^{132}Sn, ^{208}Pb)

The spin orbit interaction has been adjusted on ^{208}Pb and not on ^{16}O as for the interactions which were determined in the 70's and the 80's.

The only finite range interaction systematically applied to the study of super heavy nuclei (or even more generally to nuclear structure calculations) is the Gogny force[2].

Recently, mean-field methods based on relativistic Lagrangians (see for instance ref. [3]) have been applied to nuclear structure studies. The relativistic mean-field method introduced by Ring and coworkers is a Hartree method (the exchange terms are thus neglected). Its non relativistic limit has been shown to be very similar to a conventional mean-field method with a Skyrme force. One of the advantage of the RMF method is that the spin-orbit is introduced naturally and is not put by hand. The non relativistic reduction of the spin-orbit term has been shown to have an isospin dependence different from the conventional spin-orbit parametrization used in both Gogny and Skyrme interactions. This isospin dependence has been introduced in some new Skyrme parametrizations. However, one must note that the spin-orbit derived from a relativistic Hartree-Fock method would still be different. Unfortunately, there are presently only a very limited number of applications of the relativistic HF method.

The Gogny interaction is the only one to be used to describe at the same time mean-field and pairing correlations. This is due to the fact that a zero-range interaction like the Skyrme force has an unphysically large strength for high particle momenta and that up to now, the pairing interaction has only been introduced non relativistically in RMF calculations.

The most commonly pairing interaction used with a Skyrme force nowadays is a zero range density dependent force:

$$V_P = \frac{V_0}{2}(1 - P_\sigma)(1 - \frac{\rho(\vec{r}_1)}{\rho_c})\,\delta(\vec{r}_1 - \vec{r}_2) \quad . \tag{1}$$

V_0 and ρ_c are two parameters and $\rho(\vec{r})$ is the total local single-particle density in coordinate space. In order to avoid divergences, the definition of the force involves also an energy cut-off parameter in the valence single-particle space to limit the active pairing space above the Fermi level.

The parameter ρ_c determines the spatial dependence of pairing. In the limit where ρ_c goes to infinity, we recover the usual delta force. In this case, the matrix elements of V_P depend on properties of the wave functions in the whole volume of the nucleus. This leads to volume active pairing fields. On the other hand, for ρ_c equal to the nuclear saturation density ρ_0, the interaction is weak inside the nucleus and strongly attractive outside, leading to a surface active pairing field.

The pairing energy is a small fraction of the total energy of a nucleus (of the order of 10MeV on 1GeV). It is therefore hard to find a simple quantity which depends mainly on pairing correlations.

3 Applications

3.1 Self consistency effects

A first question that one can ask is whether self-consistency between all the nucleons is important to generate the mean field or if one can neglegt it and use simpler model, based on parametrized mean fields, like a Woods Saxon potential. An interesting answer to this question has been given by Pomorski and Berger [4]. These authors have first determined the fission barriers of actinids in a fully consistent calculation using the Gogny interaction. They have redone the calculation with the constraint that the neutron and the proton densities have the same shape, as it is the case in calculations perfomed with a parametrized mean field. Differences of the order of a few MeV between the barriers were obtained, an effect which has a large impact on the barrier penetration probabilities and thus on the fission half lifes.

On figure 1, are compared the fission barriers obtained for the Z=112, N=164 isotope with a Woods Saxon potential and with the Skyrme SLy4 interaction in a HF+BCS calculation (from ref. [8]). The structure of the curves is similar in both cases, with several minima as a function of deformation, two on the prolate side and one on the oblate side. However, the self-consistent minima are significantly deeper, leading to higher fission barriers and to nuclei which are more stable against fission.

3.2 Energy predictions

The conventional mean field interactions lead to rms deviations with respect to experimental masses of the order of 2 MeV (see ref [5]). However, quantities based on energy differences are in much better agreement with the data. On figure 2 are compared with the experimental data the Q_α energies obtained in Skyrme HFB calculations performed with the SLy4 interaction (from ref. [6]). The agreement is excellent, giving confidence in the ability of mean field calculations to predict Q_α energies in unknown regions of the nuclear chart. Similar results have been obtained with the RMF method [7] and with the Gogny interaction [9] .

3.3 Deformations

The density of single-particle levels close to the Fermi energy is very large for super-heavy nuclei. As a consequence, there are many competing shell effects, spherical and deformed. On figure 3 are plotted the difference of energies between the oblate and prolate minima obtained with the SLy4 Skyrme interaction (from ref. [8]). Three regions can be distinguished: a region of prolate deformation around N=176; a region of oblate deformation for Z larger than 114 and low values of N and a region of spherical shapes centered around the magic number Z=126. The topology of the quadrupole deformation energy surfaces can be quite complicated, as illustrated on figure 4 (from ref. [8]). The three selected isotopes are connected by alpha decays. Starting from a soft oblate minimum at Z=120, one is going to a well deformed prolate nucleus at Z=116 through a gamma soft nucleus at Z=118. This rapid change of deformation should have an effect on alpha emission probabilities. It also shows the limitation of pure mean-field calculations and that dynamical effects, mixing different quadrupole shapes, must be considered.

Results concerning the position of magic gaps in super-heavy nuclei are contradictory. An illustrative example is given on figure 5 (from ref. [9]) where the shell gaps obtained with the Gogny interaction. are plotted. These shell gaps correspond to the difference between the energies of the last occupied and of the first empty levels in a spherical configuration. There is a clear shell effect at N=184, for all number of protons, a result similar to several other mean-field calculations. For the protons, there is no such obvious magic number. The magnitude of the shell gap is strongly dependent on the neutron number. Moreover, as can be seen on figure 3, the isotopes for which there exist a gap for some neutron number may be deformed for others. In particular, the existence of a magic number at Z=114 which was predicted on the basis of macroscopic calculations is not confirmed by microscopic mean-field studies.

3.4 Alpha decay chains

On figure 6 are given the spectra of odd nuclei along the chain of alpha emission starting at one of the Z=114 isotope detected at Dubna[10]. Here also the energies of the Q_α particles are correctly predicted by the calculation. In this case, the odd nuclei are calculated fully self-consistently, a feature which requires to calculate several configurations for each isotope. This is necessary to determine which state in each nucleus is the most favorable one for the emission of the α particle, some low lying states being excluded by a spin mismatch with the daughter nucleus. A similar agreement has been obtained for the other recently detected chains.

4 Conclusions

In this talk, we have shown that self-consistent microscopic calculations of super-heavy nuclei can be systematically performed. The agreement with the experimental data is very good and gives some confidence in the results obtained in unknown areas. One of these results concern the deformation properties of super-heavy nuclei, which can vary significantly along an isotopic line. This result indicates the necessity of going beyond pure mean-field studies and to treat the dynammics with respect to shape collective degrees of freedom. An important conclusion form microscopic studies is also that the proton number 114 is not confirmed to be magic, a feature which should have consequences on the position of a possible isalnd of stability in the super-heavy region.

4.1 Acknowledgments

This research was supported in part by the Wallonie/Brussels-Poland integrated actions program. I would like to thank J.-F. Berger and S. Cwiok to have provided me several figures presented here.

References

1. E. Chabanat, P. Bonche, P. Haensel, J. Meyer and R. Schaeffer, Nucl. Phys. **A635** 231 (1998).

2. J. Dechargé and D. Gogny, Phys. Rev. **C21** 1568 (1980).
3. P. Ring, Prog. Part. Nucl. Phys. **37** 193 (1996).
4. J.F. Berger and K. Pomorski Phys. Rev. Lett. **85** 30 (2000).
5. Z. Patyk A. Baran, J.F. Berger, J. Dechargé, J. Dobaczeswki, P. Ring and A. Sobiczewski, Phys. Rev. **C59** 704 (1999).
6. S. Cwiok, W. Nazarewicz and P.-H. Heenen, Phys. Rev. Lett. **83** 1108 (1998).
7. M. Bender Phys. Rev. **C61** 031302 (2000).
8. S. Cwiok, W. Nazarewicz and P.-H. Heenen, in preparation.
9. J.F. Berger, L. Bitaud, J. Dechargé, M. Girod and K. Dietrich Nucl. Phys. **A685** 1c (2001).
10. Y. T. Oganessian et al., Phys. Rev. Lett. **83** 3154 (1999).

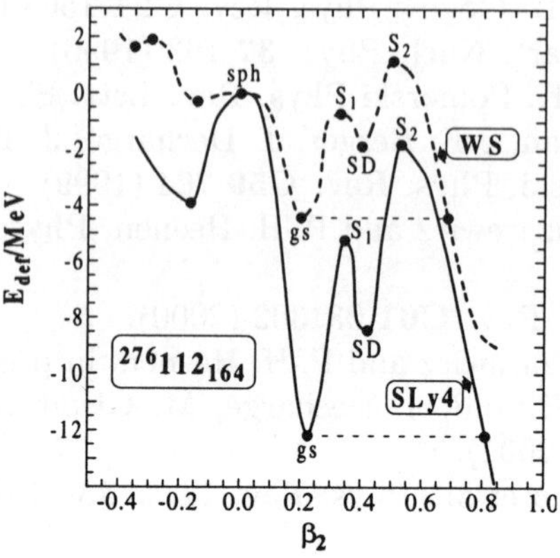

Figure 1. Figure 1: Comparison between the axial quadrupole deformation energy curves obtained in a HFBCS calculation using the Skyrme SLy4 interaction and in a Strutinsky calculation with a Woods Saxon potential.

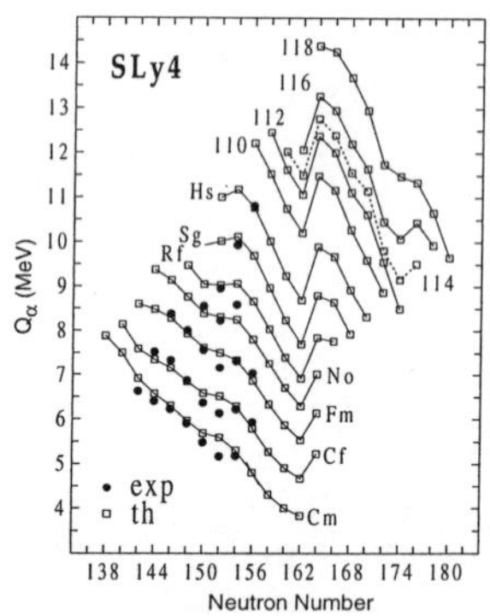

Figure 2. Figure 2: Q_α values for even-even nuclei with $96 \leq Z \leq 118$ obtained in the HFB method and the SLy4 interaction. The experimental Q_α values are plotted as filled circles (from ref [6]).

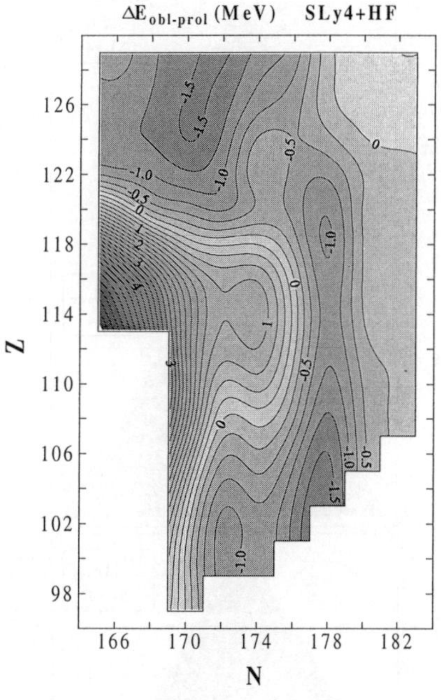

Figure 3. Figure 3: Difference between the energies of the oblate and prolate minima obtained in constrained HFBCS calculations using the Skyrme SLy4 interaction.

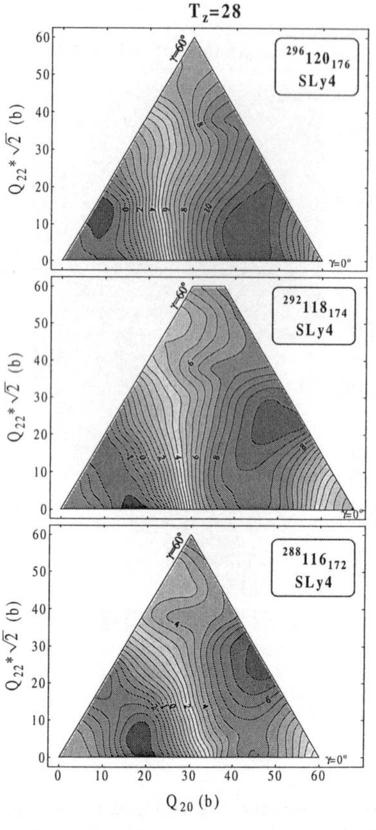

Figure 4. Figure 4: Potentiel energy surfaces obtained for three isotopes along the T_z=28 line. The HFBCS calculations have been performed with the SLy4 interaction.

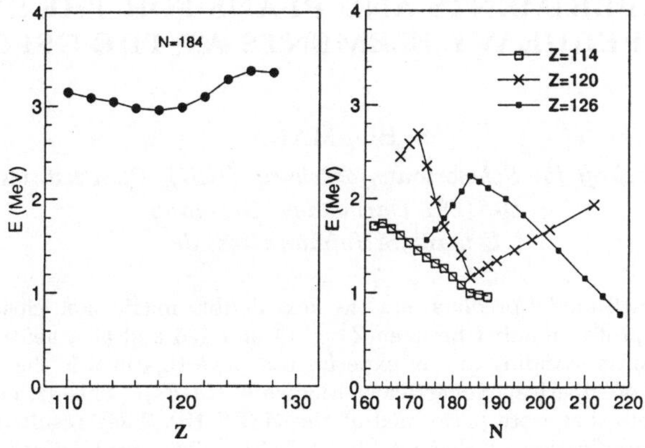

Figure 1. Left: spherical N=184 shell gap as a function of the nucleus proton number Z. Right: spherical Z=114, 120 and 126 shell gaps as functions of the nucleus neutron number N.

Figure 5. Figure 5: Shell gaps obtained for N=184 neutrons as a function of the proton number and for three values of the proton number as a function of the neutron number. The HFB calculations have been performed with the Gogny force (from ref. [9]).

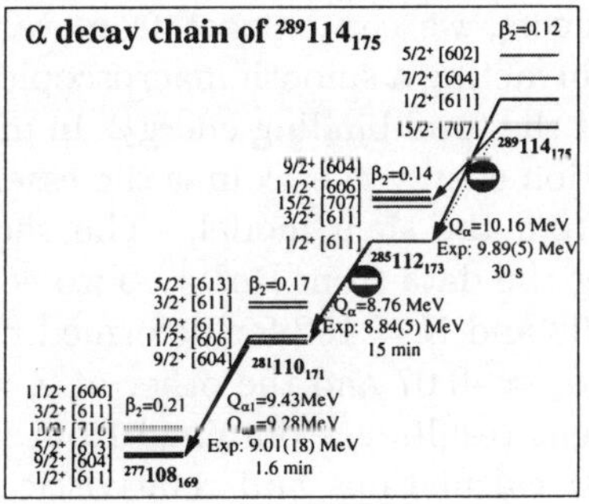

Figure 6. Figure 6: Predicted structure of the lowest one-quasiparticle excitations in the $^{289}114$ α-decay chain. Each excitation is characterized by the Nilsson quantum numbers of the odd neutron. The deformations of each isotopes is also given, together with the Q_α value for the ground-state to ground-state transition and the experimental Q_α value (from ref [6]).

RECENT EXPERIMENTS AND PLANS FOR THE SYNTHESIS OF SUPERHEAVY ELEMENTS AT THE GSI SHIP

S. HOFMANN

Gesellschaft für Schwerionenforschung (GSI), Planckstrasse 1, D-64291 Darmstadt, Germany
E-mail: S.Hofmann@gsi.de

The nuclear shell model predicts that the next doubly magic shell-closure beyond ^{208}Pb is at a proton number between Z = 114 and 126 and at a neutron number N = 184. The outstanding aim of experimental investigations is the exploration of this region of spherical 'SuperHeavy Elements' (SHEs). This article describes the experiments that were performed at the GSI SHIP. They resulted in an unambiguous identification of elements 107 to 112. They were negative so far in searching for elements 113, 116 and 118. The measured decay data are compared with theoretical predictions. Some aspects concerning the reaction mechanism are also presented.

1 Decay properties of SHEs

The basic step necessary for the determination of the stability of SuperHeavy Elements (SHE) is the calculation of the ground-state binding energy. As a signature for shell effects, we can extract from various models the shell-correction energy by subtracting a smooth macroscopic part (derived from the liquid-drop model) from the total binding energy. In macroscopic-microscopic models the shell-correction energy is of course the essential input value which is calculated directly from the shell model. The shell-correction energy is plotted in Fig. 1a using the data from Ref. [1]. Two equally deep minima are obtained, one at Z = 108 and N = 162 for deformed nuclei with deformation parameters $\beta_2 \approx 0.22$, $\beta_4 \approx -0.07$ and the other at Z = 114 and N = 184 for spherical SHEs. Different results are obtained from self-consistent Hartree-Fock-Bogoliubov (HFB) calculations and relativistic mean-field models [2,3]. They predict, as indicated by the dashed lines in Fig. 1, for the spherical nuclei shell closure at Z = 120 or 126 and N = 172, with shell strengths being also a function of the amount of nucleons of the other type.

The dominating partial half-life is shown in Fig. 1b for even-even nuclei. The two regions of deformed heavy nuclei near N = 162 and spherical SHEs merge and form an island of α emitters surrounded by fissioning nuclei. The longest half-lives are 1000 s for deformed heavy nuclei and 30 y for spherical SHEs. It is interesting to note that the longest half-lives are not reached for the doubly magic nucleus $^{298}_{184}114$, but for Z = 110 and N = 182. This

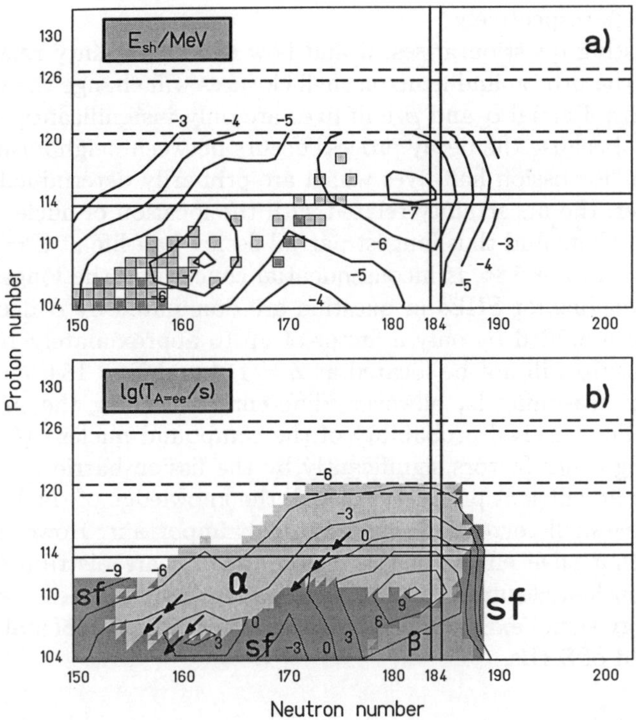

Figure 1. Ground-state shell-correction energy (E/MeV) (a) and dominating partial α, β and spontaneous fission half-life [$\log(T_{1/2}/s)$] (b) from calculations using the macroscopic-microscopic model [1,4,5]. The nuclei presently known or under investigation are marked by filled squares in (a). The arrows in (b) mark the recently measured decay chains of even-even nuclei [6,7,8].

is a result of the continuously increasing Q_α values with increasing element number. Therefore, the α decay becomes the dominant decay mode beyond element 110 with continuously decreasing half-life. The half-lives of nuclei at N = 184 and Z<110 are reduced from β^- decay.

The decay chains of recently synthesized even-even nuclei, ^{270}Hs [6], ^{270}110 [7] and ^{292}116 [8], are also drawn in the figure. In these cases the decay chains end by spontaneous fission, as theoretically predicted, at ^{262}Rf,

^{262}Sg and ^{280}110, respectively.

The interesting question arises, if and how the uncertainty related with the location of the proton and neutron shell closures will change the predicted half-life of SHEs. Partial α and β half-lives are only insignificantly modified by shell effects, because the decay process occurs between neighboring nuclei. This is different for fission half-lives which are primarily determined by shell effects. However, the uncertainty related with the location of nuclei with the strongest shell-effects and thus longest partial fission half-life at Z = 114, 120 or 126 and N = 172 or 184, is inconsequential concerning the longest 'total' half-life. The regions for SHEs in question are dominated by α decay. And α decay will be modified by only a factor of up to approximately 100, if the double shell closure will not be located at Z = 114 and N = 184.

The line of reasoning is, however, different concerning the production cross-section. The survival probability of the compound nucleus (CN) is determined among other factors significantly by the fission-barrier. Therefore, with respect to an efficient production yield, the knowledge of the location of minimal negative shell-correction energy is highly important. However, it may also turn out that shell effects in the region of SHEs are distributed across a number of subshell closures. In that case a wider region of less deep shell-correction energy would exist with corresponding modification of stability and production yield of SHEs.

2 Experimental Results

2.1 The new elements 110, 111 and 112

Element 110 was discovered in 1994 using the reaction ^{62}Ni + ^{208}Pb $\rightarrow$ ^{270}110* [9]. The main experiment was preceded by a thorough study of the excitation functions for the synthesis of ^{257}Rf and ^{265}Hs in order to determine the optimum beam energy for the production of element 110 (see Fig. 2). The data revealed that the maximum cross-section for the synthesis of element 108 was shifted to a lower excitation energy, different from the predictions of reaction theories. A total of four decay chains were observed from the decay of ^{269}110.

The isotope ^{271}110 was synthesized with a beam of the more neutron-rich isotope ^{64}Ni [10]. The important result for the further production of elements beyond meitnerium was that the cross-section was enhanced from 3.5 pb to 15 pb by increasing the neutron number of the projectile by two, which gave hope that the cross-sections could decrease less steeply with more neutron-rich projectiles. Cross-sections were measured at three projectile energies.

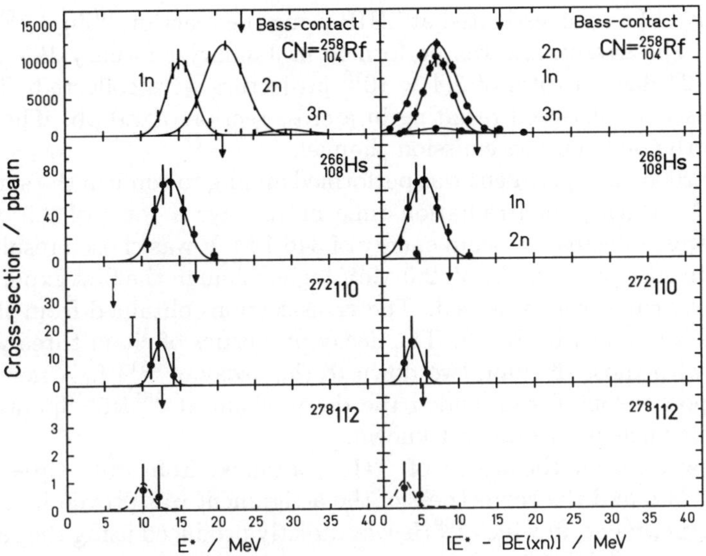

Figure 2. Measured even element excitation functions.

The values are plotted in Fig. 2.

Two more isotopes of element 110 have been reported in the literature. The first is ^{267}110, produced in the irradiation of ^{209}Bi with ^{59}Co [11,12]. The second isotope is ^{273}110, reported to be observed in the irradiation of ^{244}Pu with ^{34}S after the evaporation of five neutrons [13,14]. Both observations need further experimental clarification.

The even-even nucleus ^{270}110 was synthesized using the reaction ^{64}Ni + ^{207}Pb [7]. A total of eight α-decay chains was measured during an irradiation time of seven days. Decay data were obtained for the ground-state and a high spin K isomer, for which calculations predict spin and parity 8^+, 9^- or 10^-. The new nuclei ^{266}Hs and ^{262}Sg were identified as daughter products after α decay. Spontaneous fission of ^{262}Sg terminates the decay chain.

Element 111 was synthesized in 1994 using the reaction ^{64}Ni + ^{209}Bi $\rightarrow$ ^{273}111*. A total of three α chains of the isotope ^{272}111 were observed [15]. Another three decay chains were measured in a confirmation experiment in October 2000 [16].

Element 112 was investigated at SHIP using the reaction $^{70}Zn + ^{208}Pb \rightarrow ^{278}112^{*}$ [17]. The irradiation was performed in January-February 1996. Over a period of 24 days, a total of 3.4×10^{18} projectiles were collected. Two α-decay chains were observed resulting in a cross-section of 1.0 pb. They were assigned to the one neutron-emission channel.

In May 2000 an experiment was performed aiming to confirm the synthesis of $^{277}112$ [18]. During an irradiation time of 19 days a total of 3.5×10^{18} projectiles was collected. A beam energy of 346.1 MeV was chosen resulting in an excitation energy of 12.0 MeV, 2.0 MeV higher than in the first experiment. A third decay chain was observed. The cross-sections obtained from the two irradiations are shown in Fig. 2. The decay properties of chain three were in agreement with those of chain two down to the decay of ^{261}Rf. A new result was that spontaneous fission ended the decay chain at ^{261}Rf. Spontaneous fission of this nucleus was not yet known.

The new data on the decay of ^{261}Rf, obtained from only three decay chains of $^{277}112$, and the correctness of the assignment was proven in a recent chemistry experiment, in which ^{269}Hs was directly produced using the reaction $^{26}Mg + ^{248}Cm \rightarrow ^{274}Hs^{*}$ [6]. After separation of HsO_4 from the other reaction products, a total of about five decay chains was measured. A spontaneous fission branching of ^{261}Rf was deduced, and also the α energy of 8.52 MeV was confirmed, which had been measured in the case of chain two.

2.2 Search for element 113, 116 and 118

The experiment to search for element 113 was complicated by the fact that no excitation function for the production of odd elements from dubnium (Z=105) to element 111 was known well enough to allow for an estimate of the optimum beam energy for production of element 113. Therefore, the two reactions $^{50}Ti + ^{209}Bi \rightarrow ^{259}Db^{*}$ and $^{58}Fe + ^{209}Bi \rightarrow ^{267}Mt^{*}$ were investigated before the main experiment [19]. The result of these experiments was that the position and widths of the excitation function of the 1n channel for meitnerium and of the 1n to 3n channel for dubnium are the same as for the next lighter even element, only the cross-section value is smaller.

During a 46-days experiment in 1998 a beam dose of 7.5×10^{18} projectiles of ^{70}Zn was collected. At a mean excitation energy of 10.7 MeV, no event was measured that could be assigned to element 113. A cross-section limit of 0.6 pb resulted.

The irradiation to search for $^{290}116$, produced by radiative capture of ^{82}Se and ^{208}Pb, was completed in November-December, 1995. Cross-section limits of 5 pb were obtained at four excitation energies between 0 and 8 MeV.

The experiment precludes the fact that the radiative capture channel has an unusually high fusion cross-section at extremely low free reaction energies.

One data point was measured at an excitation energy of 10.8 MeV which is high enough for emission of one neutron. The obtained cross-section limit of 9 pb is shown in Fig. 3a.

An extrapolation of the measured cross-section data shown in Fig. 3a beyond element 112 results in values of about 10 and 1 fb for the synthesis of element 116 and 118, respectively. Although we were aware that the cross-section trend may change in the transition from deformed heavy nuclei to spherical superheavy nuclei, the announcement of the successful synthesis of $^{293}118$ at Berkeley was a surprise. A cross-section of 2.2 pb was deduced from the 3 event chains measured in the reaction $^{86}Kr + ^{208}Pb \rightarrow ^{294}118^{*}$ [20] (see Fig. 3a).

In order to confirm the data obtained in Berkeley, the same reaction was investigated at SHIP in the summer of 1999. The experiment is described in detail in [18]. No event chain was detected similar to those observed in Berkeley. The cross-section limit resulting from the SHIP experiment is 1.0 pb.

Although the Berkeley data on the synthesis of $^{293}118$ could not be proved, the negative result does not disprove those data. Several reasons could plausibly explain the difference (see [18]), among which statistical fluctuations would be the simplest interpretation.

Further attempts to prove the Berkeley data were made by repeating the experiment at the Berkeley Gas-filled Separator (BGS) itself. The experiments were made at the end of 2000 and in April-May 2001. The analysis of the data is presently under investigation.

3 Cross-sections, fusion valleys and excitation energy

The main features which determine the fusion process of heavy ions are (1) the fusion barrier and related beam energy and excitation energy, (2) the ratio of surface tension versus Coulomb repulsion which determines the fusion probability and which strongly depends from the degree of asymmetry of the reaction partners (the product Z_1Z_2 at fixed $Z_1 + Z_2$), (3) the impact parameter and related angular momentum, and (4) the ratio of neutron evaporation versus fission probability of the CN. In fusion of SHEs the product Z_1Z_2 reaches extremely large and the fission barrier extremely small values. In addition, the fission barrier itself is fragile, because it is solely built up from shell effects. For these reasons the fusion of SHEs is hampered, whereas the fusion of lighter elements proceeds unhindered through the contracting effect of surface tension.

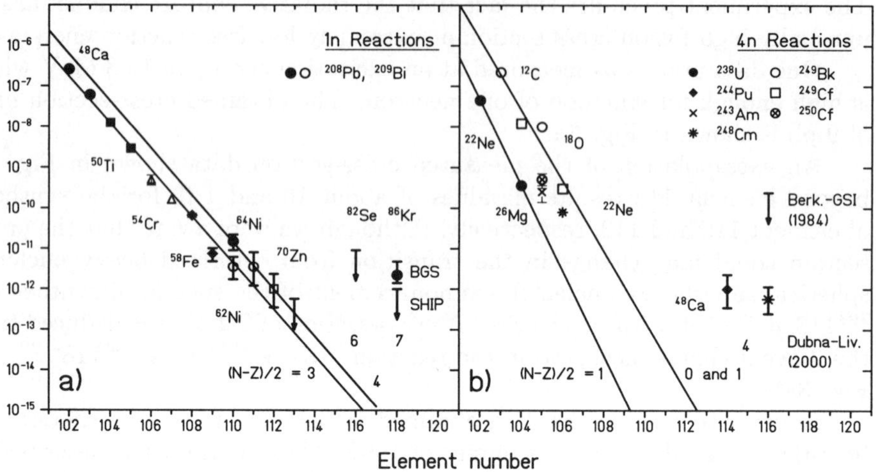

Figure 3. Measured cross-sections and cross-section limits for reactions using ^{208}Pb and ^{209}Bi targets and one neutron evaporation (left) and for reactions using actinide targets and four neutron evaporation (right).

The effect of Coulomb repulsion on the cross-section starts to act severely for fusion of elements beyond Fm. From there on a continuous decrease of cross-section was measured from microbarns for the synthesis of nobelium down to picobarns for the synthesis of element 112. The data obtained in reactions with ^{208}Pb and ^{209}Bi for the 1n evaporation channel at low excitation energies of about 10-15 MeV (therefore named *cold fusion*) and in reactions with actinide targets at excitation energies of 35-45 MeV (*hot fusion*) for the 4n channel are plotted in Fig. 3. The surprising and so far not understood data are the relatively high cross-sections for the synthesis of element 118 [20] (1n channel) and 114 and 116 [8,21] (4n channel). In both cases the obtained values of about 1 pb deviate considerably from the trend set by fusion of the lighter elements.

An explanation could be a relatively high and wide fission barrier of the CN which is created by strong shell effects in the region of spherical SHEs.

A number of excitation functions was measured for the synthesis of elements from No to 110 using Pb and Bi targets [18]. For even elements these data are shown in Fig. 2. The maximum cross-section (1n channel) was measured at beam energies well below a fusion barrier calculated in one dimension [22]. In the case of element 110 (CN = $^{272}110$) the maximum cross-section was measured at a kinetic energy of 236.2 MeV in the centre-of-mass system. At that energy projectile and target are just reaching a contact configuration in a central collision due to the repulsive Coulomb potential. However, the relatively simple fusion barrier based on the Bass model [22] is too high and a tunneling process through this barrier cannot explain for the measured cross-section.

Various processes are possible and are discussed in the literature which result in a lowering of the fusion barrier. Among these transfer of nucleons and excitation of vibrational degrees of freedom are the most important [23,24,25,26,27,28,29].

In the case of actinide targets, the target nucleus is strongly deformed and the height of the Coulomb barrier is a function of the orientation of the deformation axes. The reaction ^{48}Ca + ^{248}Cm was studied in Dubna [8], and evidence for the 4n channel was obtained at a beam energy resulting in an excitation energy of 30.4-35.8 MeV. Excitation functions were not yet measured.

It was pointed out in the literature [30] that closed shell nuclei as projectile and target are favorable for fusion of SHEs. The reason is not only a low reaction Q value and thus low excitation energy, but also that fusion of such systems is connected with a minimum of energy dissipation. The fusion path is along cold fusion valleys on the potential energy surface, where the reaction partners keep kinetic energy up to the closest possible distance. In this view the difference between *'cold'* and *'hot'* fusion is not only a result from gradual different values of the excitation energy, but there exists a qualitative difference, which is on the one side based on a well ordered fusion process along paths of minimum dissipation of energy (cold fusion), and on the other side on a process governed by the formation of a more or less energy equilibrated compound nucleus (hot fusion). This qualitative explanation is well in agreement with the results from studies of quasi-fission and compound-nucleus fission [31].

4 Summary and Outlook

Experimental work of the last two decades has shown that the cross-sections for the synthesis of the heaviest elements decrease almost continuously. However, the recent data on the synthesis of element 114 and 116 in Dubna using hot fusion and 118 in Berkeley using cold fusion seem to break this trend when the region of spherical superheavy elements is reached. Therefore a confirmation is urgently needed that the region of spherical SHEs has finally been reached and that the exploration of the 'island' has started and can be performed even on a relatively high cross-section level. However, the progress towards the exploration of the island of spherical SHEs is difficult to predict.

One can hope that, during the coming years, more data will be measured in order to promote a better understanding of the stability of the heaviest elements and the processes that lead to fusion. The microscopic description of the fusion process will be needed for an effective explanation of the measured phenomena in the case of low dissipative energies. Then, the relationships between fusion probability and stability of the fusion products may also become apparent.

An opportunity for the continuation of experiments in the region of SHEs at decreasing cross-sections will be afforded by further accelerator developments. High current beams and radioactive beams are the options for the future. At increased beam currents, values of tens of particle μA's may become possible, the cross-section level for the performance of experiments can be shifted down into the region of tens of femtobarns, and excitation functions can be measured on the level of tenths of picobarns. The high currents, in turn, require the development of a new target and improvement of the separator. The radioactive beams, not as intense as the stable beams, will allow for approaching the closed neutron shell N = 184 already at lighter elements. However, the usefulness of radioactive neutron-rich nuclei for the synthesis of superheavy elements is difficult to predict.

The half-lives of SHEs are expected to be relatively long. Based on nuclear models, which are effective predictors of half-lives in the region of the heaviest elements, values from microseconds to years have been calculated for various isotopes. This wide range of half-lives encourages the application of a wide variety of experimental methods in the investigation of SHEs, from the safe identification of short lived isotopes by recoil-separation techniques to atomic physics experiments on trapped ions, and to the investigation of chemical properties of SHEs using long-lived isotopes.

Acknowledgments

The recent experiments at SHIP were performed in collaboration with D. Ackermann, F.P. Heßberger, B. Kindler, J. Kojouharova, B. Lommel, R. Mann, G. Münzenberg, S. Reshitko, (GSI Darmstadt), A.G. Popeko, A.V. Yeremin (FLNR-JINR Dubna), S. Antalic, P. Cagarda, S. Śaro (University Bratislava), M. Leino and J.Uusitalo (University Jyväskylä).

References

1. R. Smolanczuk and A. Sobiczewski, in *Proc. of the XV. Nuclear Physics Divisional Conf. on Low Energy Nuclear Dynamics*, eds. Yu.Ts. Oganessian *et al* (World Scientific, 1995) p. 313.
2. S. Cwiok *et al*, *Nucl. Phys.* A **611,** 211 (1996).
3. K. Rutz *et al*, *Phys. Rev.* C **56,** 238 (1997).
4. R. Smolanczuk, *Phys. Rev.* C **56,** 812 (1997).
5. P. Möller *et al*, *Atomic Data and Nucl. Data Tables* **59,** 185 (1995).
6. A. Türler *et al*, private communication (2001)
7. S. Hofmann *et al*, *Eur. Phys. J.* A **10,** 5 (2001).
8. Yu.Ts. Oganessian *et al*, *Phys. Rev.* C **63,** 011301 (2000).
9. S. Hofmann *et al*, *Z. Phys.* A **350,** 277 (1995).
10. S. Hofmann, *Rep. Prog. Phys.* A **61,** 639 (1998).
11. A. Ghiorso *et al*, *Nucl. Phys.* A **583,** 861c (1995).
12. A. Ghiorso *et al*, *Phys. Rev.* C **51,** R2293 (1995).
13. Yu.A. Lazarev *et al*, *Phys. Rev. Lett.* **75,** 1903 (1995).
14. Yu.A. Lazarev *et al*, *Phys. Rev.* C **54,** 620 (1996).
15. S. Hofmann *et al*, *Z. Phys.* A **350,** 281 (1995).
16. S. Hofmann *et al*, in *GSI Report* **2001-1**, 1 (2001)
17. S. Hofmann *et al*, *Z. Phys.* A **354,** 229 (1996).
18. S. Hofmann and G. Münzenberg, *Rev. Mod. Phys.* **72,** 733 (2000).
19. S. Hofmann, *Nucl. Phys.* A **616,** 370c (1997).
20. V. Ninov *et al*, *Phys. Rev. Lett.* **83,** 1104 (1999).
21. Yu.Ts. Oganessian *et al*, *Phys. Rev. Lett.* **83,** 3154 (1999).
22. R. Bass, *Nucl. Phys.* A **231,** 45 (1974).
23. R. Smolanczuk, *Phys. Rev. Lett.* **83,** 4705 (1999).
24. Y. Aritomo *et al*, *Phys. Rev.* C **59,** 796 (1999).
25. E.A. Cherepanov, *Pramana J. Phys.* **53,** 619 (1999).
26. R. Smolanczuk, *Phys. Rev.* C **59,** 2634 (1999).
27. G.G. Adamian *et al*, *Nucl. Phys.* A **678,** 24 (2000).
28. V.Yu. Denisov and S. Hofmann, *Phys. Rev.* C **61,** 034606 (2000).
29. G. Giardina *et al*, *Eur. Phys. J.* A **8,** 205 (2000).
30. R.K. Gupta *et al*, *Z. Phys.* A **283,** 217 (1977).
31. M.G. Itkis *et al*, in *Proc. Int. Workshop on Fusion Dynamics at the Extremes*, eds. Yu.Ts Oganessian and V.I. Zagrebaev (World Scientific, 2001) p. 93.

THE STUDY OF THE CHARACTERISTICS OF NEUTRON EMISSION IN THE REACTIONS WITH ^{48}CA IONS

I. ITKIS, A. A. BOGATCHEV, M. G. ITKIS, J. KLIMAN, G. N. KNYAZHEVA, N. A. KONDRATIEV, E. M. KOZULIN, I. V. KORZYUKOV, L. KRUPA, YU. TS. OGANESSIAN, I. V. POKROVSKI, E. V. PROKHOROVA, V. M. VOSKRESSENSKI

Flerov Laboratory of Nuclear Reactions, Joint Institute for Nuclear Research, Dubna, Russia
E-mail: itkis@cv.jinr.ru

N. AMAR, J. PETER

Laboratoire de Physique Corpusculaire, Caen, France

G. GIARDINA

Department of Physics, University of Messina and INFN, Messina, Italy

F. HANAPPE, T. MATERNA

Universite Libre de Bruxelles, Belgique

P. DESESQUELLES, O. DORVAUX, N. ROWLEY, CH. SCHMITT, L. STUTTGE

Institut de Recherches Subatomiques, Strasbourg, France

The work presents the results of the study of neutron emission characteristics in the fission and quasi-fission of heavy nuclei, produced in the reactions with ^{48}Ca ions. These experiments have been performed at the U-400 accelerator of the Flerov Laboratory of Nuclear Reactions (JINR) with the use of the time-of-flight spectrometer of fission fragments CORSET and neutron multidetector DEMON. ^{208}Pb, ^{232}Th and ^{248}Cm layers 120-200 mkg/cm^2 in thickness were used as targets. The mass and energy distributions of the fission fragments, the multiplicities of pre- and post-fission neutrons at energies close to the Coulomb barrier have been measured. The analysis of neutron energy and angular distributions has shown that the total neutron multiplicity M_{tot} in the fusion-fission reactions of compound nuclei formed in the reactions with heavy ions is essentially higher than M_{tot} in the quasi-fission reactions. Note that M_{tot} monotonously increases with the growing Z of the compound nucleus.

1 Introduction

In the last ten years the multiplicity of pre-fission neutrons has been the subject of the intensive experimental and theoretical studies. Experimental data obtained using the neutron clock and thermometer method have led to the conclusion that fission is a cold slow process taking place after a delay time estimated to be of the order of a few 10^{-20} s [1,2]. However, experimental data concerns only nuclei within the mass range of about 200 atomic mass units (amu). Emission of neutrons in correlation with fission fragments in the decay of superheavy compound systems at excitation energies near to the Coulomb barrier had not been studied properly before. At the same time such investigations may be useful for an additional identification of fusion-fission and quasi-fission processes.

2 Experiment

The experiments were carried out during last year in the Flerov laboratory of nuclear reactions on the extracted beam of ^{48}Ca ions of U-400 accelerator at the energies 240, 242 MeV. ^{208}Pb, ^{232}Th and ^{248}Cm spectrometric layers 120-170 mkg/cm^2 in thickness deposited on a 40-50 mkg/cm^2 carbon backing were used as the target. The measurements were performed using a double-arm time-of-flight position-sensitive spectrometer CORSET [3,4]. This spectrometer consists of four position-sensitive microchannel plate detectors providing the registration of the reaction products within a solid angle of 360 msrad and two microchannel plate detectors with electrostatic mirrors providing a start signal for the time-of-flight measurements. The detectors were installed symmetrically relative to the beam axis at an angle of about 60^0. The neutron multidetector DEMON [5,6] was used to measure neutrons in coincidence with two coincident fragments. DEMON consists of NE213 liquid scintillator cells of 20 cm length and 16 cm diameter, optically coupled to XP4512B photomultiplier tubes. Neutrons are separated from γ-rays by pulse shape discrimination. Their energy is determined by time-of-flight measurement over a flight path of about 50-70 cm. The geometry of DEMON was close to spherical one. The discrimination thresholds were set using ^{22}Na, ^{137}Cs, ^{60}Co and ^{241}Am sources. Efficiency curves for the neutron detection as a function of the energy were obtained by Monte Carlo simulations [7]. The simulation also gives the possibility to estimate the mean interaction depth in the scintillator as a function of the neutron energy that allows obtaining more exact neutron flight path.

3 Data analysis and results

The data processing was carried out with the use of traditional techniques, assuming the process under study to be a two-body process. The energy losses of the fragments in the target, backings and start detectors were taken into account.

The decomposition of the total neutron multiplicities into its pre- and post-scission components was obtained by means of a three moving source fit. Neutrons were supposed to be emitted sequentially with Maxwellian energy distributions from the compound system and two fully accelerated fragments. The whole set of 41 DEMON detectors was taken into account in the χ^2 minimization. Figure 1 shows the experimental neutron energy and angular distributions and the results of the fitting procedure for the reaction $^{48}Ca + ^{208}Pb$, which are in good agreement with the experimental points.

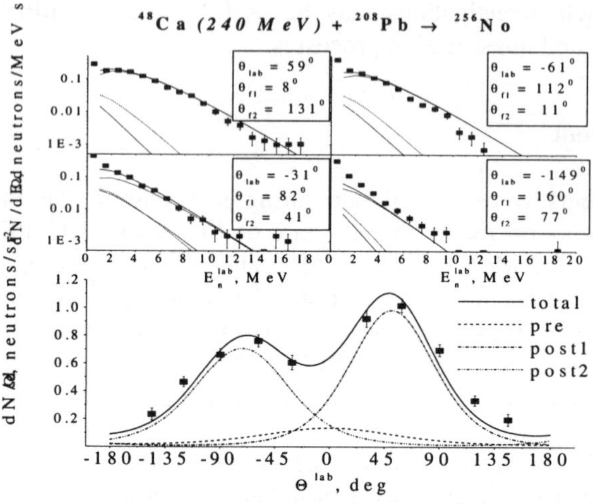

Figure 1. Neutron energy (top panels) and angular (bottom panel) distributions (squares) and the results of the fitting procedure (lines) for the reaction $^{48}Ca + ^{208}Pb$.

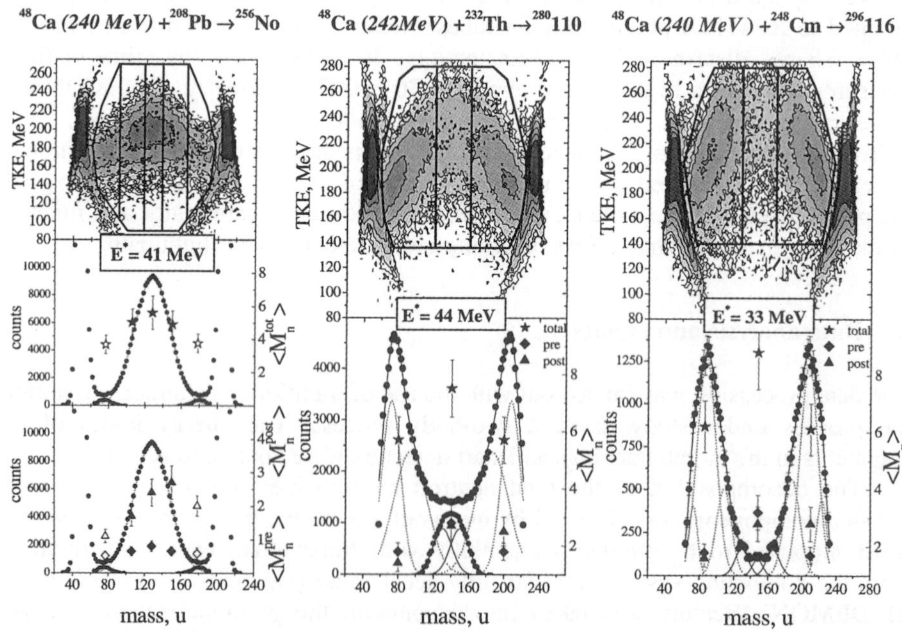

Figure 2. Two-dimensional TKE-Mass matrices (top panels) and the mass yields (solid circles) and neutron total (stars), pre- (diamonds) and single-fragment post-scission (triangles) multiplicities in dependence on the fragment mass (bottom panels).

Obtained neutron total, pre- and post-scission multiplicities are shown in Fig.2. In the case of $^{48}Ca + ^{208}Pb$, the total neutron multiplicity is nearly twice higher for the region of masses $A_{CN}/2\pm10$ amu than for the regions where quasi-fission is dominating ($A = 65\div90$ amu and complementary). Note, that the pre-scission multiplicity M_{pre} is low for all the mass ranges and increases slightly from $M_{pre}= 0.6\pm0.1$ for $A = 65\div90$ to $M_{pre}= 0.8\pm0.1$ for $A_{CN}/2\pm10$. For the heavier compound systems the total neutron multiplicity M_{tot} is also essentially higher for $A_{CN}/2\pm20$ and increases with the growing Z of compound nucleus, while for the fragment mass region where the mechanism of quasi-fission predominates M_{tot} rises insignificantly as well as M_{pre}.

4 Conclusions

The total neutron multiplicity M_{tot} in the fusion-fission reactions of compound nuclei formed in the reactions with heavy ions is considerably higher than M_{tot} in the quasi-fission reactions.

M_{tot} monotonously increases with the growing atomic number of the compound nucleus: from $M_{tot}= 5.6\pm0.8$ for the ^{256}No compound nucleus formed in the reaction $^{48}Ca + ^{208}Pb$ to $M_{tot}= 8.8\pm1.3$ for the nucleus Z=116 formed in the $^{48}Ca + ^{248}Cm$ reaction (for the fragment mass region $A_{CN}/2\pm20$ amu). At the same time for the fragment mass regions where quasi-fission is dominating the absolute values of M_{tot} are lower and the dependence on Z is rather smooth.

5 Acknowledgments

The work has been supported by the Russian Foundation for Basic Research (Grant No 99-02-17891) and by INTAS (Grant No 97-11929).

References

1. D.Hilscher and H.Rossner, *Ann. Phys. Fr.* **17** (1992) 471
2. D.J.Hinde *et al.*, *Phys.Rev.* **C 45** (1992) 1229.
3. E. M. Kozulin, N. A. Kondratjev and I. V. Pokrovski, Heavy Ion Physics, Scientific Report 1995-1996, JINR, FLNR, Dubna, 1997, p. 215.
4. N. A. Kondratiev *et al.*, Fourth Int. Conf. on *Dynamical Aspects of Nuclear Fission, (DANF'98)* (Casta-Papiernicka, Slovak Republic, October 1998), World Scientific, Singapore (1999) 431.
5. M. Moszyński *et al.*, *Nucl. Instrum. Meth.* **A 350** (1994) 226.
6. I. Tilquin *et al.*, *Nucl. Instrum. Meth.* **A 365** (1995) 446.
7. P.Desesquelles et al., *Nucl. Instrum. Meth.* **A 307** (1991) 366.

NUCLEAR FISSION AT BORDER LINES

M. G. ITKIS, A. A. BOGATCHEV, I. M. ITKIS, M. JANDEL, J. KLIMAN, G. N. KNIAJEVA, N. A. KONDRATIEV, I. V. KORZYUKOV, E. M. KOZULIN, L. KRUPA, YU. TS. OGANESSIAN, I. V. POKROVSKI, V. A. PONOMARENKO, E. V. PROKHOROVA, A. YA. RUSANOV AND V. M. VOSKRESENSKI

Flerov Laboratory of Nuclear Research, JINR, 141980 Dubna, Russia
Email:itkis@flnr.jinr.ru

F. HANAPPE AND T. MATERNA

Universite Libre de Bruxelles,1050 Bruxelles, Belgium

N. ROWLEY AND L. STUTTGE

Institut de Recherches Subatomiques, F-67037 Strasbourg Cedex, France

G. GIARDINA

Dipartimento di Fisica dell' Universita di Messina 98166 Messina, Italy

K. J. MOODY

University of California, Lawrence Livermore National Laboratory, Livermore, California 94551, USA

The process of fusion-fission of superheavy nuclei with Z=102-122 formed in the reactions with ^{22}Ne, ^{26}Mg, ^{48}Ca, ^{58}Fe and ^{86}Kr ions at energies near and below the Coulomb barrier has been studied. The experiments were carried out at the U-400 accelerator of the Flerov Laboratory of Nuclear Reactions (JINR) using a time-of-flight spectrometer of fission fragments CORSET and a neutron multi-detector DEMON. As a result of the experiments, mass and energy distributions of fission fragments, fission and quasi-fission cross sections, multiplicities of neutrons and gamma-quanta and their dependence on the mechanism of formation and decay of compound superheavy systems have been studied.

1 Introduction

Interest in the study of the fission process of superheavy nuclei interactions with heavy ions is connected first of all with the possibility of obtaining information, the most important for the problem of synthesis, on the production cross section of compound nuclei at excitation energies of ≈15-30 MeV (i.e. when the influence of shell effects on the fusion and characteristics of the decay of the composite system is considerable), which makes possible prediction on its basis of the probability of their survival after evaporating 1, 2 or 3 neutrons, i.e. in “cold” or “warm” fusion reactions. However, for this problem to be solved, there is a need for a much more penetrating insight into the fission mechanism of superheavy nuclei and for a knowledge of such fission characteristics as the fission - quasi-fission cross section ratio in relation to the ion-target entrance channel mass asymmetry and excitation energy, the multiplicity of the pre and postfission neutrons, the kinetic energy of the fragments and the peculiarities of the mass distributions of the fission and quasi-fission fragments etc. Undoubtedly all these points are of great independent interest to nuclear fission physics.

In this connection this work presents the results of the experiments on the fission of superheavy nuclei in the reactions $^{48}Ca+^{208}Pb \rightarrow ^{256}No$, $^{22}Ne+^{248}Cm \rightarrow ^{270}Sg$, $^{26}Mg+^{248}Cm \rightarrow ^{274}Hs$, $^{48}Ca+^{238}U \rightarrow ^{286}112$, $^{48}Ca+^{244}Pu \rightarrow ^{292}114$, $^{48}Ca + ^{248}Cm \rightarrow ^{296}116$, $^{58}Fe+ ^{208}Pb \rightarrow ^{266}Hs$, $^{58}Fe + ^{244}Pu \rightarrow ^{302}120$, $^{58}Fe+ ^{248}Cm \rightarrow ^{306}122$ and $^{86}Kr + ^{208}Pb \rightarrow ^{294}118$ carried out at FLNR JINR in the last year. The choice of the indicated reactions has undoubtedly been inspired by the results of the recent experiments on producing the nuclides $^{283}112$, $^{287}114$, $^{289}114$ at Dubna [1,2] and $^{293}118$ at Berkeley [3] in the same reactions.

2 Characteristics of mass and energy distributions of SHE fission fragments

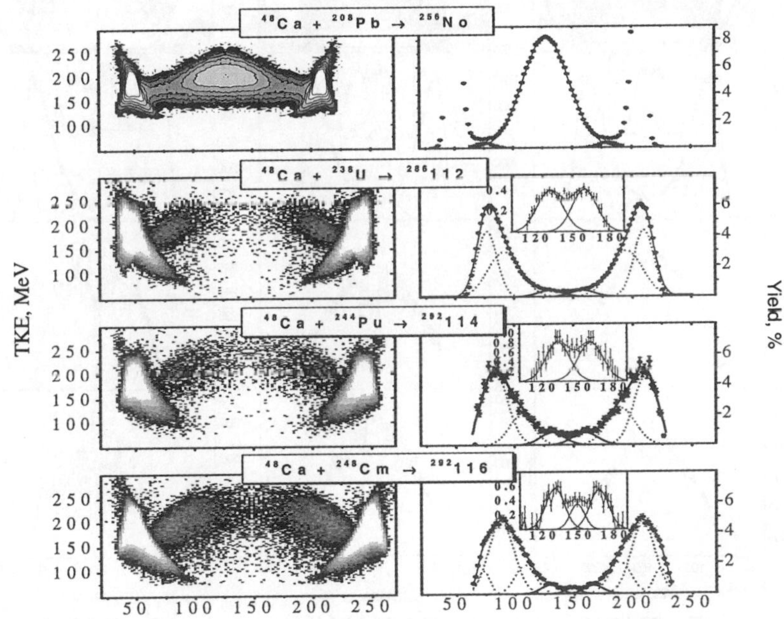

Figure 1. Two-dimensional TKE-Mass matrices (left-hand side panels) and mass yields (right-hand side panels) of fission fragments of ^{256}No, $^{286}112$, $^{292}114$ and $^{296}116$ nuclei produced in the reactions with ^{48}Ca at the excitation energy $E^* \approx 33$ MeV.

Figure 1 shows the data on mass and energy distributions of fission fragments of $^{256}102$, $^{286}112$, $^{292}114$ and $^{296}116$ nuclei produced in the reactions with ^{48}Ca at one and the same excitation energy E*≈33 MeV. The main peculiarity of the data is the sharp transition from the predominant compound nucleus fission in the case of $^{256}102$ to the quasi-fission mechanism of decay in the case of the $^{286}112$ nucleus and more heavy nuclei. It is very important to note that despite a dominating contribution of the quasi-fission process in the case of nuclei with Z=112-116, in the symmetric region of fission fragment masses (A/2 ± 20) the process of fusion-fission of compound nuclei, in our opinion, prevails.

It is demonstrated in the framings (see the right-hand panels of Fig.1) from which it is also very well seen that the mass distribution of fission fragments of compound nuclei is asymmetric in shape with the light fission fragment mass = 132-134.

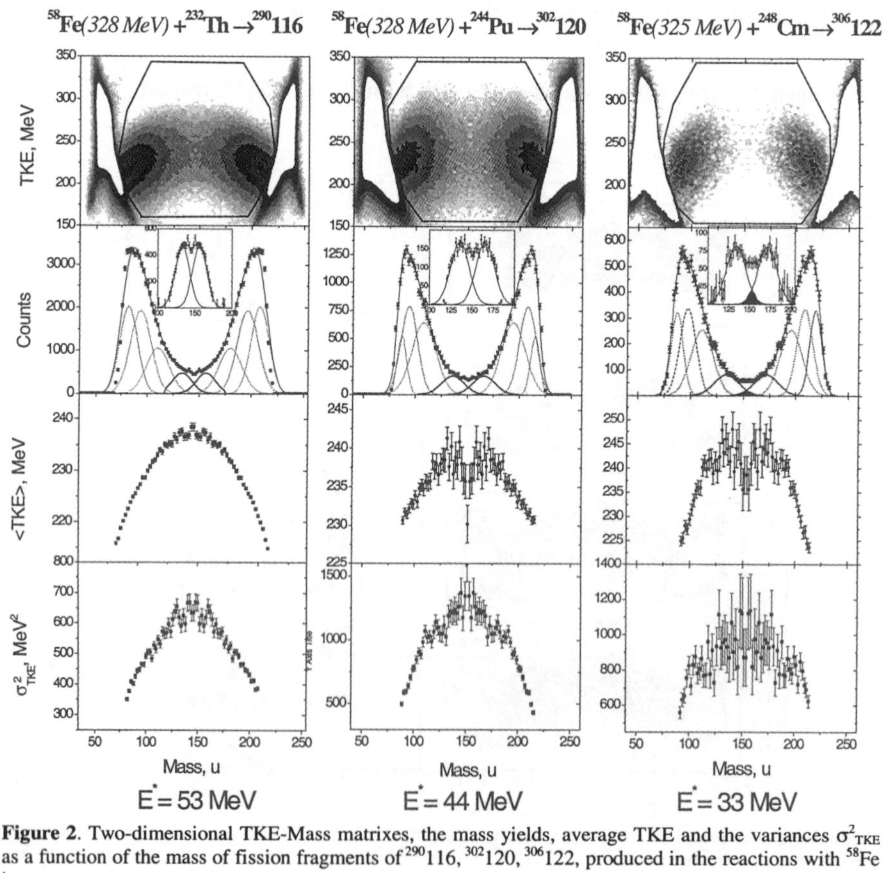

Figure 2. Two-dimensional TKE-Mass matrixes, the mass yields, average TKE and the variances σ^2_{TKE} as a function of the mass of fission fragments of $^{290}116$, $^{302}120$, $^{306}122$, produced in the reactions with ^{58}Fe ions.

Figure 2 shows similar data for the reaction between ^{58}Fe projectiles and ^{232}Th, ^{244}Pu and ^{248}Cm targets, leading to the formation of the compound system ^{290}116 and the heaviest compound systems ^{302}120 and ^{306}122 (where N=182-184), i.e., to the formation of the spherical compound nucleus, which agrees well with theoretical predictions [4]. As seen from Fig.2, in these cases we observe an even stronger manifestation of the asymmetric mass distributions of ^{306}122 and ^{302}120 fission fragments with the light fragment mass =132. The corresponding structures are also well seen in the dependence of the TKE on the mass. Only for the reaction ^{48}Ca+^{232}Th -> ^{290}116 (E*= 53 MeV) the valley in the region of M=A/2 disappears in the mass distribution as well as in the average TKE distribution which is connected with a reducing of the influence of shell effects on these characteristics. Such a decrease in the role of shell effects with increasing the excitation energy is observed also in the induced fission of actinide nuclei.

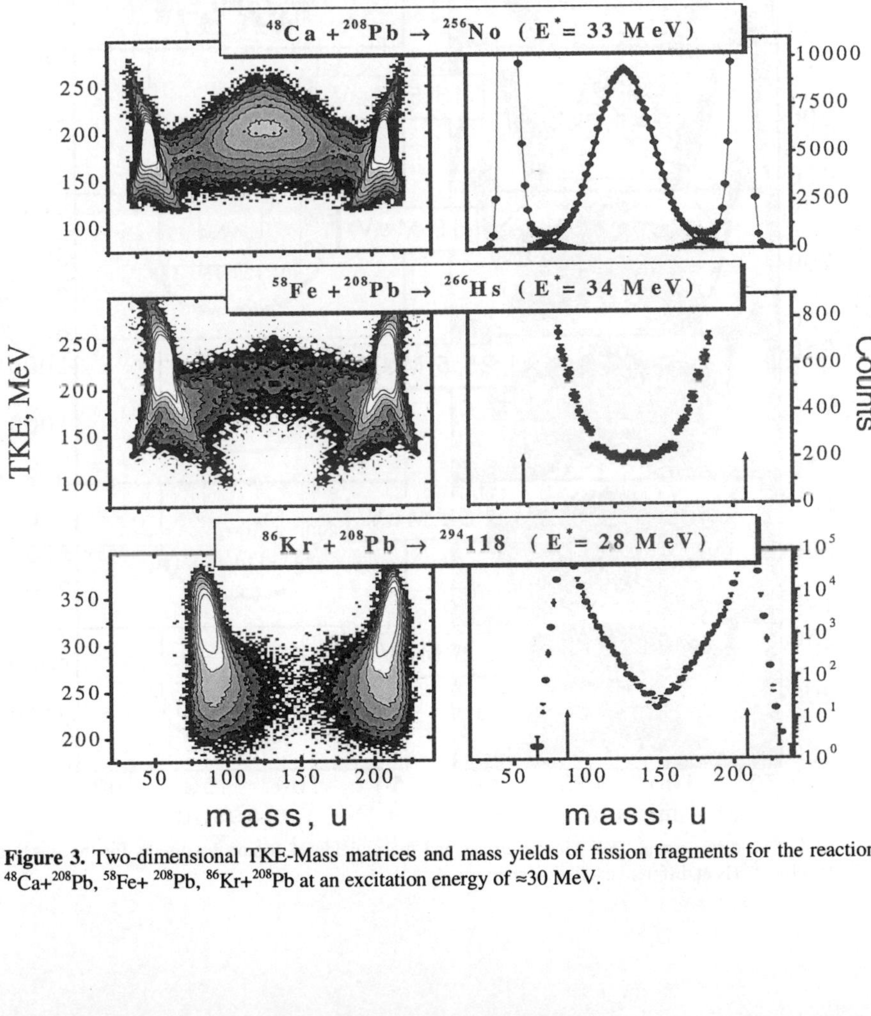

Figure 3. Two-dimensional TKE-Mass matrices and mass yields of fission fragments for the reactions ^{48}Ca+^{208}Pb, ^{58}Fe+ ^{208}Pb, ^{86}Kr+^{208}Pb at an excitation energy of ≈30 MeV.

Figure 3 shows mass and energy distributions of fission fragments for compound nuclei ^{256}No, ^{266}Hs and ^{294}118 formed in the interaction between a ^{208}Pb target and ^{48}Ca, ^{58}Fe and ^{86}Kr ions at the excitation energy of ≈ 30 MeV. It is important to note that in the case of the reaction ^{86}Kr+^{208}Pb the ratio between the fragment yields in the region of asymmetric masses and that in the region of masses A/2 exceeds by about 30 times a similar ratio for the reactions with ^{48}Ca and ^{58}Fe ions. It signifies to that fact that in the case of the reaction ^{86}Kr+^{208}Pb→^{294}118 in the region of symmetric fragment masses the mechanism of quasi-fission prevails.

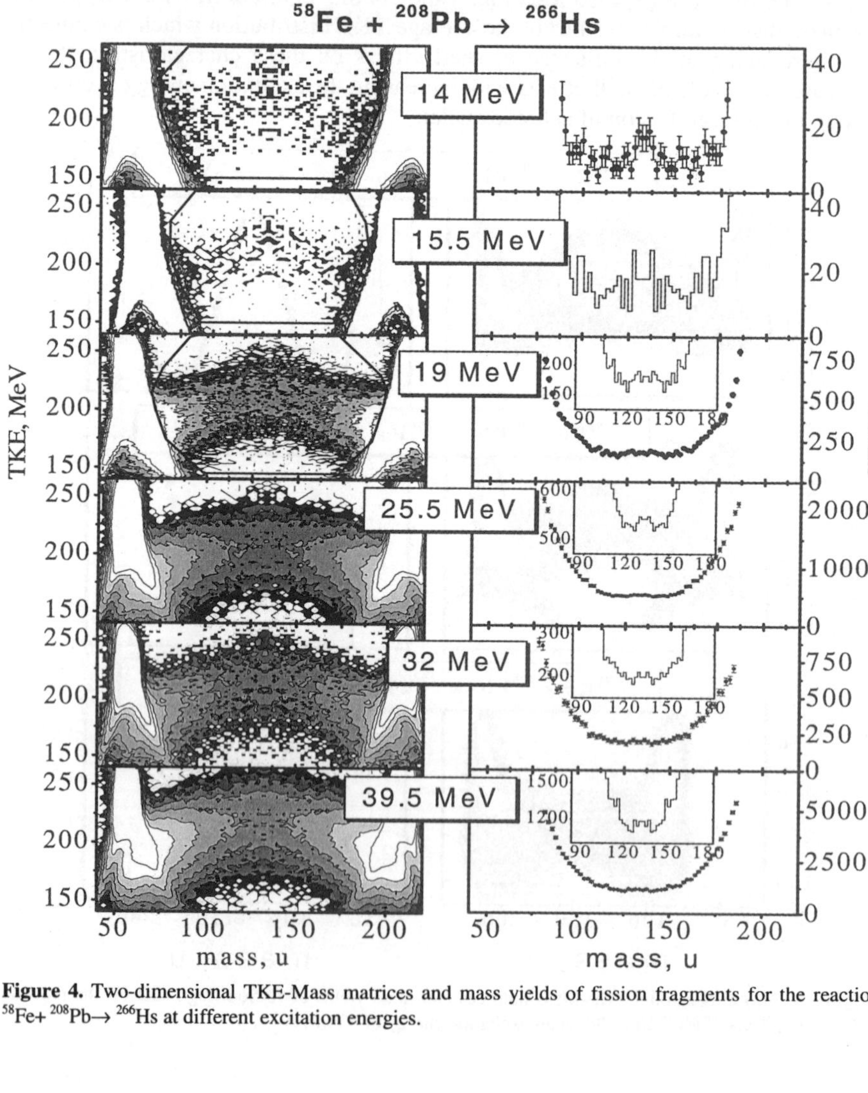

Figure 4. Two-dimensional TKE-Mass matrices and mass yields of fission fragments for the reaction ^{58}Fe+ ^{208}Pb→ ^{266}Hs at different excitation energies.

Figures 4 and 5 show similar data for the reactions ^{58}Fe+^{208}Pb→ ^{266}Hs, and ^{26}Mg+ ^{248}Cm→ ^{274}Hs obtained at different excitation energies.

Mass-energy distributions of ^{274}Hs (Fig.5) formed in the reaction ^{26}Mg+^{248}Cm differ greatly from those of the isotope ^{266}Hs (Fig.4). In this case the matrixes have triangular shapes corresponding to the classic fission, which can be described by the liquid drop model. The contribution of the quasi-fission component with a peak for the heavy fission fragment near m = 208 is not practically noticeable. The predominance of one of the mechanisms, namely, fusion-fission or quasi-fission, for one and the same element (Hs) is caused by the asymmetry of the entrance reaction channel. This situation can be explained in terms of the dinuclear system concept (DNSC) [5].

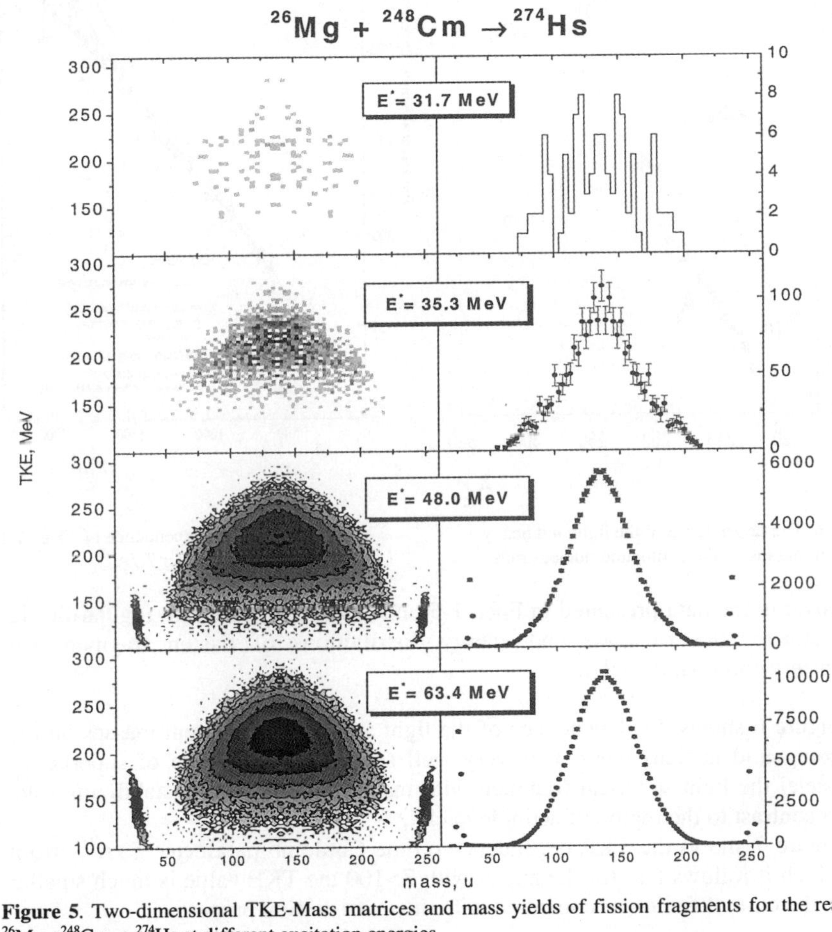

Figure 5. Two-dimensional TKE-Mass matrices and mass yields of fission fragments for the reaction ^{26}Mg +^{248}Cm→ ^{274}Hs at different excitation energies.

Besides, at a decrease in the excitation energy of ^{274}Hs down to 31.7 and 35.3 (Fig.5), there appear structures in the mass distribution, corresponding to the structures in the energy distribution. Thus, at a decrease in the excitation energy different fission modes can be observed, as is the case with the low-energy fission of ^{271}Hs and ^{270}Sg [6,7].

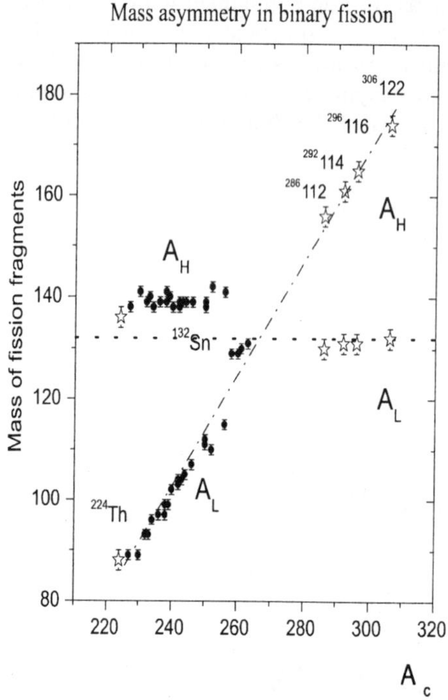

Figure 6. The dependence of the light and heavy fragment masses on the compound nucleus mass.

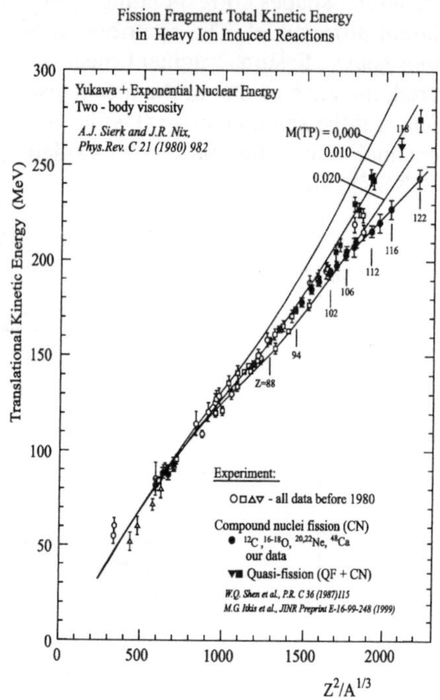

Figure 7. The dependence of TKE on the Coulomb parameter $Z^2/A^{1/3}$.

In analyzing the data presented in Figs. 1-5 one can notice two main regularities in the characteristics of mass and energy distributions of fission fragments of superheavy compound nuclei:

1. Figure 6 shows the dependence of the light and heavy fragment masses on the compound nucleus mass. It is very well seen that in the case of superheavy nuclei the light spherical fragment with mass 132-134 plays a stabilizing role, in contrast to the region of actinide nuclei.
2. Figure 7 shows the TKE dependence on the Coulomb parameter $Z^2/A^{1/3}$, from which it follows that for the nuclei with Z>100 the TKE value is much smaller in the case of fission as compared with the quasi-fission process.

3 Capture and fusion-fission cross sections

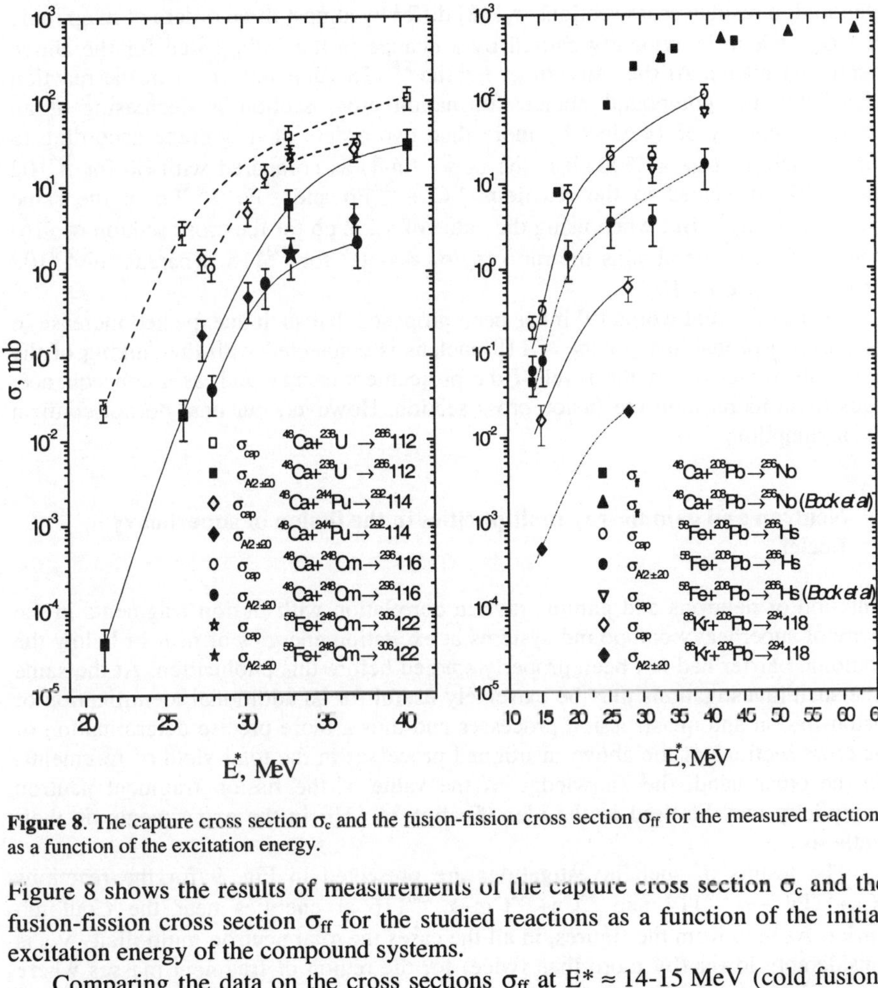

Figure 8. The capture cross section σ_c and the fusion-fission cross section σ_{ff} for the measured reactions as a function of the excitation energy.

Figure 8 shows the results of measurements of the capture cross section σ_c and the fusion-fission cross section σ_{ff} for the studied reactions as a function of the initial excitation energy of the compound systems.

Comparing the data on the cross sections σ_{ff} at $E^* \approx$ 14-15 MeV (cold fusion) for the reactions ^{58}Fe + ^{208}Pb and ^{86}Kr + ^{208}Pb, one can obtain the following ratio: $\sigma_{ff}(108)/\ \sigma_{ff}\ (118) \geq 10^{\ 2}$. In the case of the reactions from ^{48}Ca + ^{238}U to ^{58}Fe + ^{248}Cm at $E^* \approx$ 33 MeV (warm fusion) the value of Z changes by the same 10 units as in the first case, and the ratio $\sigma_{ff}\ (112)/\ \sigma_{ff}\ (122)$ is $\approx$ 4-5 which makes the use of asymmetric reactions for the synthesis of spherical superheavy nuclei quite promising.

Another interesting result is connected with the fact that the values of σ_{ff} for $^{256}102$ and $^{266}108$ at $E^* = 14\text{-}15$ MeV are quite close to each other, whereas the evaporation residue cross sections σ_{xn} [8] differ by almost three orders of magnitude (σ_{ff}/σ_{xn}) which is evidently caused by a change in the Γ_f/Γ_n value for the above mentioned nuclei. At the same time, for the $^{294}118$ nucleus formed in the reaction $^{86}Kr+^{208}Pb$, the compound nucleus formation cross section is decreasing at an excitation energy of 14 MeV by more than two orders of magnitude according to our estimations ($\sigma_{ff} \approx 500$ nb is the upper limit) as compared with σ_{ff} for $^{256}102$ and $^{268}108$ produced in the reactions $^{48}Ca+^{208}Pb$ and $^{58}Fe+^{208}Pb$ at the same excitation energy. But when using the value of ≈ 2.2 pb for the cross section $\sigma_{ev}(1n)$ from work [3], one obtains the ratio $\sigma_{xn}/\sigma_{ff} \approx 4\cdot10^{-6}$ for $^{293}118$, whereas for $^{266}108$ the ratio is $\sigma_{xn}/\sigma_{ff} \approx 10^{-6}$.

In one of recent works [9] it has been proposed that such unexpected increase in the survival probability for the $^{294}118$ nucleus is connected with the sinking of the Coulomb barrier below the level of the projectile's energy and, as a consequence, leads to an increase in the fusion cross section. However, our data do not confirm this assumption.

4 Neutron and gamma-ray multiplicities in the fission of superheavy nuclei

Emission of neutrons and gamma-rays in correlation with fission fragments in the decay of superheavy compound systems at excitation energies of near or below the Coulomb barrier had not been properly studied before this publication. At the same time such investigations may be extremely useful for an additional identification of fusion-fission and quasi-fission processes and thus a more precise determination of the cross sections of the above mentioned processes in the total yield of fragments. On the other hand, the knowledge of the value of the fission fragment neutron multiplicity may be used in the identification of SHE in the experiments on their synthesis.

The results of such investigations are presented in Fig. 9 for the reactions $^{48}Ca+^{244}Pu \rightarrow {}^{292}114$ and $^{48}Ca+^{248}Cm \rightarrow {}^{296}116$ at energies near the Coulomb barrier. As seen from the figures, in all the cases the total neutron multiplicity ν_{tot} is considerably lower (by more than twice) for the region of fragment masses where the mechanism of quasi-fission predominates as compared with the region of fragment masses where, in our opinion, the process of fusion-fission prevails (in the symmetric region of fragment masses).

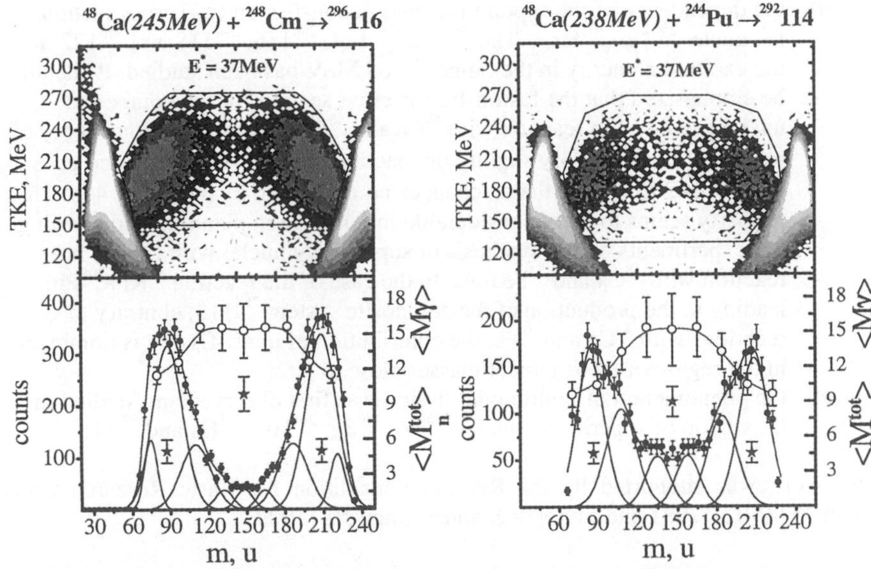

Figure 9. Two-dimensional TKE-Mass matrices (top panels) and the mass yields (the solid circles), neutron (the stars) and gamma ray (the open circles) multiplicities in dependence on the fission fragment mass (bottom panels) for the reactions $^{48}Ca+^{248}Cm \rightarrow ^{296}116$ and $^{48}Ca+^{244}Pu \rightarrow ^{292}114$.

Another important peculiarity of the obtained data is the large values of $\nu_{tot} \approx 9.2$ and 9.9 for the fission of $^{292}114$ and $^{296}116$ compound nuclei, respectively. As well as for ν_{tot} noticeable differences have been observed in the values of γ-ray multiplicities for different mechanisms of superheavy compound nucleus decay.

5 Conclusion

As a result of the experiments carried out, for the first time the properties were studied of the fission of the compound nuclei ^{256}No, ^{270}Sg, ^{266}Hs, ^{271}Hs, ^{274}Hs, $^{286}112$, $^{292}114$, $^{296}116$, $^{294}118$, $^{302}120$ and $^{306}122$, produced in reactions with ions ^{22}Ne, ^{26}Mg, ^{48}Ca, ^{58}Fe and ^{86}Kr at energies close to and below the Coulomb barrier.

On the basis of those data a number of novel important physics results were received:

a) it was found, that the mass distribution of fission fragments for compound nuclei $^{286}112$, $^{292}114$, $^{296}116$, $^{302}120$ and $^{306}122$ is asymmetric one, whose nature, in contrast to the asymmetric fission of actinides, is determined by the shell structure of the light fragment with the average mass 132-134. It was established that TKE, neutron and γ-ray multiplicities for fission and quasi-fission of superheavy nuclei are significant different;

b) the dependence of the capture (σ_c) and fusion-fission (σ_{ff}) cross sections for nuclei ^{256}No, ^{266}Hs, ^{274}Hs, ^{286}112, ^{292}114, ^{296}116, ^{294}118 and ^{306}122 on the excitation energy in the range 15–60 MeV has been studied. It should be emphasized that the fusion-fission cross section for the compound nuclei produced in reaction with ^{48}Ca and ^{58}Fe ions at excitation energy of ≈30 MeV depends only slightly on reaction partners, that is, as one goes from ^{286}112 to ^{306}122, the σ_{ff} changes no more than by the factor 4–5. This property seems to be of considerable importance in planning and carrying out experiments on the synthesis of superheavy nuclei with Z>114 in reaction with ^{48}Ca and ^{58}Fe ions. In the case of the reaction ^{86}Kr+^{208}Pb, leading to the production of the composite system ^{294}118, contrary to reactions with ^{48}Ca and ^{58}Fe, the contribution of quasi-fission is dominant in the region of the fragment masses close to A/2;

c) the phenomenon of multimodal fission was first observed and studied in the region of superheavy nuclei ^{256}No, ^{270}Sg, ^{266}Hs, ^{271}Hs and ^{274}Hs.

This work was supported by the Russian Foundation for Basic Research under Grant № 99-02-17981 and by INTAS under Grant № 11929.

References

1. Oganessian Yu.Ts. *et al, Eur. Phis. J, A* **5** (1999) 63.
2. Oganessian Yu.Ts. *et al, Nature*, 400 (1999) 242.
3. Ninov V. *et al, Phys. Rev. Lett.* 83 (1999) 1104.
4. Patyk Z., Sobiczewski A., *Nucl.Phys. A* **533** (1991) 132.
5. Antonenko A. V. *et al, Phys. Lett.* B 319 (1993) 425
6. Itkis M. G. *et al, Phys. Rev. C,* **59** (1999), 3172
7. Itkis M. G. *et al, Proc. 7-th Int. Conf. Clustering aspects of nuclear structure and dynamics (Claster'99) Rab Island, Croatia, 1999* (WS, 2000) 386.
8. Hofmann S., Münzenberg G., *Reviews of Modern Physics*, 72 (2000) №3.
9. Myers W.D. and Swiatecki W.J., *Phys. Rev. C*, **62** (2000) 044610.

INVESTIGATION OF NEUTRON AND GAMMA MULTIPLICITIES IN REACTIONS WITH HEAVY IONS LEADING TO THE PRODUCTION OF SUPERHEAVY NUCLEI CLOSE TO THE ISLAND OF STABILITY

E. M. KOZULIN, M. G. ITKIS, YU. TS. OGANESSIAN, A. A. BOGATCHEV, A.YU.CHIZHOV, I. M. ITKIS, M. JANDEL, J. KLIMAN, G. N. KNIAJEVA, N. A. KONDRATIEV, I. V. KORZYUKOV, L. KRUPA, I. V. POKROVSKI, V. A. PONOMARENKO, E. V. PROKHOROVA AND V. M. VOSKRESENSKI

Flerov Laboratory of Nuclear Reactions, Joint Institute for Nuclear Research, Dubna, 141980, Russia, E-mail: kozulin@dubna.ru

F. HANAPPE AND T. MATERNA

Universite Libre de Bruxelles,1050 Bruxelles, Belgium

A. NINANE

Universite of Louvain, Nuclear Physics Department Ch. du cyclotron, 2-1348, Louvain-la-Neuve, Belgium

N. ROWLEY, L. STUTTGE, CH. SCHMITT, O. DORVAUX AND B.GALL

Institut de Recherches Subatomiques, F-67037 Strasbourg Cedex, France

G. GIARDINA

Dipartimento di Fisica dell' Universita di Messina 98166 Messina, Italy

J. PETER, N. AMAR, J. M. GAUTHIER AND ST.GREVY

Laboratoire de Physique Corpusculaire, 14050, Caen, France

G. CHUBARIAN

TEXAS A&M Univercity, Cyclotron Institute, College Station, 77843-3366, Texas, USA

P. DESESQUELLES

Institut de Physique Nucleaire d'Orsay, F-91406, Orsay, France

V. A. RUBCHENYA, W. H. TRZASKA AND Z. RADIVOJEVIC

Department of Physics, University of Jyväskylä, FIN-40351, Finland

CH. STODEL

GANIL, 14076, Caen, France

This work is devoted to the investigation of characteristics of neutron and gamma emission in the processes of fission and quasi-fission of the compound nuclei ^{256}No, ^{270}Sg, ^{266}Hs, ^{271}Hs, ^{274}Hs, ^{286}112, ^{292}114, ^{296}116, ^{302}120 and ^{306}122, produced in reactions with ions ^{26}Mg, ^{48}Ca, ^{58}Fe at energies close to and below the Coulomb barrier (i.e. when the influence of the shell effects on the fusion and on the characteristics of the decay of the composite system is considerable). The major part of these experiments has been performed at the U-400 accelerator of the Flerov Laboratory of Nuclear Reactions (JINR, Dubna) and also at the SARA accelerator of the Institut des Sciences Nucleaires (Grenoble, France) using a time-of-flight spectrometer of fission fragments CORSET (CORrelation SET-up) [1] and a neutron multi-detector DEMON [2]. The analysis of energy and angular distributions of neutrons and gamma quanta has shown that the $\langle M^{FF}{}_{\gamma}\rangle$ value for the fusion-fission process is greater than the value of $\langle M^{QF}{}_{\gamma}\rangle$ for the quasi-fission process, and on the other hand, the total neutron multiplicity M_{tot} and total gamma-multiplicity $\langle M_{\gamma}\rangle$ monotonously increase with the atomic number Z of the compound nucleus as well as with the bombarding energy.

The information on the problem of heating the heavy nuclei is very scarce and their decay properties has been studied very poorly. Difficulties in choosing the experimental approach to observing the nuclei with a temperature of at least 1 MeV are connected with a high probability of emission or evaporation of particles, first of all neutrons, which leads to the cooling of the fissile system and formation of a composite mixture of fission fragments pertaining to different fissile nuclei at different excitation energies. It substantially hampers not only quantitative but also qualitative analysis of the experimental data. It demands the carrying out of more detailed experimental investigations of the characteristics of fission in the deep sub-barrier region at low excitation energies (12-30 MeV) of the compound nucleus. In this case the most probable source of neutrons which have the fission fragment nature is the fissile system itself in its strongly deformed state corresponding to the post-scission stage of the shape evolution, i.e. the stage of going down from the top of the potential energy barrier.

The experiment was carried out at the extracted beams of ^{26}Mg, ^{48}Ca, ^{64}Ni and ^{58}Fe ions of the FLNR JINR U-400 accelerator, and the SARA accelerator complex at ISN, Grenoble, France [3] using a set-up which included:

- the two-armed time-of-flight reaction products spectrometer CORSET built with the use of microchannel plates (MCP);
- 8-41scintillation modules of time-of-flight neutron multidetector DEMON [4];
- 4-6 detector scintillation γ-quanta multiplicity spectrometer.

As a target, 40-170 μg/cm^2 ^{208}Pb, ^{238}U, ^{248}Cm and ^{244}Pu spectrometric layers deposited on a 40- 50 μg/cm^2 carbon backing were used.

The combined use of the efficient 4π-multidetector system DEMON with the high-resolution triggering device CORSET provides good technical advantages as compared with previous neutron-multiplicity measurements and thus will allow high-quality data to be obtained. The set-up permits the study of characteristics of neutron and gamma-quanta emission with taking into account the kinematics of angular distribution of fission fragments formed at the break-up of the di-nuclear system or nuclear fission which makes it possible to obtain information on different stages of the di-nuclear system evolution.

Methods of experimental data processing have been constantly improved which is also reflected in this series of works. Methods of extracting mass and energy distributions of fission fragments are based on the use of time-of-flight method with taking into account the coordinates of fission fragments hitting the detector as well as on the assumption on the two-body nature of the process under registration. The neutron energy and angular distributions for selected masses and energies of fragments were obtained using a JUNO data visualization and analysis program. Multiple-source fitting analysis with 3 emitting neutron sources was used to get the pre- and post-scission neutron multiplicities. No pre-equilibrium emission was taken into account because of the low excitation energy of the compound nuclei (E*=20-50 MeV). Acceleration neutrons were neglected.

Figure 1 shows two-dimensional TKE-mass matrices (left-hand side panels), mass yield, total neutron multiplicities, and average γ-ray multiplicities of fission fragment of $^{256}102$, $^{280}110$, $^{286}112$, $^{292}114$ and $^{296}116$ nuclei produced in the reactions with ^{48}Ca as a function of the fission fragment mass (middle panels), pre-and post – neutron multiplicities of fission fragments of $^{256}102$, $^{280}110$, and $^{296}116$ (right-hand side panels) as a function of the fission fragment mass.

The main peculiarity of the data is the sharp transition from the predominant compound nucleus fission in the case of $^{256}102$ to the quasi-fission mechanism of decay in the case of the $^{296}116$ nucleus (Fig.1). Average total γ–ray multiplicities as a function of the fragment mass (the $^{48}Ca+^{208}Pb$ reaction) are depressed in the region of symmetric masses of fission fragments. It is perhaps connected with the fact that in that mass region we obtain fission fragments with Z=50 which is very close to the closed shell nuclei. Fig.1 also shows the average total neutron multiplicities in dependence on the fission fragment mass. One can see that the total neutron multiplicities are much greater in the case of the fusion-fission process than those in the case of quasi-fission. The right-hand side of the figure shows average multiplicities of pre- and post-fission neutrons emitted before the fission and after it.

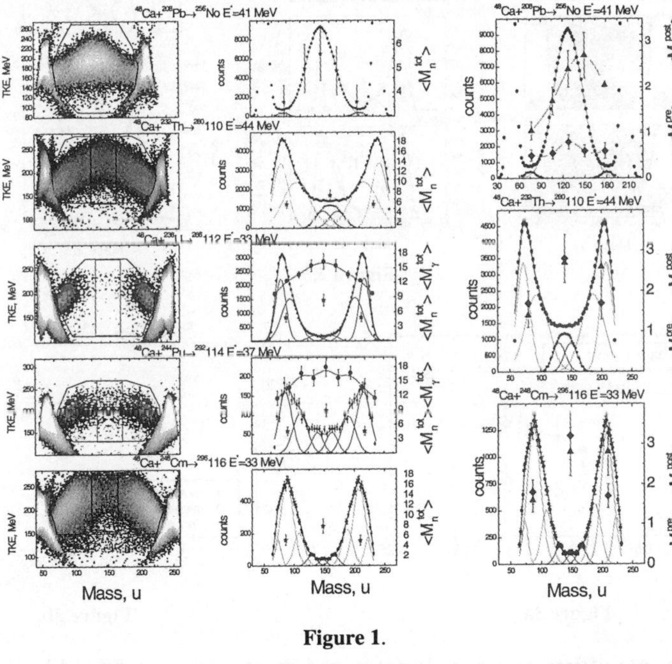

Figure 1.

Figure 2 shows two-dimensional TKE-mass matrices (left-hand side panels), mass yield, total neutron multiplicities, and average γ-ray multiplicities of fission fragments of ^{266}Hs, $^{290}116$, $^{302}120$, $^{306}122$ nuclei produced in the reactions with ^{58}Fe as a function of the fission fragment mass (middle panels), pre-and post-neutron multiplicities of fission fragments of ^{266}Hs, ^{290}116, ^{302}120 (right-hand side panels) as a function of the fission fragment mass. Figure 3a shows the total neutron multiplicities as a function of atomic number of compound nuclei produced in the reaction with ^{26}Mg, ^{48}Ca, ^{58}Fe. Average γ-ray multiplicities for fusion-fission and quasi-fission regions as a function of excitation energy are shown on Figure 3b. Preliminary experimental data are shown in the table.

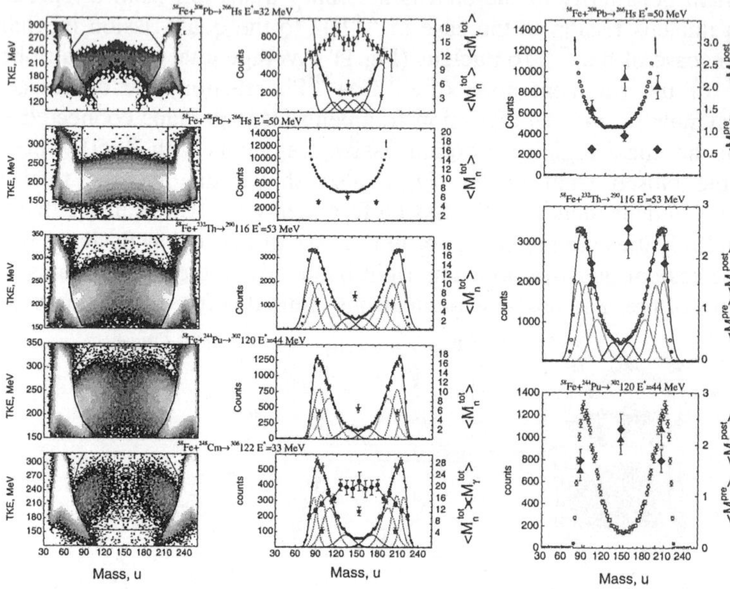

Figure 2.

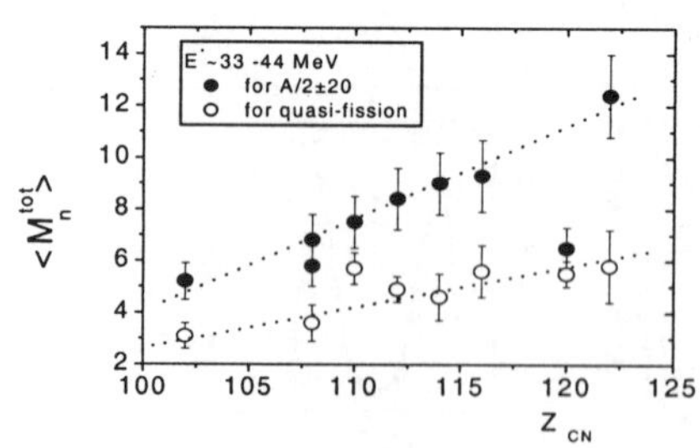

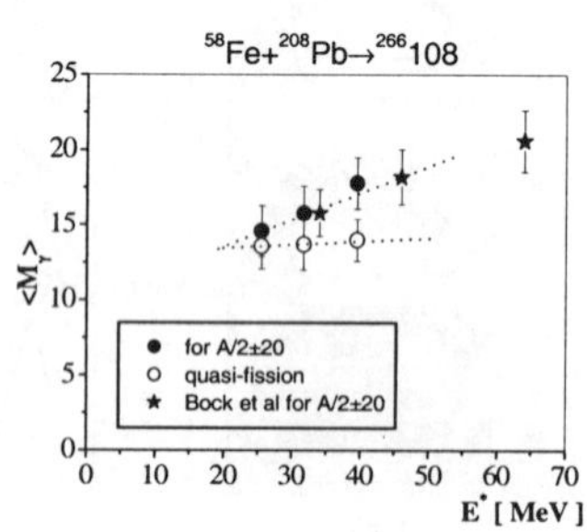

Figure 3a. **Figure 3b.**

In conclusion, we affirm that for all presented reactions a considerably larger yield of the average total number of neutrons M_{tot} emitted in the fusion-fission reactions (for the fragment mass region A/2) is observed as compared with the yield of neutrons M_{tot} in the quasifission at the same excitation energy. Note that the total neutron multiplicity M_{tot} monotonously increase with the atomic number Z of the compound nucleus as well as with the bombarding energy. At the same time in the quasi-fission reactions the absolute value of M_{tot} is considerably lower and the dependence on Z is rather smooth. The analysis of energy and angular distributions of gamma rays showed that the $\langle M^{FF}_{\gamma}\rangle$ for the fusion-fission process is greater than $\langle M^{QF}_{\gamma}\rangle$ for the quasi-fission process. Average $\langle M_{\gamma}\rangle$ increases with both the bombarding energy and the total mass of the system.

Reactions	E_{lab}	E^*	M^{tot}_n A/2±20 (FF)	M^{tot}_{pre} A/2±20 (FF)	M^{tot}_{post} A/2±20 (FF)	M^{tot}_n (QF)	M^{tot}_{pre} (QF)	M^{tot}_{post} (QF)	$\langle M_\gamma\rangle$	$\langle M_\gamma\rangle$ (FF)	$\langle M_\gamma\rangle$ (QF)
$^{48}Ca + ^{208}Pb \rightarrow ^{256}No$	230	33	5.2±0.7			3.1±0.5			13.8±1.3	14.0±1.3	9.3±1.8
$^{48}Ca + ^{208}Pb \rightarrow ^{256}No$	240	41	5.6±0.7	0.8±0.2	4.8±0.6	3.7±0.7	0.6±0.2	1.0 ±0.1 2.1±0.3			
$^{58}Fe + ^{208}Pb \rightarrow ^{266}Hs$	297	25							13.8±1.5	14.6±1.7	13.6±1.5
$^{58}Fe + ^{208}Pb \rightarrow ^{266}Hs$	305	32	5.8±0.8			3.6±0.7			14.1±1.7	15.8±1.8	13.7±1.7
$^{58}Fe + ^{208}Pb \rightarrow ^{266}Hs$	315	40							14.7±1.4	17.8±1.7	14.0±1.4
$^{58}Fe + ^{208}Pb \rightarrow ^{266}Hs$	328	50	5.2±0.7	0.9±0.2	4.4±0.6	4.1±0.5	0.6±0.2	1.5±0.2 2.0±0.3			
$^{26}Mg + ^{248}Cm \rightarrow ^{274}Hs$	143	45	6.8±1.0						16.2±1.4	16.2±1.4	
$^{48}Ca + ^{232}Th \rightarrow ^{280}110$	242	44	7.5±1.0	2.8±0.4	5.4±0.6	5.7±0.6	1.7±0.2	1.4 ±0.2 2.6 ±0.3			
$^{64}Ni + ^{208}Pb \rightarrow ^{272}110$	377	65	9.7±0.78	3.3±0.6	6.6±0.6						
$^{48}Ca + ^{238}U \rightarrow ^{286}112$	232	33	8.4±1.2			4.9±0.9			12.7±1.4	17.5±1.6	12.6±1.3
$^{48}Ca + ^{244}Pu \rightarrow ^{292}114$	233	32	9.0±1.2			4.6±0.9			10.8±1.6	14.6±1.8	10.6±1.6
$^{48}Ca + ^{248}Cm \rightarrow ^{296}116$	245	37	9.9±1.4			5.6±1.0			12.7±1.6	14.8±1.8	12.5±1.6
$^{58}Fe + ^{232}Th \rightarrow ^{290}116$	328	53	7.2±0.9	2.6±0.3	4.6±0.6	5.7±0.8	1.9±0.2	1.5±0.2 2.2±0.3			
$^{58}Fe + ^{244}Pu \rightarrow ^{302}120$	328	44	6.5±0.8	2.3±0.3	4.2±0.6	5.5±0.7	1.7±0.2	1.5±0.2 2.3±0.3			
$^{58}Fe + ^{248}Cm \rightarrow ^{306}122$	325	32	12.4±1.6			5.8±1.4			15.4±1.7	21.4±2.1	14.9±1.7

E_{lab} - beam energy in laboratory framework; E^* - excitation energy of compound nuclei; $M^{tot}_{n(FF)}$ - total neutron multiplicities for fission fragments (A/2±20), $M^{tot}_{pre(FF)}$ - total pre-neutron multiplicities for fission fragments (A/2±20); $M^{tot}_{post(FF)}$ - total post-neutron multiplicities for fission fragments (A/2±20); $M^{tot}_{pre(QF)}$ - total pre-neutron multiplicities for quasi fission fragments; $M^{tot}_{n\,(QF)}$ - total neutron multiplicities for quasi fission fragments; $M^{tot}_{pre(QF)}$ - total pre-neutron multiplicities for quasi fission fragments; $M^{tot}_{post(QF)}$ - total post-neutron multiplicities for left and right quasi fission fragments; $\langle M_\gamma\rangle$ - average total γ-ray multiplicities; $\langle M_\gamma\rangle_{(FF)}$ - average γ-ray multiplicities for fission fragments (A/2±20); $\langle M_\gamma\rangle_{(QF)}$ - average γ-ray multiplicities for quasi fission fragments.

• We thank the INTAS foundation (Grant YSC 4258) and the Organizing Committee of the Conference for the financial support. The authors are also grateful to M.Morozova for her help in the preparation of this contribution.
The work has been supported by the Russian Foundation for Basic Research (Grant No 99-02-17891) and by INTAS (Grant No97-11929).

References

1. Kondratiev N. A., Kozulin E. M., Pokrovsky I. V., Prokhorova E. V., Proc. 4 Int. Conf. DANF98, October 19-23 1998, Slovak Republik, Eds. Yu.Ts.Oganessian, J.Kliman, S.Gmuca, World Scientific, Singapore, 1999, p.431.
2. Tilquin I. et al, Nucl. Instr. and Meth. A365 (1995), pp.446-461.
3. Donadille L. et al., Nucl.Phys.A656, 1999, pp.259-283.
4. Moszyński M. et al., Nucl. Instr. Meth. **A 350** (1994) 226.

CHARGE RADIUS CHANGE IN THE HEAVY TIN ISOTOPES AROUND A=132 FROM LASER SPECTROSCOPY

F. LE BLANC, E. COTTEREAU, S. ESSABAA, J. OBERT, J. OMS, A. OUCHRIF, B. ROUSSIERE, J. SAUVAGE AND D. VERNEY

Institut de Physique Nucléaire, IN2P3-CNRS, 91406 Orsay Cedex, France

L. CABARET AND J. PINARD

Laboratoire Aimé Cotton, 91405 Orsay Cedex, France

J.E. CRAWFORD AND J.K.P. LEE

Foster Radiation Laboratory, Mc Gill University, H3A2T8 Montréal, Canada

R. HORN, G. HUBER AND J. LASSEN

Institut für Physik der Universität Mainz, 55099 Mainz, Germany

ISOLDE COLLABORATION

CERN, 1211 Genève 23, Switzerland

Laser spectroscopy measurements have been carried out on the very-neutron-rich tin isotope with the COMPLIS experimental setup. Using the $5s^25p^2\ {}^3P_0 \rightarrow 5s^25p6s^3P_1$ optical transition, hyperfine spectra of $^{126-132}Sn$ and $^{125m,127m,129m-131m}Sn$ where recorded for the first time. The variation of the mean square charge radius ($\delta\langle r^2\rangle$) between these nuclei and nuclear moments of the isomers and the odd isotopes were thus measured. An odd even staggering which inverts at A=130 is clearly observed. This indicates a small appearance of a plateau on the $\delta\langle r^2\rangle$ which should be confirmed by measuring the isotope shift beyond A=132.

1 Introduction

The doubly-magic nuclei are of great interest in nuclear physics because their properties (binding energy, radius...) is used to perform parametrization of effective interactions used for mean-field calculations [1,2]. For the last two decades, these calculations successfully described global properties of the nuclear ground states [3,4,5,6]. In the same time, the relativistic mean-field theory [7] was getting success for describing nuclear ground state properties [6,8,9]. In parallel, numerous new and accurate results were obtained making systematic data available along isotopic series from light to heavy nuclei. More recently, it appeared that the mean-field models had to be improved to correctly reproduce these precise results. This motivated many theoretical works in particular to improve the parameters of the effective interactions currently used. The goal of these theoretical studies is to define an effective interaction valid not only along the stability but also for exotic nuclei. We want thus to study the effect of the shell closure at N=82 on the mean square charge radius variation ($\delta\langle r^2\rangle$) far from stability. The most important question is : will the $\delta\langle r^2\rangle$ exhibit a slope change at ^{132}Sn as it does for ^{208}Pb ? Moreover, $\delta\langle r^2\rangle$ curves have been calculated for neuron-rich tin [10,11] and the predictions depend on the type of calculations.

Measurement of isotope shift of hyperfine structures gives a direct acces to the $\delta\langle r^2\rangle$ along isotopes series. To perform such measurements on tin isotopes, we have successfully used a technique of ion-beam implantation followed by Resonant Ionisation Spectroscopy (RIS) studies of the laser desorbed radioactive element. Such a system (COMPLIS) is installed at the ISOLDE-Booster facility. Accumulated products from the implanted ISOLDE beams are prepared as pulsed mass-separated ion beams by laser desorption and selective ionization. In this contribution, we report on recent laser spectroscopy measurements performed on these heavy tin isotopes up to A=132. From the $5s^25p^2\ ^3P_0 \rightarrow 5s^25p6s^3P_1$ optical transition, the hyperfine spectra of $^{126\text{-}132}Sn$ as well of these of $^{125m,127m,129m\text{-}131m}Sn$ where recorded for the first time. The variation of the mean square charge radius ($\delta\langle r^2\rangle$) between these nuclei and the nuclear moments of the isomers and the odd isotopes were thus measured. These results are discussed and compared with different theoretical predictions.

2 Experimental Method

The radioactive tin isotopes are produced via fission reactions on the ISOLDE UC_2 target with the 1GeV CERN PS-Booster proton beam. The ions are extracted at 60 kV and mass separated by the Ground Purpose Separator (GPS) of ISOLDE. They enter the COMPLIS beam line, are slowed to 1 kV and are thus deposited on the first atomic layers of a rotating graphite substrate. Once the amount of the collected atoms is optimum (the collection time depending on the half-life of the isotope to be studied) they are desorbed by a Nd:YAG laser and selectively ionized by a set of two pulsed, tunable dye lasers where the first excitation step at 286.3 nm ($5s^25p^2\ ^3P_0 \rightarrow 5s^25p6s^3P_1$) is obtained from frequency doubling. The ions are finally detected with time-of-flight identification using a microchannel plate detector. This experimental set up is shown on Fig. 1. The frequency scan over the hyperfine structure of a given isotope (and eventually isomer) is made as follows : after a sufficient collection time, the desorption of the tin atoms is made over the entire collection spot on the slowly rotating target at a given frequency step. After the desorption is complete, a new cycle of implantation desorption is run at an advanced frequency step. Whenever the laser frequency corresponds to a hyperfine transition, the desorbed atoms are excited and ionized by the other fixed laser frequency. The number of counted ions at the detector is directly proportional to the intensity of the hyperfine transition. With this apparatus, the efficiency we measured is of about 10^{-6} with a resolution of 170 MHz.

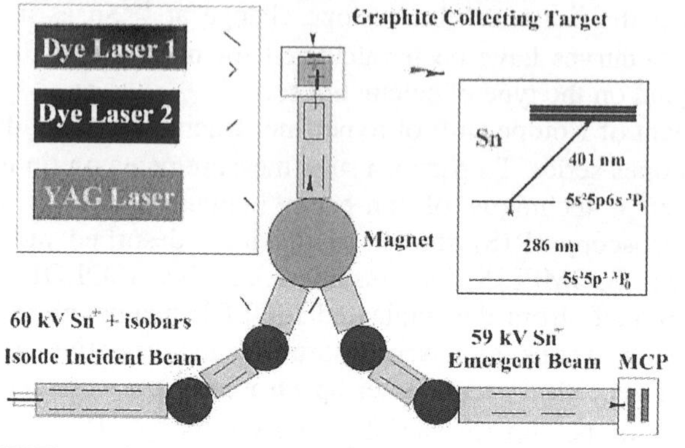

Figure 1. The COMPLIS experimental scheme used for tin

For the first experiment on tin, we measured for the first time all the isotopes and isomers from A=125 to A=132. The measured hyperfine spectra of ^{132}Sn, $^{131}Sn^{g+m}$ and $^{130}Sn^{g+m}$ are shown as examples in Fig. 2. From the displacement of the centers of gravity of the hyperfine spectra, we were able to extract the isotope shift.

3 Experimental results

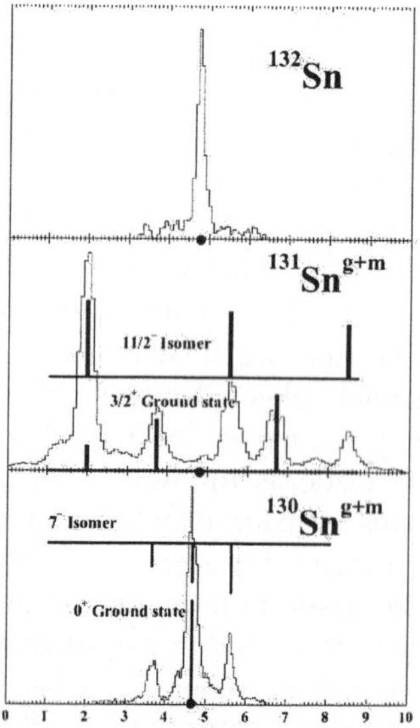

Figure 2. Hyperfine spectra of $^{130}Sn^{g+m}$, $^{131}Sn^{g+m}$, $^{131}Sn^{g+m}$

The experimental isotope shift consists of a mass shift $\delta\nu_{MS}$ and a field shift $\delta\nu_{FS}$; it is from this last contribution that $\delta<r_c^2>$ can be extracted : $\delta\nu_{FS}^{AA'} = K.F_{\lambda}.\delta < r_c^2 >^{AA'}$ [12], where K=0.975 [12] and F_λ=3.35(20) GHz/fm^2 [13]. Using formula (2) in [14] but with the correct values of F_λ and $\delta<r_c^2>^{124,\ 116}$=0.442(5) [13], one can avoid to introduce the specific mass shift which is very difficult to estimate. The updated $\delta<r_c^2>$ curve is presented in Fig. 3. The main remarkable thing that we can observe is the small appearance of a plateau at mass 131. This suggests a neutron skin at A=132 and beyond but experimental data are now needed to confirm this hypothesis. The $\delta<r^2>$ of the even isotopes can however be compared with existing mean-field calculations : some including relativistic effect (NL3, NL-Z) [10], some using skyrme forces (SkI1, SkI4, Sly6) [10] and one using the Gogny force [11] (see Fig. 4).

Most of them are not as far from the experimental values but the appearance of the plateau is badly reproduced. Once more, experimental data on ^{134}Sn are waited to definitely constrain the models.

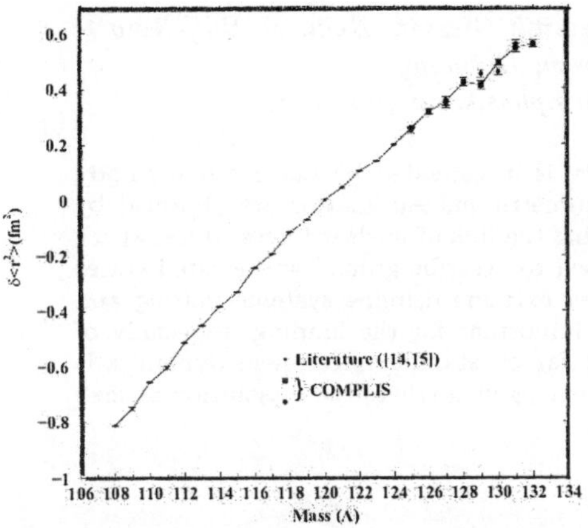

Figure 3. Charge radius change in the tin isotopes. The line relies the ground states.

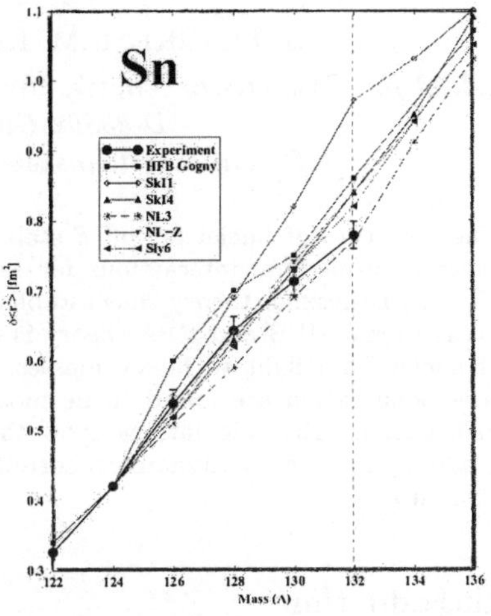

Figure 4. Comparison of $\delta<r_c^2>$ with some theoretical predictions.

Moreover, from the *A* and *B* hyperfine constants, one can extract the magnetic and quadrupole moments of each isomer and odd isotope. The data are under analysis and the extracted experimental values will be compared with calculations assuming a given nuclear orbital.

References

1. D. Vautherin and D.M. Brink, Phys. Rev. C5 (1972).
2. J. Dechargé and D. Gogny, Phys. Rev. C21 (1980) 1568.
3. P. Quentin and H. Flocard, Annu. Rev. Nucl. Sci. 28 (1978) 523.
4. M. Girod and B. Grammaticos, Phys. Rev. C27 (1983)2317.
5. M. Girod et al., Phys. Rev. Lett. 62 (1989) 2452.
6. Z. Patyk et al., Phys. Rev. C59 (1999) 704.
7. J.D. Walecka, Annu. Phys. N.Y. 83 (1974) 491.
8. P.G. Reinhard, Rep. Prog. Phys. 52 (1989) 439.
9. V. Blum et al., Phys. Rev. B323 (1994) 262.
10. P.G. Reinard, private communication (1997).
11. M. Girod, private communication (1998).
12. G. Torbohm et al., Phys. Rev A31 (1985) 2038.
13. C. Piller *et al.*, Phys. Rev. C42 (1990) 182.
14. M. Anselment *et al.*, Nucl. Phys. A451 (1986) 471.
15. J. Eberz *et al.*, Z. Phys. A326 (1987) 121.

MANY-BODY THEORY AT EXTREME ISOSPIN

H. LENSKE, C.M. KEIL AND F. HOFMANN

Institut für Theoretische Physik, Universität Giessen, Heinrich-Buff-Ring 16, D-35392 Giessen, Germany

E-mail: horst.lenske@theo.physik.uni-giessen.de

The structure of nuclei far off β-stability is investigated by nuclear many-body theory. In-medium interactions for asymmetric nuclear matter are obtained by (Dirac-) Brueckner theory thus establishing the link of nuclear forces to free space interactions. HFB and RPA theory is used to describe ground and excited states of nuclei from light to heavy masses. In extreme dripline systems pairing and core polarization are found to be most important for the binding, especially of halo nuclei. The calculations show that far off stability mean-field dynamics is gradually replaced by dynamical correlations, giving rise to the dissolution of shell structures.

1 Introduction

Nuclear many-body theory provides a very general scheme allowing to investigate the whole variety of nuclear systems from (purely theoretical) infinite nuclear matter and macroscopic objects as neutron star matter down to the microscopic level of finite nuclei. Testing the concepts and methods of nuclear many-body theory at the extremes as found in exotic nuclei is an important and challenging question for structure physics. The vast amount of new data on nuclei far off stability [1] has initiated corresponding efforts on the theoretical side, ranging from new impacts on cluster models and *ab initio* shell model calculations for light nuclei to mean-field calculations over the entire mass range. New phenomena, like halo and skin formation, have been explained or predicted by nuclear theory, documenting the progress made over the last years.

Until very recently, data were mainly available for ground state properties of exotic nuclei. At present, the situation is changing because the progress in experimental techniques allows to obtain information also on dynamical processes in neutron- and proton-rich nuclei. For theory this means to extend the methods to a new sector of phenomena, e.g. inelastic and charge exchange response functions and the fragmentation patterns of single particle strength functions in breakup and transfer reactions. The description of such phenomena clearly requires new approaches, accounting properly for new effects as the coupling to the nearby continuum and extreme isospin. Since one is entering unexplored territories models providing intrinsically an extrapolation scheme for interactions and the resulting nuclear dynamics are requested. Phenomenological approaches like Skyrme theory and relativistic mean-field theory are moderately successful in this respect, but their flexibility and generality is constraint by the assumed operator structure of the model Hamiltonian or Lagrangian and the ability to determine the model parameters by fits to data only.

An important alternative is to approach the problem microscopically. A clear advantage of such a program is that - at least in principle - investigations can be based on a systematic order-by-order hierarchy of interaction diagrams, as typical for many-body theory, thus avoiding ambiguities and taking advantage of results being tested independently in other regions of nuclear physics. Taking this as a guideline the approach presented in this contribution starts from calculating interactions in asymmetric nuclear matter by Brueckner theory. As discussed in section 2 applications to the equation of state of infinite matter, neutron star matter and neutron star structure calculations give information on the global properties of the interaction model, e.g. indicating the importance for going beyond the pure ladder approximation inherent to Brueckner theory.

Applications of the microscopic approach to finite nuclei are discussed in section 3, illustrating pairing in dripline nuclei for ^{11}Li, and core polarization far off stability is the subject of section 4 as examples for the transition into a new dynamical regime of correlation dynamics. A summary and an outlook are given in section 5.

2 Interactions in Asymmetric Nuclear Matter

Good confidence on the free NN-interaction has been obtained with field theoretical meson-exchange models like the Bonn potentials [17], showing that the ladder approximation is an appropriate scheme in free space. In a medium, however, the situation is less certain in the sense that additional classes of diagrams will contribute, as illustrated in Fig.1. The in-medium pieces introduce an additional density dependence and contribute new types of operators, not found in free space interactions [8,9]. Among those, the coupling of mesons to medium polarization modes and three body forces from intermediate excitations of nucleonic resonances (TBF) will alter also the isospin structure of interactions. Little is known about the dependence of these contributions on varying the proton-to-neutron composition.

In ref.[2] Dirac-Brueckner Hartree-Fock (DBHF) theory was used to investigate 2-body interactions in infinite matter over a large range of asymmetries. In-medium meson-nucleon coupling constants were extracted for the isoscalar

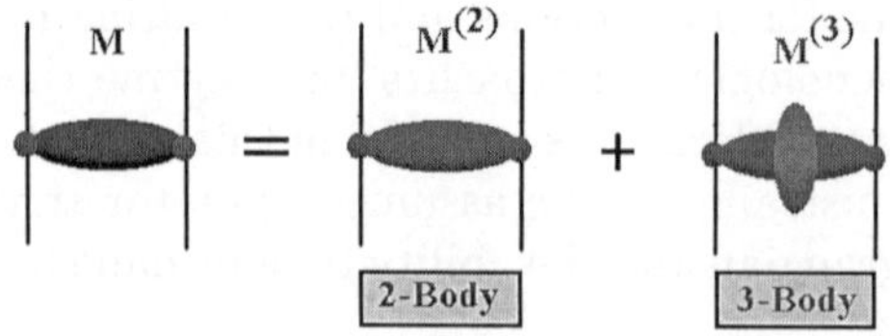

Figure 1. The in-medium NN scattering amplitude including 2- and 3-body contributions, respectively. The 3-body pieces, being of order $\sim \mathcal{O}(\rho)$, result from the coupling to nucleon resonances and polarization graphs.

σ and ω and the isovector ρ and δ mesons, respectively [2]. In all meson channels a pronounced dependence on the isoscalar bulk density is found while the dependence on the asymmetry is close to negligible. Hence, to a very good approximation in-medium strong interactions remain intrinsically independent of isospin thus conserving the fundamental isospin symmetry. Using these coupling constants in the density dependent relativistic hadron (DDRH) field theory [3] finite nuclei [4], hypernuclei [5] and neutron stars [6] are well described, thus underlining the success of a microscopic approach.

In Fig.2 the equation of state obtained in non-relativistic theory is shown, including results in ladder approximation only and with TBF. The TBF are seen to act in total *attractive* in the low density region, but turn to *repulsive* at high densities, reflecting their specific density dependence. The full D3Y interaction, including TBF, is used in the following sections.

3 Pairing at the Dripline

In nuclear matter pairing is found to be a low density phenomenon, having a maximum contribution to the energy density around $\frac{1}{3}$ of the saturation density $\rho_0 = 0.16 fm^{-3}$. Therefore, it has to be expected that pairing is enhanced in nuclei with a pronounced low-density region. This is exactly the situation in 2- or multi-nucleon halo systems, and, to a lesser extent, in weakly bound skin nuclei. A prominent case is ^{11}Li which exist as a particle-stable system mainly because of the mutual interactions among the two valence neutrons. Their low separation energy, S_{2n}=320 keV, points to the importance of continuum coupling. For a detailed description of the valence wave function the conventional BCS and HFB methods using representations in terms of mean-field wave functions are not suitable. Theoretically, continuum effects

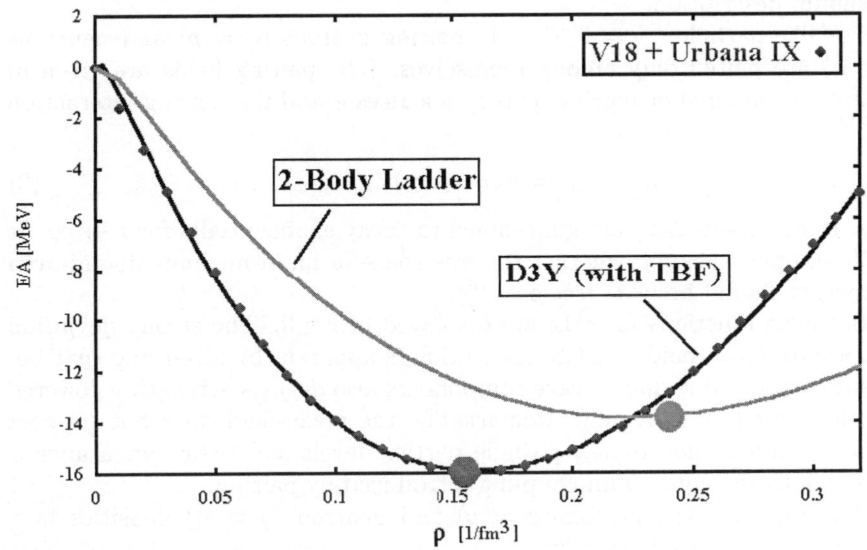

Figure 2. Equation of state of symmetric nuclear matter in non-relativistic Brueckner theory without and with 3-body interactions. The saturation point are indicated by full circles. For comparison, variational result (dots) of the Urbana group [8] are also shown.

are properly described by solving thc Gorkov-equations [14]

$$(h - e_+)\Phi_+ - \Delta\Phi_- = 0 \quad , \quad (h - e_-)\Phi_- + \Delta\Phi_+ = 0 \tag{1}$$

for the hole and particle components [15] $\Phi_\pm$, coupled by the pairing field Δ. The single particle energies are $e_\pm = \lambda \pm E$ where λ and E (≥ 0) denote the chemical potential and the quasiparticle energy, respectively. $\Phi_\pm$ will in general *not* coincide with the eigenfunctions of the mean-field Hamiltonian h, i.e. the states $\Phi_\pm$ are off the (mean-field) energy shell.

Particle-stability is given for $\lambda < 0$. Then, irrespective of the value of E, the hole-type solutions Φ_- are exponentially decaying for $r \to \infty$. For E$\leq |\lambda|$, an exponential asymptotic behaviour is also found for the particle-type components Φ_+ and a finite subset of discrete eigenvalues E is obtained. For E$> |\lambda|$, eq.(1) has to be solved with continuum wave boundary conditions for Φ_+ leading to single particle spectral functions being distributed continuously in energy. Hence the quasi-particle picture, underlying BCS theory and, to some degree, also discretized HFB theory, is replaced by a fully dynamical continuum description.

For like-particle ($S = 0, T = 1$) pairing protons ($q = p$) and neutrons ($q = n$) are paired only among themselves. The pairing fields are given in terms of the anomal or pairing density matrices κ and the pairing interaction V_{SE} [18],

$$\Delta_q(r_1, r_2) = V_{SE}(r_1, r_2)\kappa_q(r_1, r_2) \quad . \tag{2}$$

If $\lambda < 0$, $\kappa_q(r)$ and $\Delta(r)$ are guaranteed to decay exponentially for $r \to \infty$. It is still an open question whether the free space or an in-medium singlet-even interaction should be used (see e.g. [16]).

Strength functions for ^{11}Li are displayed in Fig.3. The strong deviation from a pure mean-field or BCS description is apparent by observing that besides the expected s- and p-wave components also $d_{5/2,3/2}$ strength is lowered into the bound state region. Remarkably, the mean-field does not support neither bound $2s$ nor $1d_{5/2,3/2}$ single particle levels and their appearance is solely due to the continuum coupling introduced by pairing.

Theoretically, the proton ($q = p$) and neutron ($q = n$) densities in a systems like ^{11}Li are defined by

$$\rho_q(r) = \sum_{j\ell} \frac{2j+1}{4\pi} \int_{-\infty}^{\lambda} de_- v^2_{qj\ell}(e_-)|G_{qj\ell}(r, e_-)|^2 \quad , \tag{3}$$

where $\Phi_- = v^2 G$ was used [18]. The particle numbers are given by

$$N_q = \sum_{j\ell} (2j+1) \int_{-\infty}^{\lambda} de_- v^2_{qj\ell}(e_-) = \sum_{j\ell} (2j+1) n_{qj\ell} \quad , \tag{4}$$

The neutron partial wave occupation numbers $N_{j\ell} = (2j+1)n_{j\ell}$ for ^{11}Li are displayed in Fig.3. In stable nuclei discrete levels at $2\lambda < e_- < \lambda$ will contribute to eq.(3). In extreme dripline nuclei they are missing because of the smallness of $|\lambda|$. In Fig.4 the proton and neutron ground state densities are shown, multiplied by r^2, emphasizing the differences in shape and the neutron halo component. High-energy elastic scattering of ^{11}Li on a proton target is well described with the HFB densities [19]. Measured ^{11}Li response functions were analyzed by QRPA calculations in [20].

4 Dynamical Core Polarization at the Dripline

Dynamical core polarization (DCP) is seen most clearly in nuclei with a single nucleon outside a core. Approaching the driplines the core nucleus by itself is already far off stability with a lower surface tension. Therefore, the restoring

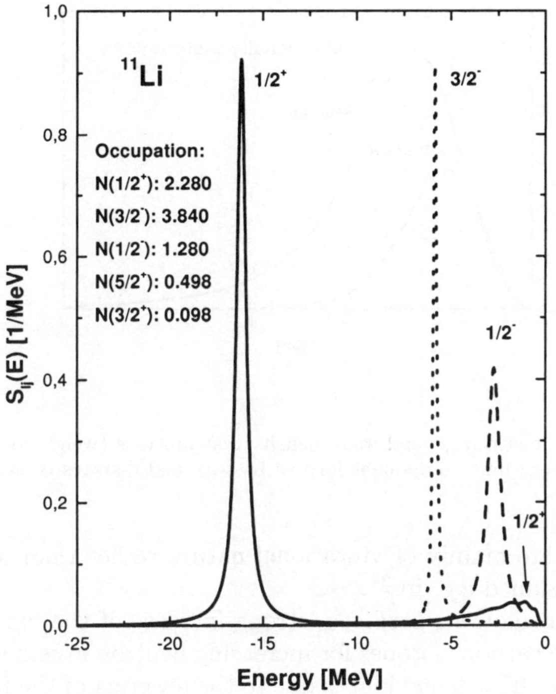

Figure 3. Single particle spectral functions for s- and p-wave neutron states in ^{11}Li. The finite widths of the states are due to continuum coupling. The partial occupation numbers $N_{j\ell}$ are also shown, including d-wave contributions.

forces against external perturbations are reduced. Such "soft core" systems are, for example, found in the neutron-rich even-mass carbon isotopes. A good indicator is the existence of low-energy 2^+ states, decreasing in energy with increasing mass. Continuum QRPA calculations with a residual interaction derived in Landau-Fermi liquid theory from the D3Y in-medium interaction reproduce the systematics of 2^+ states rather well. This leads to the con-

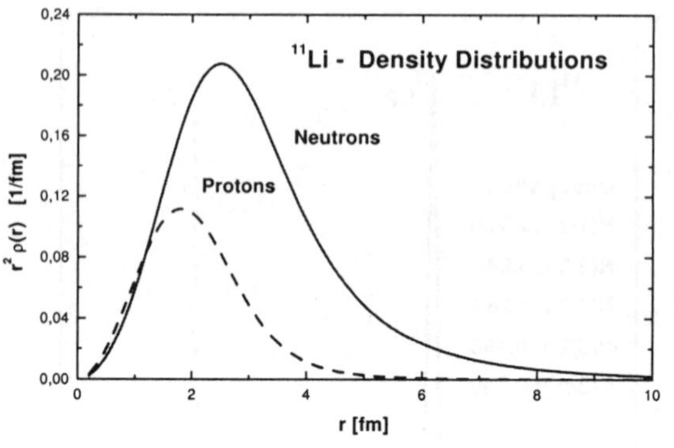

Figure 4. Proton and neutron ground state density distributions (weighted by r^2) for ^{11}Li. The pronounced neutron halo component formed by s-,p- and d-states is clearly visible.

clusion that they are mainly of vibrational nature rather than due to static deformation as assumed e.g. in [24].

The QRPA calculations predict a strong increase of the quadrupole polarizabilities in the carbon isotopes for increasing neutron excess with a maximum around ^{16,18}C. This behaviour is due to the lowering of the first 2^+ state which in ^{12}C is located at $E_x = 4.44$ MeV and moves down to $E_x = 1.66$ MeV in ^{18}C. The dipole (or electric) polarizability, however, changes only within 10% over the $^{10-22}$C isotopic chain.

In a system with core polarization the valence particle (or hole) obeys a non-static wave equation

$$(H_{MF} + \Sigma_{pol}(\varepsilon) - \varepsilon)\,\Psi = 0 \tag{5}$$

including the static (HFB) mean-field Hamiltonian H_{MF} and the non-local and energy-dependent polarization self-energy Σ_{pol} describing the rescattering of the nucleon off the core thereby exciting it into states of various multipolarities and excitation energies followed by subsequent de-excitations back to the ground state (see Fig.5).

During these processes the particle can be scattered virtually into high lying orbitals. The quantum numbers of the intermediate $2p1h$ (or $1p2h$) configurations are only constraint by the requirement that spin and parity must match those of the leading particle configuration, given by the state moving with respect to the core ground state. In nuclear matter and finite

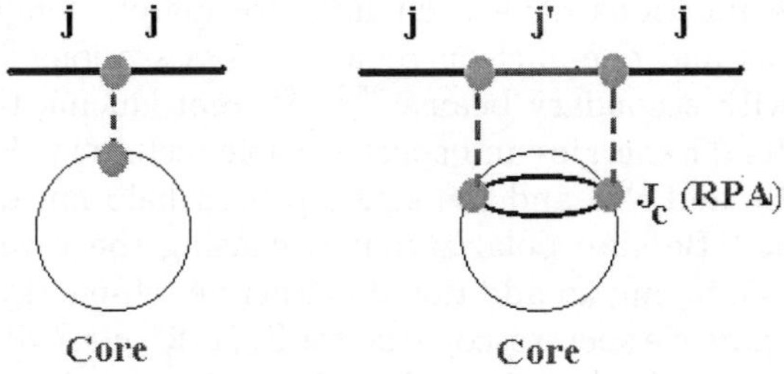

Figure 5. Diagramatic structure of mean-field (left) and core polarization interactions (right) of a nucleon in single particle state j. Interactions (meson exchange) are indicated by dashed lines. Core polarization leads to intermediate states $(j'J_c)$ with a (QRPA) core excitation J_c and a single particle state j'.

Table 1. Energies, spins and ground state spectroscopic factors from core polarization calculations.

Nucleus	j^π	*Energy keV*	$S(j^\pi, g.s.)$
^8B	$3/2^-$	130	0.73
^{11}Be	$1/2^+$	510	0.74
^{17}C	$5/2^+$	760	0.43
^{19}C	$1/2^+$	263	0.41

nuclei these processes are known to give rise to a depletion of momentum space ground state occupation probabilities [22,23] from the step function momentum distribution of a Fermi gas of quasiparticles interacting only by a static mean-field. In practice, the core excitations are calculated in QRPA theory thus extending the static HFB picture in a consistent way to a dynamical theory.

Dynamical single particle self-energies Σ_{pol} affect separation energies and wave functions. In dripline nuclei these effects are strongly enhanced. In fact, dynamical self-energies provide the main source of binding for ^{19}C$(1/2^+, g.s.)$ [18].

Theoretically, the DCP description amounts to use a multi-configuration ground state containing a single particle component - reminiscent of the static mean-field configuration - and a multitude of configurations where the valence particle is rescattered into other orbits by interactions with the core [15,21]. Coupling to the lowest 2^+ and 3^- states only as advocated e.g. in [25] cannot account for the complexity of the process.

The DCP wave functions have been used to analyze longitudinal momentum distributions and one-nucleon removal cross sections in relativistic breakup reactions with secondary beams [26,27,28], reproducing the data very satisfactorily. Results for energies and spectroscopic factors in the single neutron halo nuclei ^{11}Be and ^{19}C and the single proton halo nucleus ^{8}B [26] are shown in Tab.1. In ^{11}Be core polarization is causing the reversal of $1/2^+$ and $1/2^-$ states by supplying an additional attractive self-energy in the $1/2^+$ channel. The single particle spectroscopic factor $S_n(1/2^*, g.s.)$=0.75 is in good agreement with recent transfer and breakup data. A more dramatic effect is found in ^{19}C and also ^{17}C where the g.s. components are strongly suppressed as seen from the small spectroscopic factors in Tab.1. Hence, the valence nucleon is no longer attached to a definite mean-field orbital but exists in a wave packet-like state spread over a certain range of shell model states. In other words, mean-field dynamics have ceased to be the dominant source of binding. Rather, the ^{17}C and ^{19}C results indicate a new type of binding mechanism in dripline nuclei where shell structures are dissolved. The mean-field has lost its dominant role and binding is produced by dynamical valence-core interactions.

5 Summary and Outlook

Asymmetric nuclear matter and nuclei far off stability were described by microscopic approaches. An important advantage over empirical models is the clean diagrammatic structure allowing to extend the theory in a well defined way by inclusion of an increasingly larger class of diagrams. This was illustrated on the example of 3-body contributions to in-medium interactions. Pairing and core polarization far off stability were discussed in some detail. The results show that close to the dripline nuclear dynamics change from mean-field to correlation dynamics. As an important consequence shell structures are dissolved. This reflects that the traditional understanding of nuclear structure as being dominated by a static potential plus some residual interactions cannot account for nuclear structure far off stability.

Acknowledgments

This work was supported in part by DFG, project Le 439/3, European Graduate School "Complex Systems of Hadrons and Nuclei", Giessen-Copenhagen, GSI Darmstadt and BMBF.

References

1. P.G. Hansen, A.S. Jensen, B. Jonson, *Ann. Rev. Phys. Sci. (N.Y.)* **45**, 591 (1995).
2. F. de Jong, H. Lenske, *Phys. Rev.* C **57**, 3099 (1998).
3. H. Lenske, F. Fuchs, *Phys. Lett.* B **345**, 355 (1995); C. Fuchs, H. Lenske, H. Wolter, *Phys. Rev.* C **52**, 3043 (1995).
4. F. Hofmann, C. Keil, H. Lenske, *Phys. Rev.* C , (publication scheduled for Aug. 2001); nucl-th/0007050.
5. C. Keil, F. Hofmann, H. Lenske, *Phys. Rev.* C **61**, 06401 (2000).
6. F. Hofmann, C. Keil, H. Lenske, *Phys. Rev.* C **64**, 025804 (2001); nucl-th/0008038.
7. P. Ring *et al.*, *Ann. Phys. (N.Y.)* **198**, 132 (1990).
8. V.J. Pandharipande *et al*, *Phys. Rev.* C **58**, 1804 (1998).
9. S.A. Coon, M.T. Peña, *Phys. Rev.* C **48**, 2559 (1993); S.A. Coon, M.T. Peña, D.O. Riska,*Phys. Rev.* C **52**, 2925 (1995).
10. H. Lenske, G. Schrieder, *Euro. Phys. J.* A **2**, 41 (1998); C. Gund *et al.*, *Euro. Phys. J.* A **10**, 85 (2001), see also nucl-ex/0010005.
11. F. Hofmann, H. Lenske, *Phys. Rev.* C **57**, 183 (1998).
12. E. Chabanat *et al.*, *Nucl. Phys.* A **627**, 273 (1997).
13. F.T. Baker *et al.*, *Phys. Rep.* **289**, 235 (1997).
14. L.P. Gorkov, *Sov. Phys.* JETP **9**, 1364 (1959); see also [16].
15. H. Lenske, *J. Phys.* G **24**, 1429 (1998).
16. For a review see e.g. W. Nazarewicz, *Prog. Part. Nucl. Phys.* **46**, (2001); P.Ring, *ibid.*.
17. R. Machleidt, *Adv. Nucl. Phys. (N.Y.)* **19**, 189 (1989).
18. H. Lenske *et al.*, *Prog. Part. Nucl. Phys.* **46**, 187 (2001).
19. P. Egelhoff, *Prog. Part. Nucl. Phys.* **46**, (2001).
20. M. Zinser, H. Emling, H. Lenske et al., *Nucl. Phys.* A **619**, 151 (1997).
21. F.J. Eckle *et al.*, *Phys. Rev.* C **39**, 1662 (1989) and *Nucl. Phys.* A **506**, 199 (1990).
22. F. de Jong, H. Lenske, *Phys. Rev.* C **54**, 1488 (1996).
23. J. Lehr *et al.*, *Phys. Lett.* B **483**, 324 (2000).
24. F. Nunes *et al.*, *Nucl. Phys.* A **596**, 171 (1996).
25. N. Vinh-Mau, *Nucl. Phys.* A **592**, 33 (1995).
26. M.H. Smedberg *et al.*, *Phys. Lett.* B **452**, 1 (1999).
27. T. Baumann *et al.*, *Phys. Lett.* B **439**, 256 (1998).
28. D. Cortina-Gil *et al.*, *Euro. Phys. J.* A **10**, 49 (2001).

STUDY OF β–ν ANGULAR CORRELATIONS IN NUCLEAR β DECAY

E. LIÉNARD, G. BAN, G. DARIUS, P. DELAHAYE, D. DURAND,
A. M. VINODKUMAR, F. MAUGER, O. NAVILIAT
LPC CAEN, IN2P3-CNRS, ISMRA, F-14050 Caen Cedex, France
E-mail : lienard@caelav.in2p3.fr

P. VAN HOVE
GANIL, CEA/DSM-CNRS/IN2P3, BP 5027, F-14076 Caen Cedex 5, France

New developments in ion cooling and ion bunching allow the trapping of radioactive ions from a low energy beam, in a very short time-scale and with very good efficiency. In the present work, we intend to use a RFQ device to inject $^6He^+$ ions in a transparent Paul trap in order to study the β decay spectrum. The ion beam will be delivered by the future low energy beam line of the SPIRAL facility at GANIL. The cooling of the beam before injection into the trap will be achieved by the use of a H_2 buffer gas in the RFQ, for which simulations give a 10% transmission. In the Paul trap, the ions will be in vacuum and nearly at rest, allowing a careful measurement of the β - recoil ion coincidence spectrum. These data will be sensitive to the angular correlation parameter "a", which depends on the coupling constants of the weak hamiltonian. It should be equal to –1/3 if the interaction is only of Axial vector type (V-A theory). Deviation from this value would imply a new Tensor-like interaction leading to new physics beyond the Standard Model.

1 Motivation

For allowed β transitions, the weak interaction hamiltonian may be written in terms of five Lorentz invariants referred to as Scalar-, Vector-, Tensor-, Axial and Pseudoscalar-interactions (*S, V, T, A, P*) [1]. The relative strength of these interactions are characterized by the coupling constants C_i and C_i'; the second ones take into account the parity violation in β decay. Data suggest a strong dominance of the *V*- and *A*-couplings. In the Standard Model, $C_V = C_V'$ and $C_A = C_A'$ (maximum parity violation), $C_S = C_T = C_P = 0$, C_i and C_i' are real (time–reversal invariance).

Any deviation from this pattern would imply new physics beyond the Standard Model. In the case of unpolarized nuclei, one can show that the angular distribution of the events depends on a coefficient 'a' related to the coupling constants [2]. Choosing a pure transition, of Fermi (F) or Gamow-Teller (GT) type, gives access separately to the contributions of the exotic terms C_S or C_T respectively. The value of 'a' influences the distribution of the events within the limits imposed by the three-bodies kinematics. The recoil nucleus energy is revealed to be an observable very sensitive to this parameter. The present best value on a_{GT} (-0.3343±0.0030) was extracted from the energy spectrum shape of the recoil ions following the decay of

^{6}He nuclei [3]. This value leads to an upper limit of 13% at the 95 % C.L. on $|C_T/C_A|$ [4]. Recently, new methods were developed to avoid the direct detection of the daughter nuclei which have a very low energy (~ 100 eV – 1 keV). They are based on the study of the recoil effects on the energy of γ's or delayed particles emitted by the daughter nucleus. For the scalar interaction, the present best upper limit of 12% on $|C_S/C_V|$ was obtained using such a method. The experiment was performed with ^{32}Ar nuclei decaying to a particle unstable excited state of ^{32}Cl [5]. Several projects are in progress around the world and we can expect that the present limit on the scalar interaction will be still improved in a near future.

We propose to search for Tensor type contribution, using a new method. Simulations have shown that the sensitivity to 'a' was maximal for an angle between the β and the recoil ion greater than 150° [6]. To select this region we must measure β-recoil ion coincidences, and this imposes to perform the experiment in a matter free environment such as the one ensured by a Paul trap [7]. This device gives the opportunity to perform the measurements under favourable conditions because the ions are in vacuum, well localized and nearly at rest. In this context the ^{6}He seems to be an excellent candidate, mainly for the following reasons : this is a pure decay from g.s. to g.s. (measurement of the exact kinematics), with a high Q-value ("high" maximum recoil energy of 1.4 keV), long lifetime (preparation of the beam) and possible high production rate. To be trapped, the ions coming from the production source must be nearly thermalized and this will be performed by a special beam handling system called a RFQ cooler-buncher [8]. This ion cooling device is able to slow down ions from few keV to less than 1 eV in some hundreds of μs while keeping a reasonable beam emittance.

2 The RFQ cooler - buncher

The injection of ^{6}He from the future low energy beam line of SPIRAL (GANIL) to the Paul trap requires specific attention since ions are trapped with nearly thermal energy and have then to be decelerated by several tens of keV. The assumed characteristics of the beam (10^9/s $^6He^+$ at 30 keV, emittance : 80π.mm.mrad) impose the use of a cooling device in order to avoid a prohibitive spreading of its emittance. A RFQ cooler-buncher is now under operation at LPC Caen [9].

At the entrance of the system, the ions are slowed down to 100 eV by a focalization electrode which avoids a complete loss of the beam when it is decelerated. The RFQ itself consists in 4 segmented rods connected at RF potentials which ensure a radial confinement of the beam. DC potentials set on the segments create a well in the axial direction, and allow the bunching of the ions at the end of the device. Finally the cooling of the ions is achieved by their collisions with a buffer gas which is introduced in the RFQ.

Such devices are also in operation at JYFL (Finland) and ISOLTRAP (CERN) [10]. They work well with a transmission efficiency close to 100 % when the mass of the gas is much lower than the ion mass. Unfortunately, in our case, the ion is already very light (^{6}He) and the only possibility for the buffer gas is hydrogen. Microscopic Monte Carlo simulations give a transmission efficiency of 10 %, with a mean transit time of the order of 200 μs [11]. Figure 1 shows the results of the simulated cooling process, for $^6He^+$ on H_2 buffer gas, with the incoming beam parameters given above. The smooth operation of the RFQ in normal conditions (N_2 ions with He buffer gas) has already been tested. The tests with H_2 buffer gas and lighter ions are presently in progress.

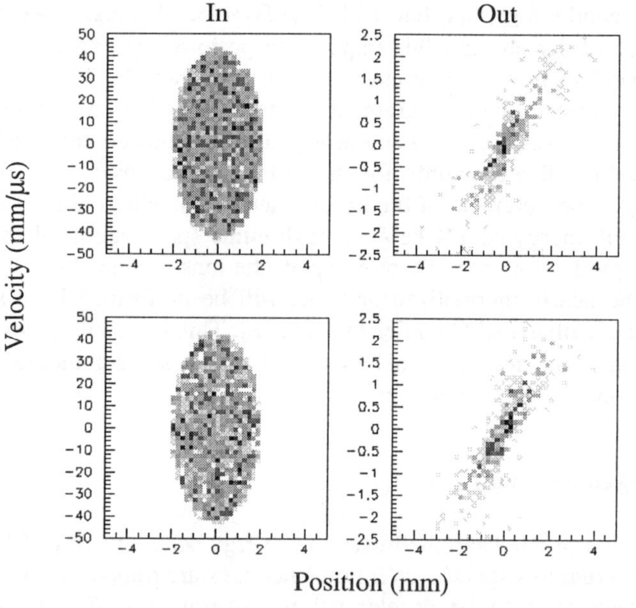

Figure 1. Results of the simulated cooling process. The velocity distribution of the simulated events is presented versus the positions in X (top) and Y (bottom) directions. The pictures on the left (resp. right) show the transverse phase space of the beam at the entrance (resp. exit) of the cooler. A clear reduction of the velocity distributions is observed.

3 The transparent Paul trap and the detection set-up

The transparent Paul trap consists of two end cap electrodes and a transparent ring made of thin wires (∅ ~100 μm). Such a frame minimizes the amount of material, and then offers a large solid angle for a detection set-up, which can be installed all around the trap. Simulations have been performed and allowed to infer that :

a) The injection efficiency from the buncher is of the order of 10 % ; b) The ion cloud has a width of the order of 2 to 5 mm. The effects of this width on the parameters of the experiment are being studied. Several trap sizes are tested to reduce the width as much as possible ; c) The fraction of β scattering on ring wires (∅ 20 μm) is less than 1 %.

The RF applied on the trap will perturb the trajectory of the ion. In order to obtain good reconstruction of the kinematics, two methods are investigated. Firstly, the RF potentials may be shut down when the relativistic electron is detected. This cut should be performed within 100 ns. Secondly, the phase of the RF may be recorded allowing an off-line reconstruction of the ion trajectory with the help of simulations.

The experimental set-up will consist of two detectors placed back to back and at distances according to the sensitivity requirements of the experiment as discussed in section 1. The three parameters of the experiment easily reachable are the energies of the β and the recoil ion, and the relative angle between them. The β will be detected with a telescope made of a position sensitive silicon detector followed by a thick lithium drifted silicon detector. The ion will be detected by a micro-channel plate, its energy being obtained by a time-of-flight referenced to the fast β signal. A delay-lines anode will be used to get the detection position of the ion. Some tests on the detection set-up have already been performed. For the ion detector, a spatial resolution of about 100–200 μm has been reached. Presently the response functions of the silicon detectors are studied, as well as the efficiency of the micro-channel plate to low energy ions.

In principle, the tests and simulations of the whole set-up should be achieved by the end of year 2002. At the same time, the low energy beam line at GANIL should be built. First on line measurements are expected during 2003.

References

1. Boothroyd A.I. et al., *Phys. Rev. C* **29** (1984) pp. 603-617.
2. Jackson J.D., Treiman S.B. and Wyld H.W., *Nucl. Phys.* **4** (1957) pp.206–212.
3. Johnson C.H. et al.., *Phys. Rev.* **132** (1963) pp. 1149–1165.
4. Glück F., *Nucl. Phys. A* **628** (1998) pp. 493–502.
5. Adelberger E., *Phys. Rev. Lett.* **83** (1999) pp. 1299–1302.
6. Ban G. et al., *LPC Caen report* LPCC-99-16 (1999) unpublished.
7. Zhao W. Z. et al., *Hyp. Int.* **108** (1997) pp. 483–495.
8. Lunney M.D. and Moore R.B., *Int. J. Mass Spec.* **190** (2000) pp. 153–160.
9. Delahaye P. et al., *Hyp. Int.* (2000) in press.
10. Kellerbauer A. et al., *CERN-EP/2000-048.*
11. Szerypo J. et al., *LPC Caen report* LPCC-99-18 (1999) unpublished.

SEARCH FOR A LONG-LIFETIME COMPONENT IN THE FISSION OF ^{235}U INDUCED BY ~30 MEV ALPHA-PARTICLES

F.MALAGUTI, L.PATRIZII, V.TOGO, A.UGUZZONI
Physics Department and INFN, Bologna, Italy

P.OLIVO
Matemates Department, Bologna University

D.EREMENKO, O.FOTINA, S.PLATONOV, O.A.YUMINOV
Institute of Nuclear Physics, Moscow State University, Moscow, Russia

G.GIARDINA
Physics Department and INFN, Messina, Italy

Glass track detectors and the crystal blocking method have been used to measure time delays associated with the production of fission fragments in the reaction induced by 30 MeV α particles on a 235UO_2 crystal. Tracks have been read by an automatic, computer driven, microscope. Preliminary results indicate a possible long-lifetime component, as suggested by recent theories.

The existence of a LLC (long-lifetime component) in the fission of Uranium-like nuclei was simultaneously suggested, some time ago, both by crystal-blocking experiments and by theoretical considerations [1,2]. It is our opinion that, before getting involved in such a deep discussion on the origin and properties of LLC, one should at first confirm its existence with experimental data of much better quality than those presently available.

The reason for the poor quality of present data is that the most popular estimator of time delay, $\Delta\chi_{min}$ (the difference between the relative yields of the fission fragments at the minimum of blocking patterns affected and unaffected by the displacement of the fissioning nucleus from the lattice site) is made ambiguous by the presence of a peak in the center of the blocking dip, a rather surprising feature at first sight. The second widely used estimator, the volume Ω of the dip, has a more sound physical basis than $\Delta\chi_{min}$ but its use is limited by its big sensitivity to the "random yield" , that is difficult to measure using present position-sensitive detectors. The reason is that they cover too small solid angles, insufficient to "average off" the spatial fluctuations of the "random" yield in a crystal, due to channeling, see [3].

In this paper we present preliminary results of crystal-blocking investigations about the induced fission time in the $^{235}U + \alpha$ reaction with beam energy varying from 26 to 31 MeV. Experiments were performed at the cyclotron of Moscow State University using a thick enriched UO_2 single crystal target. The angular distributions of outgoing fission fragments were measured by glass track detectors in the vicinities of two different <111> crystallographic axes and <110> crystallographic ax. The angles of these axes with the backward direction of bombarding particles were -10°, 25° and 61°, respectively.

We therefore demonstrate how the use of an "old-fashioned" detector (glass plates for nuclear tracks), if coupled with an automatic setup for reading and classifying track images (ELBEK , Image Analyzer System) [3] , and sustained by the power of present day "micro" computers, is able to produce interesting data., if their analysis is of comparable accuracy.

To get the required precision, data were collected on an event-by-event basis and receive individual treatment, postponing their classification in histograms as late as possible. The microscope is able to analyze individual tracks, performing many measurements on each one, and writing the results, event by event, in a data base. The data reduction program will accept only those events that satisfy the conditions imposed by the operator. In our case, each event was described by the coordinates X and Y of the track, and by its eccentricity E and area A, after having "fitted" an ellipse on the track shape .

The speed of the microscope + computer assembly allows to read some 10^4 tracks per hour, therefore the blocking dip can easily be observed up to some 150 mrad from the axis, with a much better definition of the random yield, because now fluctuations do to channcling arc largely "averaged off", so we can define the random yield simply as the number of events observed within some solid angle, divided by the solid angle itself.

We performed many measurements at 5 different beam energies ranging from 25.9 to 31.2 MeV, and at 3 lab. angles, -10, 25 and 71 degrees from 2 different crystal axes, <111> and <110>.

The large amount of data is being analyzed now, and will take some time.
Anyway we extracted a self-contained data subset, just to have a feeling of what results one can expect. A preliminary analysis of those data is displayed in Figure 1., and can be qualitatively "read" as follows :

- Both dips correspond to the same crystal axis <110> and lab angle 25 deg., but to 2 different beam energies, 31.2 and 27.1 MeV.
- Contrary to naive expectations (more beam energy => more excitation energy in the compound nucleus => shorter lifetime), the 31.2 dip is clearly more "filled-in" than the 27.1 one, and seems to show a peak in the origin, therefore at 31.2 MeV some fraction of the reaction flux passes through a new open channel, that was closed at 27.1 Mev and has a large time delay at 31.2 MeV.

Therefore, the LLC seems to be present in this reaction.

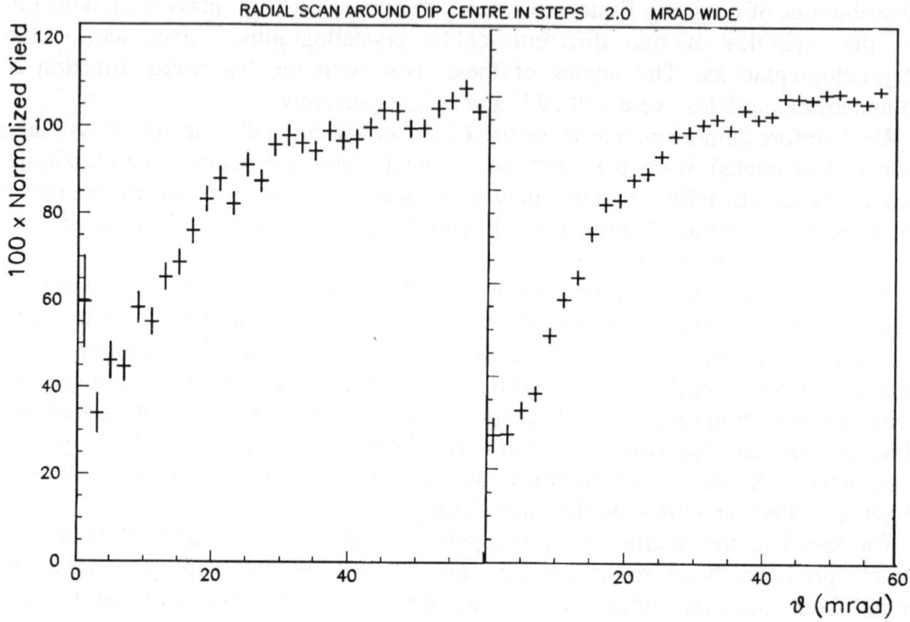

Figure 1 Radial scans across two blocking dips produced at ϑ_{lab}= 25 degrees around the <110> axis of a $^{235}UO_2$ crystal by α-particles with kinetic energies 31.2 MeV (left) and 27.1 MeV (right). The corresponding dip volumes are 686 ± 38 and 806 ± 19 microsterad, respectively.

References

1. J.S.Forster et al., Nucl.Phys.A, **464** , 497 (1987)
2. O.A.Yuminov et al., J.Phys.G, **21**, 1243 (1995)
3. See, for instance, "Handbook of modern ion beam materials analysis", J.R.Tesmer and M.Nastasi editors, Materials Research Society, Pittsburgh, Pennsylvania 15237 USA, ch. 10
4. A.Noll et al., Nuclear tracks and radiation Measurements, **25**, 265 (1988)

PROBING CORRELATIONS IN MANY-BODY HALOES

F.M. MARQUÉS MORENO[abc]
Laboratoire de Physique Corpusculaire, IN2P3-CNRS,
ISMRa et Université de Caen, F-14050 Caen cedex, France
E-mail: Marques@caelav.in2p3.fr

An experimental programme aiming to probe correlations in two-neutron halo nuclei is presented. Three different techniques have been applied to the three well-established systems, ^{6}He, ^{11}Li and ^{14}Be: intensity interferometry following the dissociation, radiative capture of high-energy protons, and the study of three-particle correlations in the β-n-n decay. The results of the different experiments, as well as the combination of them, provide an important step forward in the knowledge of the correlations within these Borromean systems, as for example the relative distance between neutrons in the halo.

1 Correlations in nuclei

The quest for the drip-lines, which define the limits of binding for nuclear systems, has only been attained for light nuclei. As such, these nuclei are unique in displaying the manner in which nucleons bind in an A nucleon system from the most neutron deficient to the most neutron rich. The strong binding of the α particle leads to many clustering phenomena beyond $A = 4$ [1], and it appears that even more-than-two-body correlations can play a dominant role, responsible for example of the mass gaps at $A = 5, 8$ [2]. Indeed, α particles and nucleons cannot form two-body systems (^{5}He, ^{5}Li and ^{8}Be are unbound), but the presence of a third particle makes the bridge of the gaps possible (^{6}He, ^{9}Be and ^{12}C*, all with a three-body structure) [2].

The study of these three-body correlations in nuclei finds an ideal case in two-neutron halo systems [3], also known as Borromean since the two-body subsystems (core-n and n-n) are unbound [4]. The three such nuclei known to date, ^{6}He, ^{11}Li and ^{14}Be, exhibit a three-body cluster structure in the ground state, as they are very weakly bound and the two valence neutrons extend far beyond the core of the nucleus to form the halo. However, even in the

[a] The dissociation experiment was performed at GANIL by: LPC-Caen, Chalmers-Göteborg, ULB-Bruxelles, Univ. Aarhus, CSIC-Madrid, Univ. Surrey, Univ. Oxford, Univ. Birmingham, IRES-Strasbourg, GANIL-Caen, IPN-Orsay, Univ. Messina.

[b] The radiative capture experiment was performed at GANIL by: LPC-Caen, KVI-Groningen, IFIN-Bucharest, Univ. Surrey, Univ. Birmingham, ULB-Bruxelles, IFIC-Valencia, NPI-Řež, GANIL-Caen.

[c] The β-decay experiment was performed at GANIL by: LPC-Caen, IFIN-Bucharest, GANIL-Caen.

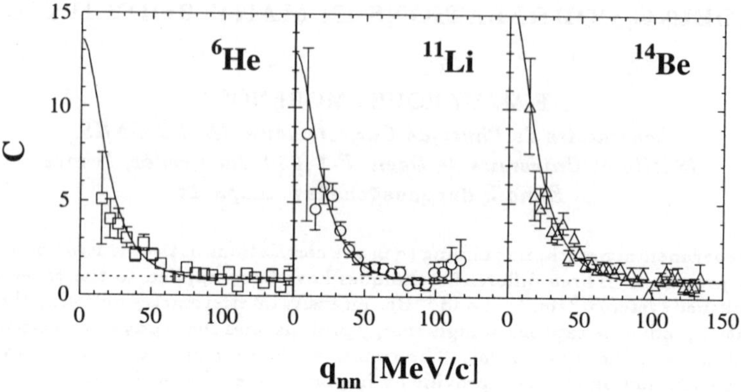

Figure 1: Two-neutron correlation functions following the breakup of ^{6}He, ^{11}Li and ^{14}Be into core+n+n by a Pb target. The solid lines correspond to a fit as a function of the n-n distance [5].

case of the most studied of these, ^{6}He and ^{11}Li, little is known regarding the correlations.

We have undertaken a series of experiments aiming to probe correlations in two-neutron halo nuclei. Three different techniques have been applied. We will present first the results of n-n interferometry analyses following the dissociation into core+n+n, in which the n-n final-state interaction (FSI) is exploited in order to measure their relative distance at emission. Then, radiative capture of high-energy protons on ^{6}He will be proposed as a new technique to investigate clustering in exotic nuclei, with the advantage of measuring photons which are not subject to FSI. Finally, we will apply the interferometry analyses developed in the dissociation experiments to an spontaneous process, the β-n-n decay of ^{11}Li. A summary and outlook will be given.

2 The dissociation experiment: intensity interferometry

When neutrons are emitted in close proximity in space-time, the wave function of relative motion is modified by the known FSI and quantum statistical symmetries [6]. Following on from the pioneering application to stellar interferometry [7], two-neutron intensity interferometry, and in particular the correlation function [8], was introduced to relate this modification to the space-time separ-

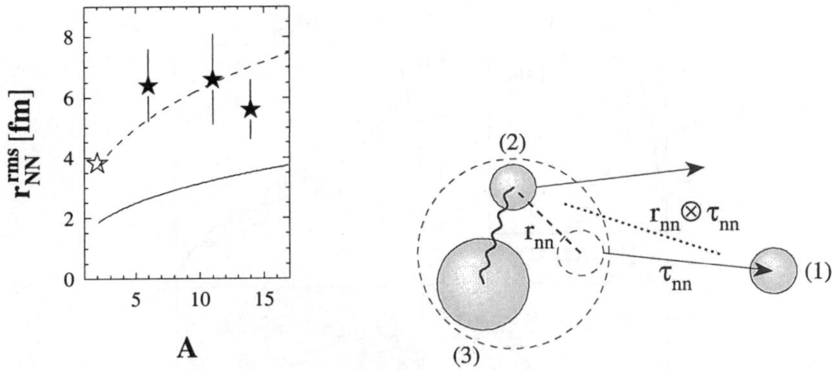

Figure 2: Left: n-n distances (closed symbols) obtained from the correlation functions of the three Borromean systems. The solid line corresponds to the distance between two nucleons within a liquid drop (multiplied by 2 as dashed line) and the open symbol to the n-p distance in the deuteron. Right: schematic view of the sequential breakup of a two-neutron halo nucleus, whereby dissociation passes through a resonance in the core-n system; the effective source size is a convolution of the n-n separation $r_{\rm nn}$ and the delay $\tau_{\rm nn}$.

ation of the particles at emission as a function of their four-momenta:

$$C_{\rm nn}(p_1, p_2) = \frac{d^2n/dp_1dp_2}{(dn/dp_1)\,(dn/dp_2)} \tag{1}$$

The numerator is the measured two-particle distribution and the denominator the product of the independent single-particle distributions [5,8], which are commonly projected onto one dimension, the relative momentum $q = |\vec{p}_1 - \vec{p}_2|$.

The technique has been applied to extract the intrinsic n-n correlation functions from the dissociation of ^{6}He, ^{11}Li and ^{14}Be secondary beams at 50, 30 and 35 MeV/N, respectively, by a Pb target (Fig. 1). The core fragment was measured with Si-CsI telescopes and the neutrons with the 100 elements of the DéMoN array [9,10]. The analysis [5], following the formalism of Ref. [11] and assuming that the neutrons were emitted simultaneously after dissociation in the Coulomb field of the target, lead to n-n separations of the order of $r_{\rm nn}^{\rm rms} \sim 6$ fm (Fig. 2, left).

The same analysis has been applied to dissociation of ^{11}Li and ^{14}Be by a C target, in order to investigate the influence of the reaction mechanism. Results suggestive of somewhat larger separations, $r_{\rm nn}^{\rm rms} \sim 8$ fm, were obtained [13]. This raised the question as to whether simultaneous emission can be assumed a priori. The simple three-body nature of the system breaking up suggests

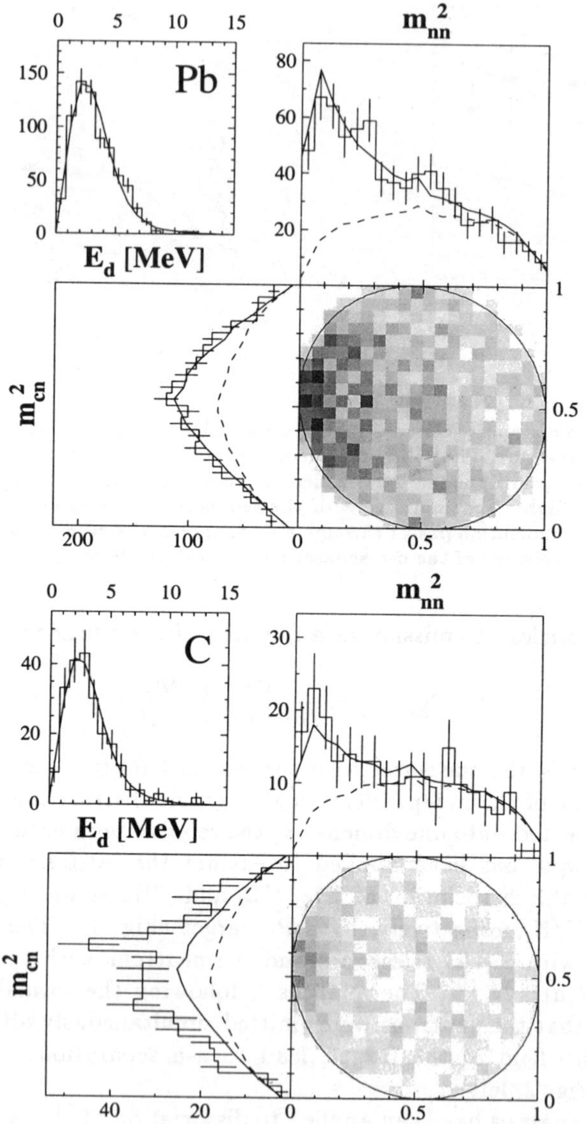

Figure 3: Dalitz plots (core-n versus n-n), and the projections onto both axes, for the data from the dissociation of ^{14}Be by Pb (upper) and C (lower panels). The lines are the results of the phase-space model simulations with/without (solid/dashed) n-n FSI. The insets correspond to the ^{14}Be decay energy [12].

that any delay in the emission of one of the neutrons will arise from core-n FSI/resonances in the exit channel (Fig. 2, right).

These resonances have been investigated with a new technique combining n-n interferometry and Dalitz plots, which are extensively used in particle physics for the search of resonances in three-particle decays [14]. The comparison of the core-n vs n-n invariant mass plots with phase-space simulations (Fig. 3) indicate that the n-n interaction dominates dissociation by Pb, while in dissociation by C the n-n interaction effects decrease at the same time that core-n resonances appear [12]. This new technique allows us to determine not only the n-n distance in the halo, but also the energy of the core-n resonances and the delay they introduce in the emission of the second neutron (their lifetime) [12].

3 The radiative capture experiment: an electromagnetic probe

We have investigated radiative capture of protons as a probe of clustering in nuclei far from stability. In brief, the photon emitted in the collision of a two-neutron halo nucleus, at tens of MeV/N, with a liquid Hydrogen target is measured with an electromagnetic calorimeter. The photon energy (tens of MeV), not affected by FSI, will be different depending on the subsystem inside the halo nucleus that captured the proton (at these energies the proton wavelength is $\lesssim 1$ fm).

The first such experiment was performed with a ^{6}He beam, produced at 40 MeV/N by fragmentation [15]. The cross-sections of the different quasi-free capture (QFC) processes allowed, ^{4,5}He(p,γ)^{5,6}Li and 1,2n(p,γ)^{2,3}H, were known [16] or deduced from photodisintegration [17]. The three-cluster configuration of ^{6}He was thus expected to be probed by the relative strength of the different processes. Photons were measured with the 74 BaF_2 crystals of the Château-de-Cristal array, and the capture system was detected in coincidence with the SPEG spectrometer [17].

Capture into ^{7}Li is observed as the strongest channel (Fig. 4). In addition, QFC was investigated by searching for γ-rays in coincidence with fragments lighter than ^{7}Li. The corresponding energy spectra (Fig. 5a,c,e) do indeed exhibit peaks below 42 MeV. The characteristics of these fragment-γ coincidences are very well described by QFC on the ^{5}He cluster, the α core and one halo neutron [17].

There were additional QFC channels, 2n(p,γ)t and t(p,γ)α, that could have been observed with finite efficiency in this experiment but were not [15]. These two processes lead to t-γ coincidences, which have not been observed. Within the picture of QFC on clusters, this is simply explained by the small contributions to the ^{6}He ground state of the α-2n [4] and t-t [18] configurations.

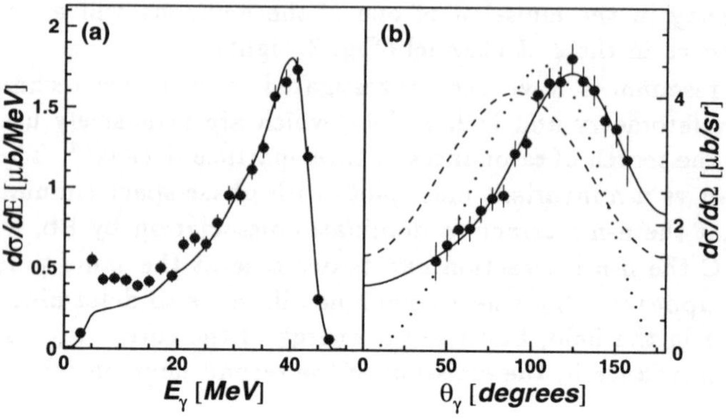

Figure 4: Energy (a) and angular distributions (b) in the ^{6}He+p c.m. for photons in coincidence with ^{7}Li. The solid line in (a) is the response of the Château to $E_\gamma = 42$ MeV. The lines in (b) correspond to a classical electrodynamics calculation (dotted), a microscopic cluster model (dashed), both normalized to the data, and to a Legendre polynomial fit [15] (solid).

This suggests that the dominant configuration is α-n-n in which the n-n separation is relatively large, in agreement with Ref. [5]. It would be of interest to search for larger distance correlations between the neutrons using lower energy protons.

4 The β-decay experiment: correlations in unbound states

In the dissociation experiment we have seen that the reaction mechanism (related to the different targets used) plays a role in the interpretation of the correlations observed between core-n-n. A possibility of suppressing these effects is the study of a spontaneous process in which two neutrons are emitted. This appears in the "n-n radioactivity" of ^{11}Li, in which the Q_β window is wide enough (20.8 MeV) to open the decay into excited states in ^{11}Be which lay above the 1 and 2 neutron thresholds [19].

The ^{11}Li beam, at 30 MeV/N, was stopped in a plastic scintillator at the center of the neutron array TONNERRE [20,21]. In the beam-off window, the β decay was triggered by the signal induced in the scintillator by the electron, which served as start of the neutron time-of-flight; the arrival of neutrons at the TONNERRE elements provided the stop [20]. The granularity of the array provided for the two momenta of the neutrons. The process taking place at rest, the momentum of the ^{9}Be recoil could be reconstructed by momentum conservation.

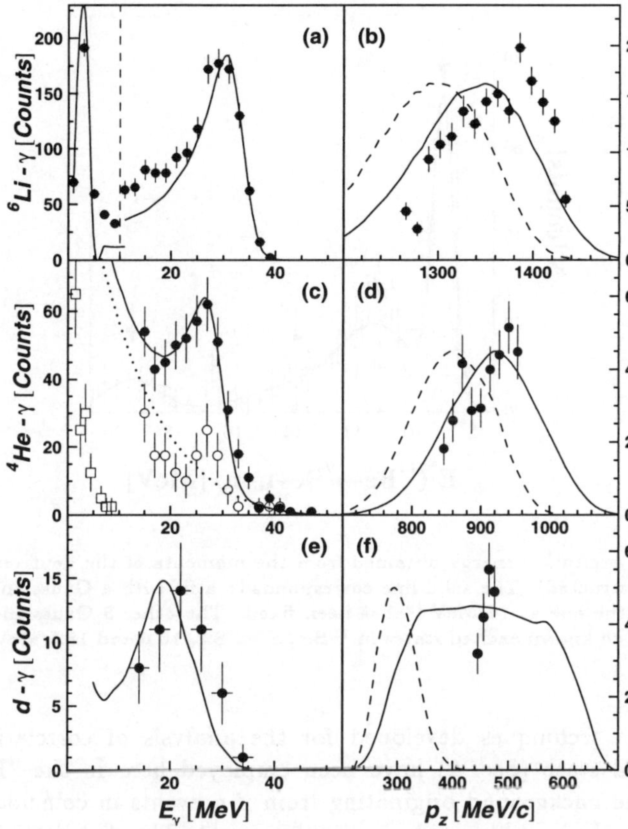

Figure 5: **γ-ray energy spectrum in the ^{6}He+p c.m. and momentum distribution of the coincident fragment for ^{6}Li (upper), α particles (middle) and deuterons (lower panels). The lines correspond to calculations of QFC on the ^{5}He cluster, the α core and one halo neutron, respectively, on the right with/without (solid/dashed) fragment FSI [17]. The distribution in (a) was divided by 3 below 10 MeV, and the open symbols in (c) are from an analysis investigating the role of the neutron background[17].**

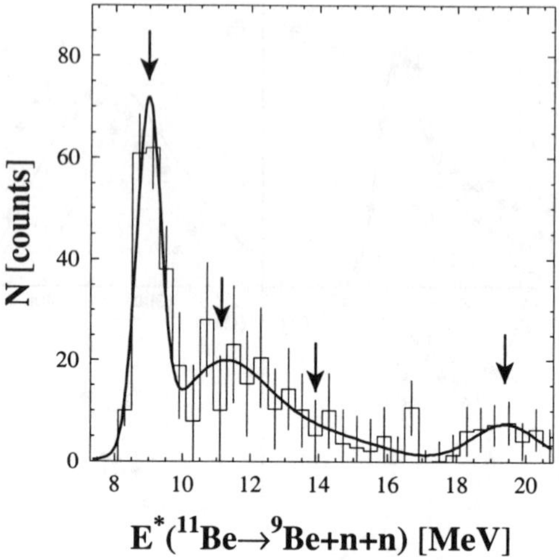

Figure 6: ^{11}Be excitation energy obtained from the momenta of the neutrons (measured) and ^{9}Be (reconstructed). The solid line corresponds to a fit with 4 Gaussians, where only the position of the one at 14 MeV [19] has been fixed. The other 3 Gaussians so obtained correspond well to known excited states in ^{11}Be ($E^* = 8.8$, 10.6 and 18.5 MeV).

The same techniques developed for the analysis of correlations in the ^{9}Li+n+n final state (Sec. 2) have been employed here in the ^{9}Be+n+n final state. The background originating from β-n events in coincidence with a cosmic ray has been subtracted via simulations [19]. The distribution of energy shared by the three particles, shown in Fig. 6 beyond the ^{9}Be+n+n threshold in ^{11}Be, exhibits peaks, indicating that the β-n-n decay does not proceed into the continuum. In addition, these peaks correspond well to known excited states of ^{11}Be.

The preliminary results from the ^{9}Be-n-n correlation analysis suggest that ^{10}Be resonances are not formed, and thus the two neutrons leave the ^{9}Be fragment simultaneously. Neutron-neutron interferometry, however, leads to a n-n distance of the order of 4 fm, smaller than the one obtained for the halo of ^{11}Li in Sec. 2. This can be explained if the nucleon participating to the β decay is one of the halo neutrons, as then the n-n distance measured in the ^{9}Be-n-n final state corresponds to the distance between one neutron in the halo and one in the core of ^{11}Li. Improved statistics would allow these analyses to be applied to each unbound state (Fig. 6) populated in the β decay.

5 Conclusions and outlook

Two-neutron haloes represent the most exotic manifestation of clustering in nuclei. As such, they are unique in displaying how three-body correlations act in the formation and stability of A nucleon systems. Recent developments in the production of secondary beams, together with the use of powerful multi-detector arrays (DéMoN, Château-de-Cristal, TONNERRE) and original analysis techniques (neutron interferometry, Dalitz plots), have provided new steps forward in the knowledge of the correlations within these systems. Results have been presented on the correlations inside ^{6}He, ^{11}Li and ^{14}Be. Among them, the access to the n-n distance in the halo, which plays a major role in the binding of these core-n-n nuclei, and the probe of spatial configurations with an electromagnetic probe.

The short-term outlook consists of the measurement of multi-particle correlations in the heavy Helium isotopes ^{6,8}He, where the simultaneous study of both nuclei will open up new possibilities in exploring many-body correlations in light nuclei. These include the study of unbound systems, like ^{5}H (t+n+n), formed by proton stripping from ^{6}He, and ^{4}n, formed by dissociation of ^{8}He into α particles. At longer term, the extension of the radiative-capture programme to lower energies, the improved measurement of the β-n-n decay of ^{11}Li, and the application of these techniques to the proton-rich side of the nuclear chart as a probe of 2p emission, appear as interesting possibilities.

Acknowledgments

I am grateful to my colleagues at LPC, specially to Nigel Orr, who leads the experimental programme briefly described here, and to the persons in charge of the data analysis of these experiments, involving complex multi-detector arrays: DéMoN plus CHARISSA (Marc Labiche), Château-de-Cristal plus SPEG (Emmanuel Sauvan), and TONNERRE (Cozmin Timiş).

References

1. M. Freer *et al.*, Phys. Rev. Lett. **82**, 1383 (1999).
2. H. Oberhummer, A. Csótó, H. Schlattl, Nucl. Phys. **A689**, 269c (2001).
3. P.G. Hansen, B. Jonson, Europhys. Lett. **4**, 409 (1987).
4. M.V. Zhukov *et al.*, Phys. Rep. **231**, 151 (1993).
5. F.M. Marqués *et al.*, Phys. Lett. B **476**, 219 (2000).
6. D.H. Boal, C.K. Gelbke, B.K. Jennings, Rev. Mod. Phys. **62**, 553 (1990).
7. R. Hanbury-Brown, R.Q. Twiss, Philos. Mag. 45 (1954) 663.
8. G.I. Kopylov, Phys. Lett. **B50**, 472 (1974).

9. M. Labiche *et al.*, Phys. Rev. Lett. **86**, 600 (2001).
10. K. Jones *et al.*, in preparation.
11. R. Lednicky, L. Lyuboshits, Sov. J. Nucl. Phys. **35**, 770 (1982).
12. F.M. Marqués *et al.*, nucl-ex/0101004.
13. F.M. Marqués *et al.*, in preparation.
14. See, for example, D.H. Perkins, *Introduction to High Energy Physics*, Addison-Wesley Publishing Company, 122 (1987).
15. E. Sauvan, PhD Thesis, Université de Caen, LPCC T-00-01 (2000).
16. M. Hoefman *et al.*, Phys. Rev. Lett. **85**, 1404 (2000).
17. E. Sauvan *et al.*, Phys. Rev. Lett. **87**, 042501 (2001).
18. K. Arai, Y. Suzuki, R.G. Lovas, Phys. Rev. C **59**, 1432 (1999).
19. C. Timiş, PhD Thesis, Université de Caen (2001).
20. A. Buţă *et al.*, Nucl. Instr. and Meth. A **455**, 412 (2000).
21. J.C. Angélique, contribution to these proceedings.

HIGH SPIN STATES POPULATED VIA PROJECTILE FRAGMENTATION IN VERY NEUTRON-RICH NUCLEI AROUND MASS 180

P. MAYET, J. GERL, CH. SCHLEGEL
GSI, Planckstrasse 1, D-64291 Darmstadt, Germany
E-mail: p.mayet@gsi.de

ZS. PODOLYÁK, P.H. REGAN, P.M. WALKER, M. CAAMAÑO
Dept. of Physics, University of Surrey, Guildford, GU2 7XH, UK

M. PFÜTZNER
Institute of Experimental Physics, Warsaw University, Pl-00681 Warsaw, Poland

M. HELLSTRÖM , M.N. MINEVA
Div. of Cosmic and Subatomic Physics, Lund University, SE-22100 Lund, Sweden

FOR THE GSI ISOMER COLLABORATION

Projectile fragmentation associated with γ-ray spectroscopy has been used to populate and study isomeric states of very neutron-rich nuclei in the transitional region with A~190-200. This experiment has proven the ability of such a technique to populate medium to high spin states and γ-ray transitions have been observed for the first time in a number of neutron-rich nuclei, including ^{190}W, ^{195}Os, ^{197}Ir, ^{202}Pt.

1 Introduction

The NPBL Conference is aimed at showing the exciting challenges of nuclear physics at the rise of the new century. The complex transitional region is one of these challenges. It extends from the W isotopes, which have well-developed quadrupole deformation but also strong hexadecapole distortions and some degree of axial asymmetry, through the less deformed and highly asymmetric Os isotopes to the weakly deformed and nearly γ unstable Pt nuclei (see Fig. 1). The Hf-W-Os region is also a well established testing ground for high K isomeric states [1]. The decay of these K-isomers populates lower energy excited states and the nuclear structure of corresponding nuclei can be studied.

Until now, our knowledge of the neutron-rich side of the valley of stability, in this region, was limited due to the difficulty in producing efficiently these species in order to perform further detailed spectroscopic studies. Such nuclei are not accessible by fusion-evaporation reactions using stable beam/target

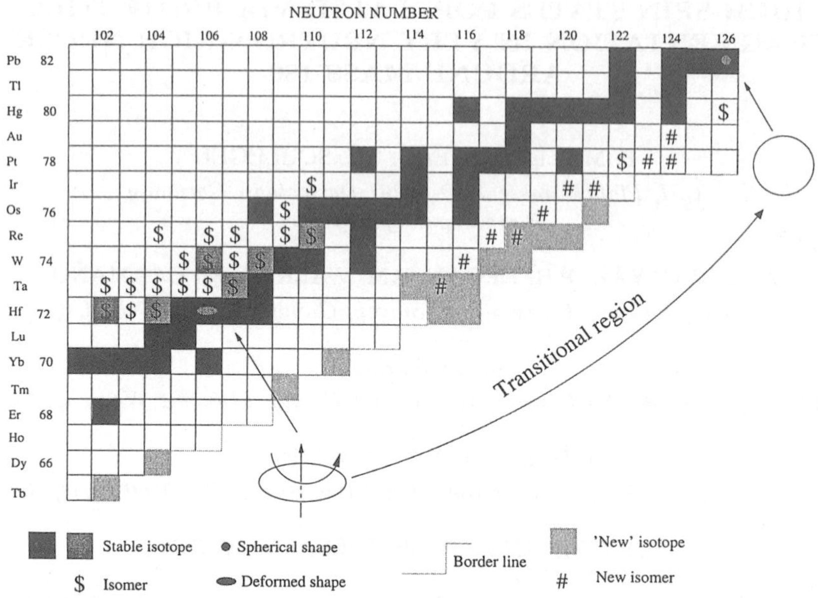

Figure 1. The nuclidic chart in the transitional region. Two key features appear : the transition between very prolate deformed shapes and spherical shapes and the existence of K-isomers on the neutron-rich side of the valley of stability

combinations and they are too heavy to be produced by fission. Deep inelastic reactions have been used to access high spin states in nuclei with a few neutrons more than the most neutron-rich stable isotopes [2] but this technique suffers from a lack of selectivity.

Projectile fragmentation has proven to be an efficient method in populating heavy neutron-rich isotopes [3] with cross-sections sufficient to perform γ-ray spectroscopy. We report here about the results of an experiment done at the GSI FRagment Separator [4] (FRS) using projectile fragmentation of a relativistic ^{208}Pb beam to populate K-isomers with lifetimes ranging from nano- to milliseconds in the A$\sim$180-200 region and to study the successive γ decay cascades.

2 Experimental set-up

The nuclei of interest were populated following the fragmentation of a ^{208}Pb beam at 1 GeV/nucleon impinging on a 1.6 g/cm^2 ^{9}Be target. Fragments were separated through the FRagment Separator (FRS) at GSI and identified event by event by standard time of flight, energy loss and position tracking techniques.

Ions were stopped at the exit of the spectrometer in a 4 mm thick aluminium catcher surrounded by an array of four Segmented Clover detectors. This array recorded γ rays in coincidence with implanted ions within time ranges of 0 to 8 μs and 0 to 80 μs using TDCs and TACs respectively. The absolute photopeak efficiency of the array was about 6% at 1.3 MeV and 17% at 300 keV for the central position of the catcher.

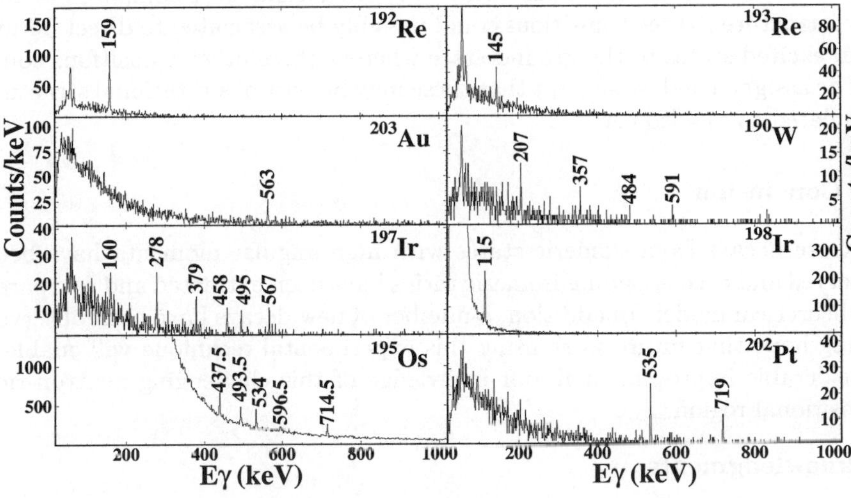

Figure 2. A selection of spectra from isomeric decays observed for the first time in the current work.

3 Results

Several settings of the FRS were used to reach different areas of the nuclidic chart during the course of the experiment. The first setting was employed to check the ability of projectile fragmentation to populate high spin states. Therefore previously reported K=35/2 isomers in ^{175}Hf, ^{179}W and ^{181}Re have been identified, representing the highest spin states observed so far in a fragmentation reaction. The isomeric ratios predicted by the abrasion-ablation model [5] have been confronted with the experimental measurements [6]. It has been found that the theoretical ratios are in relatively good agreement with the experimental results as long as only yrast states are considered while it is observed that the population probability for an isomer is generally reduced if it is located far from the yrast line. Finally the predictions of this model are found to give an upper limit for the measured isomeric ratios.

In more exotic settings, new decays have been observed in ^{190}W [7], ^{192}Re, ^{193}Re, ^{195}Os, ^{197}Ir, ^{198}Ir, ^{201}Pt, ^{202}Pt and ^{203}Au (see Fig. 2), providing the first spectroscopic information on the structure of these nuclei. In particular, the odd-even ^{197}Ir nucleus exhibits an isomer with half-life of 27.5 ± 2.5 μs decaying into a series of γ-ray transitions which were never observed so far. The partial level scheme of this nucleus was previously reported by J. Cizewski et al. [8] using transfer reactions. The γ-ray transitions were compared to this level structure : three transitions could possibly be associated to direct decays from excited states to the ground-state whereas three other transitions could not be assigned and we suggest that these may be part of a rotational structure populated by the isomer.

4 Conclusion

Discrete decays from isomeric states with high angular momenta have been observed and corresponding isomeric ratios have been extracted and compared to theoretical model. In addition, a number of new decays have been observed giving hope that future work using this experimental technique will enable a considerable improvement in our knowledge of this challenging neutron-rich transitional region.

Acknowledgments

This work was supported by GSI, EPSRC(UK), CEA/DSM/DAPNIA (France), the EU Access to Large Scale Facilities Program and the Polish Committee of Scientific Research under grant KBN 2 P03B 036 15.

References

1. P.M. Walker and G. Dracoulis, *Nature*, Volume **399**, 35 (1999).
2. C. Wheldon *et al., Phys. Lett.* B **425**, 239 (1998).
3. J. Benlliure *et al., Nucl. Phys.* A **660**, 87 (1999).
4. H. Geissel *et al., Nucl. Instr. and Meth.* B **70**, 286 (1992).
5. M. de Jong *et al., Nucl. Phys.* A **613**, 435 (1997).
6. M. Pfützner *et al., submitted to Phys. Rev.* C.
7. Z. Podolyák *et al., Phys. Lett.* B **491**, 225 (2000).
8. J.A. Cizewski *et al., Phys. Rev.* C **27**, 1040 (1983).

BREAKUP PROCESSES OF ^{6,7}Li PROJECTILES FROM A ^{208}Pb TARGET AT COULOMB BARRIER ENERGIES

M. MAZZOCCO, C. SIGNORINI, L. STROE, A. ANDRIGHETTO, A. VITTURI, D. FABRIS, M. LUNARDON, G. NEBBIA, G. VIESTI
Physics Department and INFN, via Marzolo 8, Padova, I-35131, Italy
e-mail:mazzocco@pd.infn.it

G. F. PRETE, M. CINAUSERO, E. FIORETTO
INFN, Laboratori Nazionali di Legnaro, Legnaro, I-35020, Italy

F. SORAMEL
Physics Department and INFN, Udine, I-33100, Italy

A. BRONDI, G. LA RANA, R. MORO, E. VERDACI
Physics Department and INFN, Napoli, I-80125, Italy

N. GELLI, F. LUCARELLI
Physics Department and INFN, Firenze, Italy

Large α production cross sections were measured for the systems ^{6,7}Li+^{208}Pb; they were identified with the total breakup cross sections. A theoretical description based on coupled channels method, also including coupling to a discretized continuum, underestimates the experimental data. These large cross sections, also found for the systems ^{6}He,^{9}Be+^{209}Bi, originate from the low binding energies of these light nuclei. The total reaction cross sections for these systems, deduced from the elastic scattering, coincide with the fusion + breakup ones.

The breakup process (BU) of loosely bound light nuclei at Coulomb barrier energies presently attracts a great interest in view of future research programs with radioactive ion beams (RIBs). Many planned light RIBs are loosely bound and, in addition, some of them have a halo structure intimately connected to a very small binding energy of the last nucleon. The BU process is expected to be very relevant for such nuclei and, as a consequence, should influence (renormalize) all open processes in this energy regime.

In this respect several light stable nuclei are interesting due to their low binding energies. Among these there are ^{6}Li, ^{7}Li and ^{9}Be. The nucleus ^{6}Li, with S_α=1.475 MeV, has the lowest binding energy among all stable nuclei; it is therefore particularly interesting since it constitutes a natural link between stable beams and RIBs. Though the breakup process should have smaller strength if compared with RIBs/halo nuclei like ^{11}Be (S_n=0.501 MeV) or ^{11}Li (S_{2n}=0.290 MeV), nevertheless, it can be studied with much higher statistical accuracy for ^{6}Li beam since its intensities are 3-4 orders of magnitude higher than RIBs ones.

The system ^{6}Li+^{208}Pb has been recently studied at Coulomb barrier energies independently by two different groups [1,2]. Both groups report large BU cross sections but differing, at some energies, even more than a factor two.

In the present paper we will present some points complementary to our work [1] and will take the opportunity to discuss about the discrepancies with ref. [2].

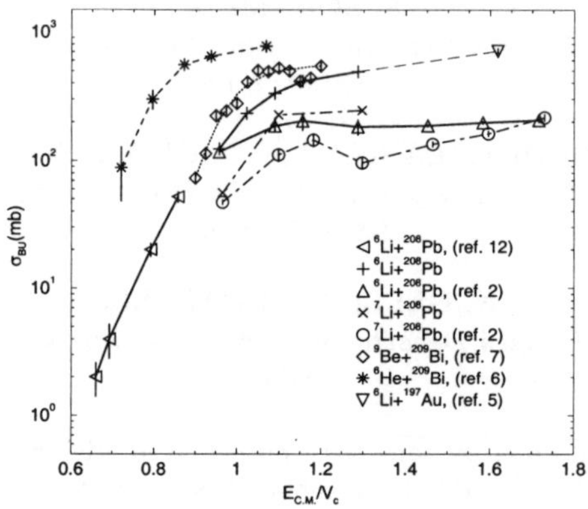

Figure 1: Experimental breakup cross sections for the systems ^{6}He+^{209}Bi, ^{6,7}Li+^{208}Pb, ^{9}Be+^{209}Bi and ^{6}Li+^{197}Au.

The system $^{6(7)}$Li +^{208}Pb was studied at the Tandem Van de Graaff accelerator of the Laboratori Nazionali di Legnaro (LNL) at 29, 31, 33, 35, 39 (29, 33, 39) MeV, the same energies at which the elastic scattering cross sections were already measured [3]. The target was an enriched ^{208}Pb 150 μg/cm^2 foil backed by a 10 μg/cm^2 carbon foil. The detector system consisted essentially of Si/ΔE (300μm) + CsI/E (5mm) telescopes each one spanning $\sim 20°(\theta)\times 17°(\phi)$ and arranged in a 4π geometry around the target. The telescopes are a portion ("ball") of the experimental setup named 8πLP of the LNL [4], and they cover an angular range from 34° to 163°. The α inclusive production cross sections, identified with the BU cross section, were deduced from the analysis of the single spectra taken with ΔE-detectors, where the α particles could be clearly identified.

The measured cross sections for both ^{6,7}Li projectiles are reported in fig.1 together with the results from ref. 2. The cross sections are in agreement within the experimental errors only at 29 MeV, and the discrepancy gets larger with increasing energy: At 39 MeV the two ^{6}Li cross sections differ by a factor 2.6! Table 1 gives the experimental and theoretical results of [1,2].

E(MeV)	^{6}Li,exp	^{7}Li,exp	^{6}Li,th	^{6}Li,exp	^{7}Li,exp	^{6}Li,th
29	123 ± 1	56 ± 1	50.9	117 ± 1	47 ± 2	50.3
31	230 ± 1		85.6			
33	332 ± 1	226 ± 3	121.6	186 ± 19	111 ± 13	70.3
35	413 ± 3		161.3	203 ± 31	145 ± 17*	79.8
39	490 ± 3	245 ± 3	180.6	183 ± 25	96 ± 10	99.9

Table 1: Experimental and theoretical BU cross sections (mb) for the systems ^{6,7}Li+^{208}Pb. The first column is the beam energy and last three are from ref. [2]. *This point was taken at 35.5 MeV.

Concerning the possible origins of this discrepancy, we would like to point out that:
a) both experiments normalize the cross sections to the same elastic data of ref. 3 (practically each spectrum is self normalizing);
b) in ref. 2 the target is self supporting, while in ref. 1 it has a carbon backing (potentially higher background ?);
c) ref. 2 gets the cross sections from the ΔE-E spectra (potentially lower efficiency ?), ref. 1 from the ΔE single spectra (potentially higher background?);
d) ref. 1 data have a much higher statistical accuracy (see table 1);
e) ref. 1 data match much better one additional point at 48 MeV, corresponding to E_{cm}/V_c=1.62 [5] (see fig.1). This point has been deduced from ref. 5.

Considering all the above mentioned points we are more inclined to believe in our results with better statistics even if with potentially higher background. Anyhow, a re-measurement of the cross sections seems necessary. Despite the existing discrepancies, the BU cross sections are considerably large as also found for the other systems reported in fig. 1: ^{6}He+^{209}Bi [6] and ^{9}Be+^{209}Bi [7]. This confirms the expectations of this work since ^{6}He has S_{2n}=0.901 MeV and ^{9}Be has S_n=1.665 MeV.

In order to understand our results we have made some exploratory calculations with the CC code FRESCO [8]. In the calculations we have considered ^{6}Li excitations to two resonant (unbound) states (sequential BU) and added excitations to continuum states above the BU threshold; this last process is expected to occur with considerable strength in loosely bound nuclei. We remind that ^{6}Li has no bound excited states. Transfer processes were also included with the DWBA formalism.

Fig.3a shows the BU cross sections calculated at E_{lab}=33 MeV for the excitation of the ^{6}Li continuum discretized in 10 energy bins up to 6.5 MeV. The strength is peaked at 3.5 to 4.0 MeV; it would be very interesting to have an experimental check of this prediction. Fig.3b shows both Coulomb and nuclear sequential BU obtained from the calculations; these two components are peaked at forward and backward angles respectively. The resulting angular distribution modulation cannot be verified experimentally due to the low angular resolution of the detecting system.

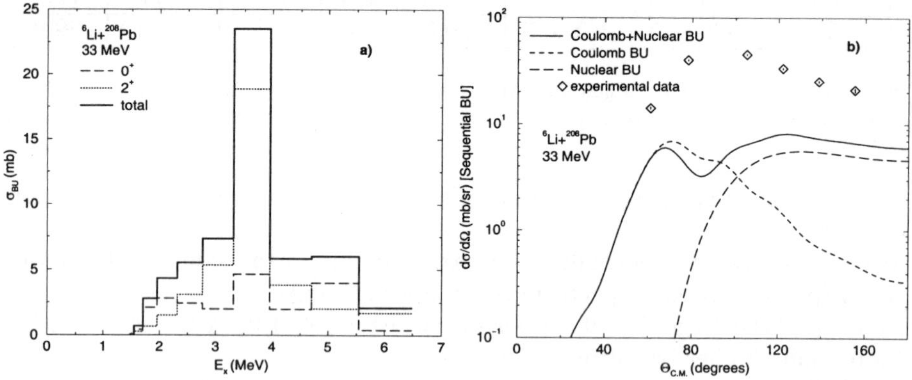

Figure 2: a) Calculated cross sections for the ^{6}Li continuum excitation via monupole and quadrupole transitions. b) Angular distributions of the various theoretical components of the sequential BU and experimental ones at 33 MeV.

The experimental angular distributions peaked around the grazing angles tell us that the two colliding nuclei interact mainly at surface.

The theory underestimates the experimental data, but the inclusion of the BU to ^{6}Li continuum is important since it has a cross section similar to the sequential BU. More elaborate theoretical treatment of the process is needed.

In ref. 2 a similar theoretical description of this process within the code FRESCO is utilized. The two theoretical approaches give the same result at 29 MeV, but differ of about a factor 2 at higher energies. This underlines once more the present limits of the theoretical descriptions based on a standard code, in particular when one considers the continuum BU.

In fig.4 we compare the reaction cross sections deduced from the elastic scattering data ([6], [9], [3]) for the systems ^{6}He+^{209}Bi, ^{9}Be+^{209}Bi and ^{6}Li+^{208}Pb to the sum of the BU + fusion experimental cross sections [10], [11]

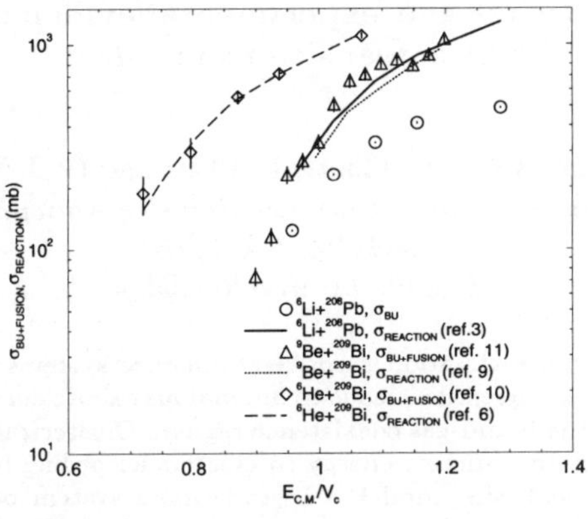

Figure 3: Reaction cross sections (continuum lines) compared with the sum of the BU + fusion cross section. In the case of ^{6}Li the fusion cross section is not yet measured. V_C= 19.9, 38.4, 29.5 MeV for ^{6}He, ^{9}Be, ^{6}Li respectively.

(the last one is not known for ^{6}Li). It is remarkable how these two sets of data well overlap for ^{6}He and ^{9}Be. This clearly confirms that these two channels are the most relevant at the barrier.

From all these data it is clear that higher quality experimental data and more accurate theoretical calculations are needed in order to understand the interplay of BU and fusion processes at the Coulomb barrier and the importance of the continuum BU.

References

1. C. Signorini et al., *Eur.Phys.J.* **A10**(2001)248.
2. G. R. Kelly et al., *Phys.Rev.* **C63**(2000)024601.
3. N. Keeley et al., *Nucl.Phys.* **A571**(1994)326.
4. G. F. Prete et al., *Nucl.Instr.Meth.Phys.Res.* **A222**(1999)263.
5. R. Kanungo et al., *Nucl.Phys* **A599**(1996)579.
6. E. F. Aguilera et a., *Phys.Rev.* **C63**(2001)061603.
7. C. Signorini et al., *Proc. Intern. Conf. BO2000*, Bologna, Italy, May 2000, eds.T.Vretenaar et al., World Scientific, Singapore in press.
8. I. J. Thompson, *Comput.Phys.Rep.* **7**(1988)167.
9. C. Signorini et al., *Phys.Rev* **C61**(2000)061602.
10. J. J. Kolata et al., *Phys.Rev.Lett.* **81**(1998)4580.
11. C. Signorini et al., *Eur.Phys.J.* **A5**(1999)7.
12. E. Speth et al., *Phys.Rev.Lett.* **24**(1970)1493.

Theoretical approaches and experimental evidence for liquid-vapor phase transitions in nuclei

L. G. Moretto, J. B. Elliott, L. Phair and G. J. Wozniak
Nuclear Science Division, Lawrence Berkeley National Laboratory, Berkeley, CA 94720
E-mail: LGMoretto@lbl.gov

The leptodermous approximation is applied to nuclear systems for $T > 0$. The introduction of surface corrections leads to anomalous caloric curves and to negative heat capacities in the liquid-gas coexistence region. Clusterization in the vapor is described by associating surface energy to clusters according to Fisher's formula. The three-dimensional Ising model, a leptodermous system par excellence, does obey rigorously Fisher's scaling up to the critical point. The ISiS multifragmentation data follows the same scaling, thus providing the strongest evidence yet of liquid-vapor coexistence.

1 Introduction

Nuclei are leptodermous, mesoscopic clusters. Their thin skin leads naturally to an expansion of their energy in powers of $A^{-1/3}$. The leading term, proportional to the number of nucleons A, is the bulk term; the second, proportional to $A^{2/3}$, is the surface term, the next, proportional to $A^{1/3}$, is the curvature term, etc. This is the basis of the liquid drop model, which by merely correcting the bulk terms with a surface term, manages to reproduce the binding energies of nuclei to within 1%. There is ample reason to believe that a similar leptodermous thermodynamical treatment of nuclear systems at $T > 0$ will lead to an equivalently good reproduction of nuclear properties.

In particular, the appearance of a vapor phase at $T > 0$ opens two new aspects: 1) the phase transition between a liquid-like nuclear droplet and its equilibrium vapor, with the attendant caloric curve; 2) the properties of the vapor phase, especially the extent to which nucleons are aggregated into clusters.

In the first part of this paper we will derive analytically the caloric curve and the (negative) heat capacity for a drop undergoing an isobaric phase transition. In the second part we will show that the clusterization in the three-dimensional Ising model can be accounted for in terms of the leptodermous expansion. The rigorous obeyance of the mass and temperature scaling law (Fisher's scaling) by the Ising model is exactly paralleled by the nuclear systems in multifragmentation, thus providing the best evidence yet for liquid-vapor coexistence.

2 The isobaric phase transition of a liquid drop

The thermodynamical equilibrium properties of first order phase transitions are completely describable in terms of the thermodynamic state variables associated with the individual separate phases. For this reason, in contrast with continuous phase transitions, first order phase transitions are "trivial," and interesting only in so far as they herald the appearance of a hitherto unknown or undescribed phase.

Renewed attention to phase transitions has been generated by studies of models with well defined Hamiltonians with either short range interactions (e.g. the Ising model [1] or the lattice gas model [2]) or incorporating long range interactions such as gravitation or electro-magnetic interactions. Features expected to disappear in the thermodynamic limit, if such a limit exists, were noticed and were claimed to be essential, characteristic indicators of phase transitions in mesoscopic systems [3]. For instance, first order phase transitions were associated with anomalous convex intruders in the entropy versus energy curves, resulting in backbendings in the caloric curve, and in negative heat capacities [3].

These anomalies have been attributed to a variety of causes, the foremost of which are surface effects, and long range forces [3]. Unfortunately however, the numerical nature of the calculations tends to make the identification of the causes of negative heat capacities rather problematic.

In the context of nuclear physics, microcanonical models of nuclear multifragmentation have associated the anomalies of a convex intruder with the onset of multifragmentation [3]. Furthermore, lattice gas models in the isobaric regime have also shown negative specific heats in the coexistence region, where multifragmentation also appears [2]. The question of whether the two transitions are related and possibly coincident with the liquid-vapor transition is still very much open.

Recently, the claim has been made of an empirical observation of these anomalies, such as negative heat capacities in nuclear systems [4]. These negative heat capacities have been inferred from the study of fluctuations in multifragmenting nuclear systems.

In particular, it would be highly desirable to ground any evidence for these anomalies, theoretical or otherwise, on thermodynamics itself, minimally modified to allow for the possible role of surface effects related to the finiteness of the system.

In this section we illustrate analytically how effects such as negative heat capacities can arise within a standard thermodynamic treatment. We consider the evaporation of a drop of ordinary liquid. Our only concern with mesoscopicity is the explicit treatment of the surface of the drop.

No qualitative difference in the picture results by considering a drop of finite radius r. The only difference is that the overall free energy of the drop (solid curves) as we shall see below, is higher than that of the bulk (dashed curves) and the equilibrium vapor pressure is correspondingly higher.

The state of equilibrium between a liquid and its vapor can be described in the simplest way by the Clapeyron Equation

$$\frac{dp}{dT} = \frac{\Delta H_m}{\Delta V_m T} : \tag{1}$$

where, p and T are the pressure and temperature, ΔH_m is the molar enthalpy of vaporization and ΔV_m is the difference of the molar volumes of vapor, V_m^v, and liquid, V_m^l.

Specialization to the case of a drop of radius r can be achieved by modifying the enthalpy to account for the surface energy [5]

$$\Delta H_m = \Delta H_m^0 - c_s S_m^l = \Delta H_m^0 - \frac{3c_s V_m^l}{r} \tag{2}$$

where ΔH_m^0 is the "bulk" molar enthalpy, S_m^l and V_m^l are the surface and volume of the drop and c_s is the surface energy coefficient.

Neglecting V_m^l compared to V_m^v and considering the vapor ideal, i.e. $V_m^v = T/p$, we can integrate Eq. (1), assuming also ΔH_m to be constant. We obtain

$$p = p_0 \exp\left(-\frac{\Delta H_m^0}{T} + \frac{3c_s V_m^l}{rT}\right) \tag{3}$$

or

$$p = p_{\text{bulk}} \exp\left(\frac{3c_s V_m^l}{rT}\right). \tag{4}$$

This equation contains *all* the thermodynamical information necessary to characterize the phase coexistence of the liquid drop of radius r with its vapor. The salient feature is the rise of the vapor pressure with decreasing radius. Fig. 1 gives a map of the function $p' = p'(T', r')$, in terms of the scaled variables

$$p' = \frac{p}{p_0},\ T' = \frac{T}{\Delta H_m^0},\ r' = \frac{\Delta H_m^0}{3c_s V_m^l} r. \tag{5}$$

For any given radius r, the function $p = p(T, r)$ describes the equilibrium condition between the drop and its vapor. In other words, it is the phase diagram of the drop.

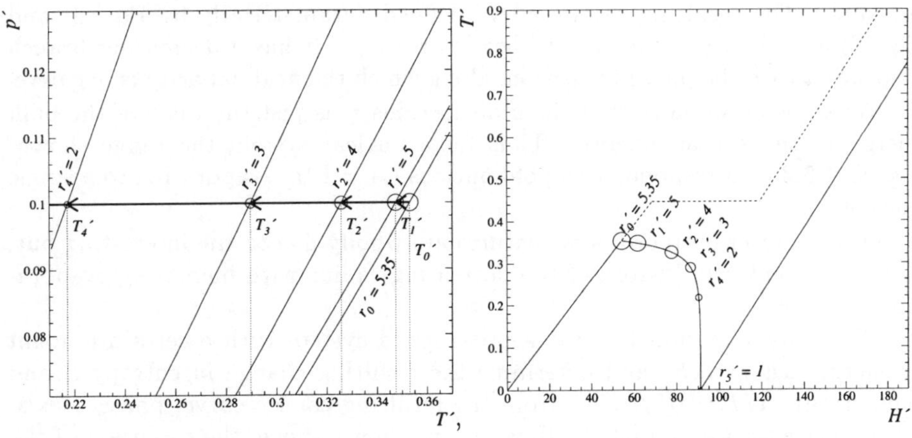

Figure 1: Left: Saturated vapor pressure as a function of the temperature for different droplet radii. The size of the open circles is proportional to the droplet radius. Right: The history dependent caloric curve of an evaporating drop at constant pressure. Dashed lines represent bulk behavior, solid line shows the drop's caloric curve. The scaled enthalpy is $H' = H(\Delta H_m^0/3c_s V_m^l)^3/(4\pi/V_m^l)$.

Let us now introduce some history and construct a caloric curve for a drop of radius r_0 at constant pressure p_0. As the drop is heated, and before the vapor can appear, the temperature increases according to $\Delta H = C_p^l \Delta T$, with C_p^l is the liquid's heat capacity and is approximately constant. When T reaches the value T_0 at which the vapor pressure $p(r_0) = p_0$, the vapor first appears and it expands against the container. The heat of vaporization is absorbed at a rate $H_m(r_0)$. However, as it evaporates, the drop sees its radius decreasing from its initial value, chosen to be $r_0' = 5.35$. At constant temperature the vapor pressure would rise, but, at constant pressure, as we are now operating, the temperature *decreases* as shown in Fig. 1, as the system absorbs its heat of vaporization, so that,

$$\Delta H = \int_r^{r_0} \Delta H_m \frac{dV}{V_m^l} = \frac{4\pi}{V_m^l} \left(\frac{3c_s V_m^l}{\Delta H_m^0}\right)^3 \left[\frac{1}{3}(r_0'^3 - r'^3) - \frac{1}{2}(r_0'^2 - r'^2)\right] \quad (6)$$

and

$$T' = T_0' \left(\frac{1 - \frac{1}{r'}}{1 - \frac{1}{r_0'}}\right). \quad (7)$$

After the drop has totally evaporated, the vapor can increase its temperature according to $\Delta H = C_p^v \Delta T$ where C_p^v is the vapor heat capacity at constant pressure. The resulting caloric curve defined parametrically by Eq. (6) and Eq. (7) and shown in Fig. 1 is rather interesting. It has a decreasing branch associated with the phase transition, along which the heat capacity is negative!

As an aside we note that the scaled radius r' is just the ratio of the bulk energy to the surface energy. Thus for a nuclear system the range shown: $1 \leq r'_0 \leq 5.35$, corresponds to a gold nucleus ($A = 197$) evaporating to a single nucleon.

These rather extraordinary features are wholly due to the interesting but, in a way, accidental history of the decreasing radius with increasing evaporation.

Typically, experiments heat a preassigned system with a certain amount of energy ΔE or ΔH, and determine the resulting change in entropy S and temperature $1/T = \partial S/\partial E|_V$; from the resulting caloric curve, phase coexistence diagrams are extracted. However, as shown above, the evolution of the system occurring during heating introduces complications in the construction of a phase diagram from such a caloric curve.

To avoid this problem an experimentalist or theorist would have to keep the radius constant or correct for its change while determining the vapor pressure as a function of the temperature, thereby eliminating the accidental aspects associated with the evolution of the system.

The results obtained here are firmly grounded on thermodynamics with a straightforward accounting of finiteness through the surface correction. They are exact in the limit in which the liquid drop model holds, namely, down to nuclei/clusters containing 20 or so constituents. They are completely general, as they do not depend on specific details of the system but rather on its gross properties. In fact, they should be used as the paragon for lattice gas models and the like. In the limit in which these models represent liquid vapor coexistence, they must reproduce the present results.

Even more importantly, this approach obviates the need for repeating numerical calculations for each individual system or drop size. All that is required is to determine the bulk energy (enthalpy) and the surface energy coefficient of a given phase once and for all.

It is worth repeating that, once the constraint of constant pressure is enforced, the results described here are entirely general, as they apply to any small system undergoing solid-vapor or liquid vapor transitions.

3 The liquid to vapor phase transition in excited nuclei

Nuclear multifragmentation, the break up of a nuclear system into several intermediate sized pieces, has been frequently discussed in terms of equilibrium statistical mechanics, and its possible association with a phase transition [6]. However, ample uncertainty remains regarding its nature, in particular whether multifragmentation is a phase transition and if so whether it is associated with the liquid to vapor phase transition.

This study will show that: 1) high quality experimental data contain the unequivocal signature of a liquid to vapor phase transition through their strict adherence to Fisher's droplet model; 2) the two phase coexistence line is observed over a large energy/temperature interval extending up to and including the critical point; 3) several critical exponents, as well as the critical temperature, the surface energy coefficient and the compressibility factor can be directly extracted; 4) the nuclear phase diagram can be constructed with the available data; 5) the nuclear liquid at break-up is a slightly super-saturated vapor, for which the pressure and density can be determined as a function of the temperature.

Studies of multifragmentation have shown two general, empirical properties of fragment multiplicities called reducibility and thermal scaling [5].

Reducibility refers to the observation that for each energy bin the fragment multiplicities are distributed according to a binomial or Poissonian law. As such, their multiplicity distributions can be *reduced* to a one-fragment production probability according to a binomial or Poissonian distribution.

Thermal scaling refers to the feature that the average fragment yield $\langle N \rangle$ behaves with temperature T as a Boltzmann factor: $\langle N \rangle \propto \exp(-B/T)$. Thus a plot of $\ln \langle N \rangle$ vs. $1/T$, an Arrhenius plot, should be linear. The slope B in such a plot is the one-fragment production "barrier".

Both the features of reducibility and thermal scaling are inherent to any statistical model, in particular to the clusterization of droplets from a vapor as described by Fisher's droplet model [7]. The abundance of a cluster of size A is given by:

$$n_A = q_0 A^{-\tau} \exp\left(\frac{A\Delta\mu}{T} + \frac{c_0 A^\sigma}{T_c} - \frac{c_0 A^\sigma}{T}\right) \tag{8}$$

$$= q_0 A^{-\tau} \exp\left(\frac{A\Delta\mu}{T} - \frac{c_0 \varepsilon A^\sigma}{T}\right), \tag{9}$$

where n_A is the number of droplets (or fragments) of mass A, normalized to the size of the system A_0; q_0 is a normalization constant depending only on the value of τ; τ is the topological critical exponent; $\Delta\mu = \mu - \mu_l$, and μ and μ_l are the actual and liquid chemical potentials respectively; $c_0 \varepsilon A^\sigma$ is the surface free energy of a droplet of size A; c_0 is the zero temperature surface energy coefficient; σ is the critical exponent related to the ratio of the dimensionality

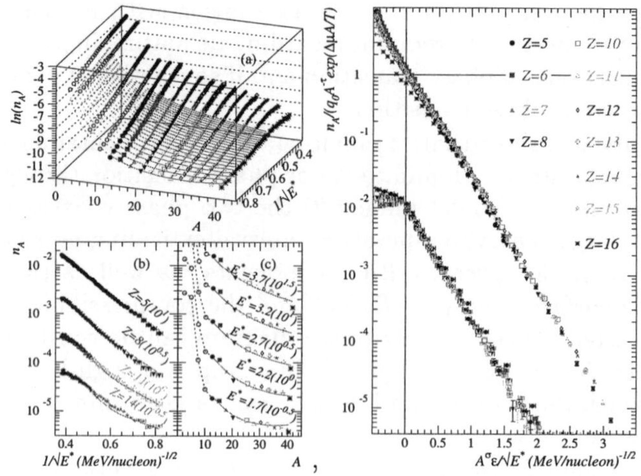

Figure 2: Left: (a) The ISiS fragment yield surface: natural log of the fragment yield vs. fragment mass and inverse temperature. (b) Arrhenius plots for representative charges. (c) Fragment mass yields for various values of E^*. Solid curves are from a fit to Fisher's droplet model; dashed curves show the extrapolation of the fit results. Error bars (statistical) are smaller than the size of the points. Right: The scaled fragment yield distribution versus the scaled temperature. ISiS data (upper) from a wide range in charge collapse to a single curve over more than five orders of magnitude. Also shown are similarly scaled data from a three-dimensional Ising model calculation (lower); the quantity $(n_A/q_0A^{-\tau})/100$ is plotted against the quantity $3.5(A^\sigma\varepsilon/T)$ for clusters of size $A = 5, 10, \ldots, 95, 100$.

of the surface to that of the volume; and $\varepsilon = (T_c - T)/T_c$ is the control parameter, a measure of the distance from the critical point, T_c.

Equation (8) shows clearly that thermal scaling is contained in this description. In particular, at coexistence ($\Delta\mu = 0$), Eq. (8) leads to a "barrier" $B = c_0 A^\sigma$.

Recently, gold multifragmentation data from the ISiS Collaboration was shown to exhibit reducibility and thermal scaling in the fragment production probabilities[8]. Since this behavior is inherent to Fisher's model, it is interesting to determine if Fisher's model describes the ISiS data set. In order to find if this is the case, the ISiS charge yields from 8 GeV/c π^-+ ^{197}Au fragmentation data (see Fig 2a) were fit to Eq. (9).

The total number of fragments of a given size N_A was normalized to the size of the fragmenting system A_0 thus $n_A = N_A/A_0$. Here $\sqrt{E^*}$ was used in lieu of T, assuming the system behaves as a degenerate Fermi gas.

The parameters of Fisher's model, e.g. τ, σ, and E_c^* (in lieu of T_c) were used as fit parameters. The distance from equilibrium $\Delta\mu$ was parameterized

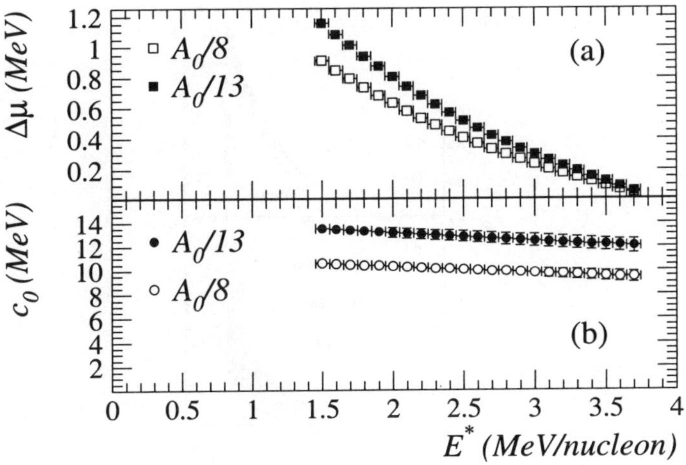

Figure 3: The behavior of $\Delta\mu$, and of the surface energy coefficient as a function of excitation energy up to $E_c^* = 3.7$ MeV/nucleon. See text for details.

by a polynomial of degree four in E^* and the coefficients of that polynomial were used as fit parameters. The surface energy coefficient c_0 was parameterized by a polynomial of degree one and the coefficients of that polynomial were used as fit parameters. The level density parameter was *absorbed* into the fit parameters for $\Delta\mu$ and c_0.

Data for $E^* \leq E_c^*$ and for $5 \leq Z \leq 16$ were included in the fit to Eq. (9) and the parameters were allowed to float to minimize χ_ν^2. Fisher's parameterization of the surface energy $c_0\varepsilon$ is invalid for $T > T_c$, thus excitation energies greater than E_c^* were not considered. Fisher's model expresses the mass/energy of a fragment in terms of bulk and surface energies. This approximation is known to fail for the lightest of nuclei where structure details (shell effects) dominate the mass. For this reason and the fact that for the lightest fragments equilibrium and nonequilibrium production cannot be clearly differentiated, fragments with $Z < 5$ were not considered in the fit. Fragments with $Z > 16$ were not elementally resolved, and were also excluded.

The behavior of the data for the (n_A, A, T) surface is reproduced over a wide range in excitation energy and fragment charge. This is also shown in Fig. 2. In this figure the results of the analysis are presented in terms of Arrhenius plots (Fig. 2b), or of the fragment yield distribution plots (Fig. 2c). A powerful method to observe the results of this analysis directly is to scale the data according to Eq. (9) using the parameters resulting from the fitting procedure. Figure 2 shows such a result. The fragment mass yield dis-

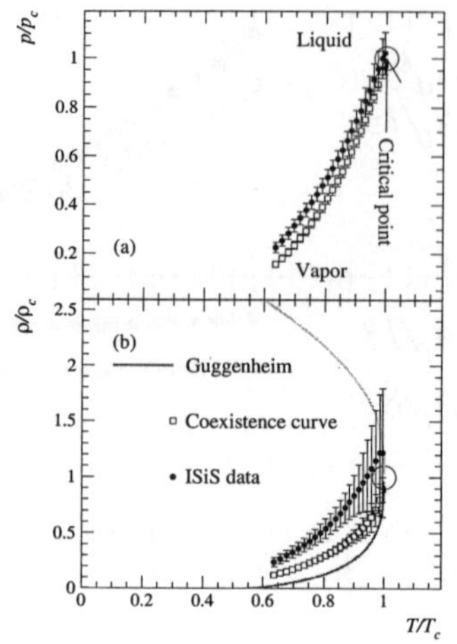

Figure 4: (a) The reduced pressure-temperature plot and (b) the reduced density-temperature plot. Filled symbols show the path of the ISiS data, open symbols show the coexistence curve of charged finite nuclear matter.

tribution is scaled by the Fisher's power law pre-factor and the bulk term: $n_A/q_0 A^{-\tau}\exp(\Delta\mu A/T)$. This quantity is then plotted against the temperature scaled by Fisher's parameterization of the surface energy: $A^{\sigma}\varepsilon/\sqrt{E^*}$. The scaled data collapse over five orders of magnitude onto a single curve, which is precisely the behavior predicted by Fisher's droplet model [7]. This curve is equivalent to a liquid-vapor coexistence line, as will be shown below, and provides the best, most direct evidence yet for a liquid to vapor phase transition in excited nuclei.

To illustrate the generality of this type of scaling, Fig. 2 shows the scaled cluster distributions from a three-dimensional Ising model calculation [9]. This system is know to model liquid-vapor coexistence up to the critical temperature. The perfect scaling of the cluster yields according to Eq. (9) demonstrates liquid-vapor-like coexistence up to T_c.

The value of τ extracted from this analysis, $\tau = 2.00 \pm 0.01$, is in the range predicted by Fisher's model. Similarly, $\sigma = 0.64 \pm 0.05$ is close to the value expected for a three dimensional system, $\sim 2/3$. The values of $\Delta\mu$ and c_0 and their energy dependence are shown in Fig. 3. The positive value of $\Delta\mu$ for $E^* < E_c^*$ indicates that the system behaves as a slightly supersaturated vapor. Values for $\Delta\mu$ and c_0 can be determined by making a choice of the level density parameter. For $a = A_0/8$, $c_0 \approx 10.0 \pm 0.3$ MeV while $a = A_0/13$ gives $c_0 \approx 12.8 \pm 0.4$ MeV; both values are in the range of the value of the surface energy coefficient of the semi-empirical mass formula (~ 17 MeV), although the linear extrapolation all the way down to $T = 0$ implied by Eq. (9) is questionable. The values of the critical exponents determined here are in agreement with those determined for the EOS multifragmentation data [10] and the value of the excitation energy at the critical point $E_c^* = 3.70 \pm 0.05$ MeV/nucleon is in the neighborhood of the value observed in the EOS analysis ($E_c^* = 4.75$ MeV/nucleon)[10]. The EOS analysis relied on the assumption that $\Delta\mu \approx 0$. No such assumption was made in this work. In fact the present effort tests that assumption and finds it to be approximately valid below the critical point since the observed $\Delta\mu$ values are at most ~ 1 MeV (assuming $a = A_0/13$).

Using the value of E_c^* determined here and a level density parameter of $A_0/13$, the critical temperature can be estimated to be approximately 6.9 MeV, which is comparable to theoretical estimates for small nuclear systems[11]. In most theoretical calculations the Coulomb force and small size of the system drastically reduces the value of T_c. It is also known that both T_c and ρ_c scale as a function of system size in many systems[12].

The coexistence line and actual position of the fragmenting system in the pressure-density-temperature (p, ρ, T) diagram can be determined from this analysis.

Fisher's theory assumes that the non-ideal vapor can be approximated by an ideal gas of clusters. Accordingly, the quantity n_A is proportional to the partial pressure of a fragment of mass A and the total pressure due to all of the fragments is the sum of their partial pressures:

$$\frac{p}{T} = \sum_{A=1}^{\infty} n_A. \tag{10}$$

The reduced pressure is then given by:

$$\frac{p}{p_c} = \frac{T \sum_{A=1}^{\infty} n_A(T)}{T_c \sum_{A=1}^{\infty} n_A(T_c)}. \tag{11}$$

This is a transformation of the information given in Fig. 3 onto a more familiar frame of reference.

When values of n_A corresponding to $\Delta\mu = 0$ are used in these equations, the coexistence curve is obtained. In other words, this analysis provides simultaneously the coexistence line and the actual line. Both are shown in Fig 4a.

The system's density can be found via

$$\rho = \sum_{A=1}^{\infty} A n_A, \tag{12}$$

and the reduced density from

$$\frac{\rho}{\rho_c} = \frac{\sum_{A=1}^{\infty} An_A(T)}{\sum_{A=1}^{\infty} An_A(T_c)}. \tag{13}$$

As before, both the actual curve and the coexistence curve can be determined. Both are shown in Fig. 4b with the coexistence curve of Guggenheim [13].

In conclusion we have shown that the data of the ISiS Collaboration contain the unequivocal signature of a liquid to vapor phase transition via their strict adherence to Fisher's droplet model. Through Fisher's scaling of the fragment yield distribution (Fig. 2) the two phase coexistence line has been determined over a large energy/temperature interval extending up to and including the critical point. The critical exponents τ and σ as well as the approximate value of T_c, the surface energy coefficient c_0 and the compressibility factor have been extracted and agree with accepted values. A portion of the nuclear phase diagram has been constructed with the available data. Finally, the nuclear liquid at break-up was observed to be a slightly super-saturated vapor for which p/p_c and ρ/ρ_c were determined as a function of T/T_c.

4 Acknowledgements

The authors would like to thank Prof. Cathy Mader for her input and invaluable efforts with the Ising model calculations. This work was supported by the the US Department of Energy

References

1. J. S. Wang and R. H. Swendsen, Physica A **167**, 565 (1990).
2. Ph. Chomaz, *et al.*, Phys. Rev. Lett. **85**, 3587 (2000).
3. D. H. E. Gross, Phys. Rep. **279**, 119 (1997).
4. M. D'Agostino *et al.*, Phys. Lett. B, **473**, 219 (2000).
5. L. G. Moretto, *et al.*, Phys. Rep. **287**, 249 (1997).
6. P. J. Siemens, Nature **305**, 410 (1983).
7. M. E. Fisher, Physics **3**, 255 (1967).
8. L. Beaulieu *et al.*, Phys. Rev. C **63**, 031302 (2001).
9. C. M. Mader *et al*, nucl-th/0103030 (2000).
10. J. B. Elliott *et al.*, Phys. Rev. Lett. **85**, 1194 (2000).
11. H. R. Jaqaman *et al.*, Phys. Rev. C **29** (1984).
12. N. B. Wilding, Phys. Rev. E **52**, 602 (1995).
13. E. A. Guggenheim, J. Chem. Phys., **13**, 253 (1945).

ROLE OF THE PRIMORDIAL DICLUSTERIC CONFIGURATION IN THE ASYMMETRIC MODE OF FISSION

G. MOUZE
Faculté des Sciences, 06108 Nice cedex 2, France

R.A. RICCI
INFN, Laboratori Nazionali di Legnaro e Dipartimento di Fisica, Università di Padova, Padova, Italy.

For actinide nuclei, as well as for superheavy nuclei, low-energy asymmetric fission results from clusterization phenomena. The role of a primordial diclusteric configuration, first suggested by the mean-mass laws of Wahl, is confirmed by a new relation describing both actinide and superheavy nuclei and independent of the mean-mass concept. New properties of the asymmetric mode of the superheavy nuclei are put into evidence.

1. Introduction

Recently, M.G. Itkis and an international team working in Dubna made the interesting observation that superheavy nuclei formed in heavy-ion reactions fission asymmetrically at low excitation energy of the compound nucleus [1]. The aim of the present paper is to show that these results furnish a new opportunity of testing the cluster model of asymmetric fission [2,3].

2. Brief outline of the cluster model of asymmetric fission.

This model is based on the mean-mass laws of A.C. Wahl [4] revised by Flynn et al. [5]: In going from $A_F = {}^{235}U + n_{th}$ to $A_F = {}^{252}Cf$ (s.f.), the mean value of the mass number A_L of the light fragments increases linearly as a function of A_F, whereas that of the heavy fragments, $\overline{A_H}$, remains constant and equal to about 139.

This model assumes that the light fragment is nothing else than the *primordial cluster*, of mass number A_c (made of $N_p + N_n$ valence nucleons of the ^{208}Pb core of the fissioning nucleus) *dressed* with a variable number N_p' of protons and N_n' of neutrons picked-up by the cluster inside the ^{208}Pb core (i.e. , in the valence shells of the deep-lying ^{132}Sn core). It further assumes that the heavy fragment is nothing else than the rump of this hold-up, and that the asymmetry results from the following properties: $\underline{N} = N'_p + N'_n \leq 76$; $N'_p \leq 32$; $N'_n \leq 44$ (however, for heavier nuclei, e.g. for ^{258}Fm, deep collective pick-up processes, for which $\underline{N}$ can be greater than 76, become possible [3]).

This situation suggests to call now "dressed cluster" (D.C.) and "rump" (R) respectivelly, what A.C. Wahl calls light fragment and heavy fragment, and to replace his concept of mean mass, which depends too much on the particular form of the mass distributions, by that of maximum value of A_{DC} and of minimum value of A_R. This choice allows a new representation of the main features of the cluster model (fig.1). One sees that, if A_F increases, the dressed cluster can become heavier than the rump: it happens for $A_F = 264$, or $A_C = 56$; indeed, for the asymmetric mode of ^{258}Fm, the gap between light and heavy mass distributions is still equal to 6 mass units, whereas there is no gap more for ^{270}Sg [3].

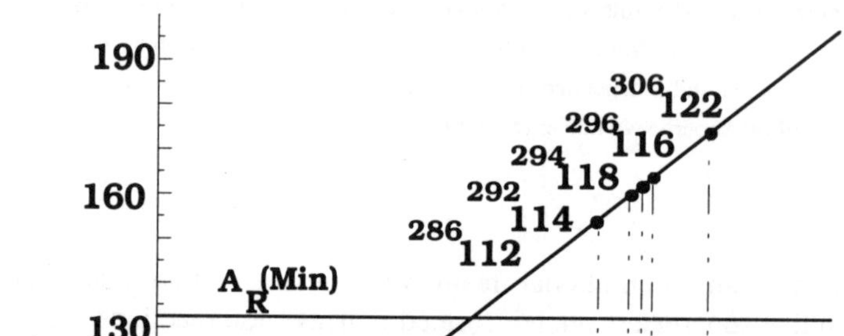

Fig.1 : Revised Flynn's plot valid for all asymmetrically fissioning nuclei.
A_{DC}(Max), the mass number of the heaviest dressed cluster, and A_R (Min), the mass number of the lightest rump, are plotted versus A_C, the mass number of the primordial cluster.

3. Properties of the asymmetric mode of superheavy nuclei according to the cluster model

Extrapolation of the A_{DC} MAX function (fig.1) to the region of the superheavy nuclei studied by Itkis et al.[1] shows that the calculated values coincide precisely with A-values corresponding to a maximum of the experimental mass distributions. In our opinion, it is a consequence of the fact that these mass distributions are not only shifted towards A = 132 and A = A_F - 132, but also that their form is much narrower than that of the actinide nuclei. We show elsewhere [6] that this situation might mean that the asymmetric mode of the superheavy nuclei constitutes a *cleaner mode of fission*. Furthermore, the predilection for the mass splits 132/ (A_F - 132) (e.g. 132/162 for $^{294}118$, having ^{86}Kr as primordial cluster) means that the transfer of as much as 76 nucleons from core to cluster is considerably favoured. This situation might mean that the smallness of the energy-releases of the rearrangement reactions (these Q_2 values are much smaller than those of actinide nuclei) is counterbalanced by *reaction-entropy effects*, in probable relation with the great number of transferred nucleons. This point, and other properties, such as the multimodality, will be discussed elsewhere [6].

4. Acknowledgements

We express our gratitude to Professor Yu. Ts. Oganessian and Professor M.G. Itkis for the communication of their recent experimental results.

References

1. M. G. Itkis et al., *Proc. Fusion Dynamics at the Extremes*, Dubna, May 25-27, 2000, preprint.
2. G. Mouze and C. Ythier, *Nuovo Cimento* A,103 (1990) 617.
3. G. Mouze, *J. Radioanal. Nucl. Chem.* 246 (2000) 191.
4. A. C. Wahl, *Proc. IAEA Symp. Phys. Chem. Fission*, Salzburg, 1965, vol.1, p.317, IAEA, Vienna.
5. R.K. Flynn et al., *Phys. Rev.* C 6 (1972) 2211.
6. G. Mouze, to be published.

THE ASTROPHYSICAL S-FACTOR FOR PROTON CAPTURE ON ^{13}C FROM THE ASYMPTOTIC NORMALIZATION COEFFICIENTS FOR PROTON REMOVAL FROM ^{14}N

A.M. MUKHAMEDZHANOV[1], A. AZHARI[1], V. BURJAN[2], F. CARSTOIU[3], C.A. GAGLIARDI[1], V. KROHA[2], A. SATTAROV[1], X. TANG[1], L. TRACHE[1], R.E. TRIBBLE[1]

[1] *Cyclotron Institute, Texas A&M University, College Station, Texas 77843 USA*
[2] *Institute for Nuclear Physics, Prague-Řež, Czech Republic*
[3] *Institute for Atomic Physics, Bucharest, Romania*

We reanalyze the astrophysical S-factor for $^{13}C(p,\gamma)^{14}N$ capture using the R-matrix approach. The low-energy S factor is dominanted by two resonant and nonresonant captures. The normalization of the nonresonant term for the transition to a particular bound state of the final nucleus is expressed in terms of the corresponding asymptotic normalization coefficient (ANC). In the analysis we use the ANC's inferred from $^{14}N(^{13}C,^{14}N)^{14}C$ and $^{13}C(^{3}He,d)^{14}N$ reactions. The fit of the calculated S factor to the experimental one turns out to be sensitive to the ANC's values and is a test of the ANC's determined by us. The total calculated S-factor at zero energy, $S(0) = 7.65$ keVb, is in an excellent agreement with the value of 7.64 keVb determined earlier [1].

1 Introduction

The $^{13}C(p,\gamma)^{14}N$ reaction is one of the important reactions of the CNO cycle. This reaction is a predecessor of the slowest CNO cycle reaction $^{14}N(p,\gamma)^{15}O$ which defines the rate of energy production in the CNO cycle [2]. In addition, the $^{13}C(p,\gamma)^{14}N$ radiative capture turns out to be very important for the nucleosynthesis of the elements via the s-processes starting from the ^{13}C nuclei via the helium burning reaction $^{13}C(\alpha,n)^{16}O$ [3]. There are measurements of $^{13}C(p,\gamma)^{14}N$ process made to as low as $E = 93$ keV energy in the c.m. reported in literature (see [1] and references therein). The value of $S(0) = 5.5$ keVb recommended in [4] was based on the measurements prior to that of King et al. [1], which are the most accurate and thorough measurements of the S-factor for $^{13}C(p,\gamma)^{14}N$ and include the captures to the all first six excited states of ^{14}N. The extrapolation of the total S-factor down to stellar energies gave $S(0) = 7.64$ keVb and $S(25\,\text{keV}) = 7.7 \pm 1.0$ keVb [1] which are higher than the recommended values in Ref. 4.

The reaction populates the first seven bound states of ^{14}N proceeding through resonant and nonresonant captures. In this work we reanalyzed the experimental data derived from Ref. 1 within the framework of the R-matrix approach. The R-matrix method allows one to treat the resonant and nonresonant contributions in a consistent way, thereby avoiding double counting. The nonresonant R-matrix amplitudes are parametrized in terms of the asymptotic normalization coefficients (ANC's) measured by us [5,6]. The only uncertainty in the nonresonant capture amplitudes comes from the dependence on the channel radius.

2 *R*-matrix approach and ANC

In the *R*-matrix approach the radiative capture amplitude is given by the sum of the resonant and nonresonat (direct) terms. The normalization of the direct term is expressed in terms of the corresponding ANC rather than in terms of the model dependent reduced width amplitude. When calculating the direct part in the *R*-matrix approach, the only fitting parameter is the channel radius r_0. The ANC's for the proton removal from the ground and first six excited states of ^{14}N have been determined from the proton transfer reactions ^{14}N(^{13}C, ^{14}N)^{13}C [5] and ^{13}C(^{3}He, d)^{14}N [6]. The parameters of the first three resonances in ^{14}N were taken from Ref. 1.

3 Analysis and Results

The nonresonant transitions to the ground and all six excited bound states are dominated by $E1$ transitions [1]. The second resonance can decay into the ground and the first five excited states in ^{14}N contributing to the corresponding resonance captures. In our fit, as in Ref. 1, we took into account the contribution of the third resonance only through the radiative capture to the ground state. A crucial fitting parameter in the R-matrix approach is the channel radius a. We find that for all the transitions to the ground and first five excited states $a = 5.0$ fm provides the best fit. The results of the fit for the astrophysical factors for captures to the ground, first excited state and the total S-factor are shown in Figs 1 and 2.

The total S-factor that we obtain at zero energy is $S(0) = 7.65$ keVb supporting the higher value found in Ref. 1. We also present the total resonant and nonresonant S-factors. At zero energy the nonresonant S-factor is $S^{(NR)}(0) = 1.490$ keVb which is 35% of the resonance S-factor $S^{(R)} = 4.28$ keVb. The interference term is $S^{(INT)}(0) = 1.88$ keVb. At $E = 25$ keV the total calculated S-factor is $S(25) = 8.067$ keVb.

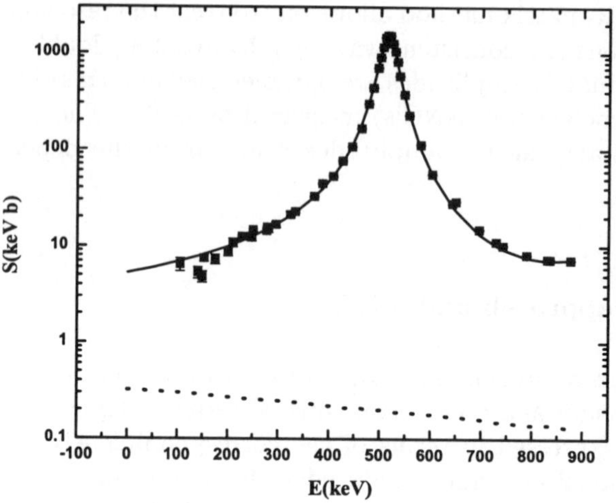

Figure 1. The $^{13}C(p,\gamma)^{14}N$ astrophysical S-factor for capture to the ground state in ^{14}N. The solid boxes are the data points[1]. The solid line represents the calculated S-factor which includes the resonant and nonresonant capture terms and their interference. The dotted line is the nonresonant capture S-factor only.

4 Summary

We reanalyzed the experimental data for the radiative capture $^{13}C(p,\gamma)^{14}N$ at energies $E < 900$ keV (in the c.m.) [1] using the R-matrix method, which turns out to be a convenient way to parametrize nonresonant capture amplitudes in terms of ANC's rather than the spectroscopic factors. Our analysis upholds the high value of the total $S(0)$-factor determined previously in Ref. 1. Hence we support the reaction rate which is higher by ≈ 1.3 in the hydrostatic hydrogen burning temperature range than the reaction rate given in the tabulation of Caughlan and Fowler [7]. Then the high reaction rate results in fewer ^{13}C available for the $^{13}C(\alpha,n)^{16}O$ reaction than was assumed before. It, correspondingly, generates fewer neutrons for the subsequent production of heavy nuclei via the s-process [3].

Acknowledgments

This work was supported in part by the U.S. Department of Energy (DE-FG03-93ER40773), the National Science Foundation (INT 9909787), the

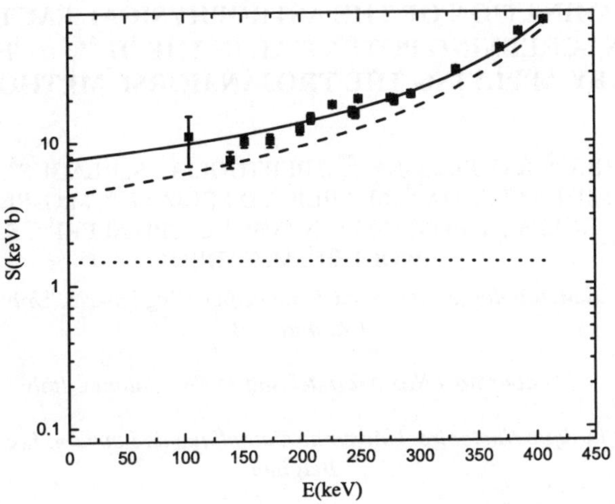

Figure 2. The total low-energy astrophysical S-factor for $^{13}C(p,\gamma)^{14}N$ capture to the ground and first six excited states in ^{14}N. The solid boxes are the data points[1]. The solid, dashed and dotted lines are the calculated total, total resonant and total nonresonant S-factors, correspondingly.

Czech Academy of Sciences and by the Robert A. Welch Foundation.

References

1. J. D. King *et al.*, Nucl. Phys. **A 567**, 354 (1994).
2. C. Rolfs, W. S. Rodney, Caudrons in Cosmos. (The University of Chicago Press, 1988).
3. R. K. Ulrich, in Essays in nuclear astrophysics, (Cambridge University Press, 1982), p. 301.
4. W. A. Fowler *et al.*, Ann. Rev. Astron. Astrophys. **5**, 525 (1967).
5. L. Trache *et al.*, Phys. Rev. C **58**, 2715 (1998).
6. P. Bem *et al.*, Phys. Rev. C **62**, 024320 (2000).
7. G. R. Coughlan and W. A. Fowler, Data Nucl. Data Tables **40**, 283 (1988).

DETERMINATION OF THE ASTROPHYSICAL FACTOR AND ELECTRON SCREENING POTENTIAL IN THE ^{2}H(^{6}Li,α)^{4}He REACTION BY APPLYING THE TROJAN HORSE METHOD

A. MUSUMARRA[1,2], R.G. PIZZONE[1,2], I. BERTUGLIA[2], S. BLAGUS[6], M. BOGOVAC[6], P. FIGUERA[2], M. LATTUADA[2,5], M. MILIN[6], Đ.MILJANIĆ[6], M.G. PELLEGRITI[1,2], D. RENDIĆ[6], C. ROLFS[4], S. ROMANO[1,2], N.SOIĆ[6], C. SPITALERI[1,2], S. TYPEL[7], H.H. WOLTER[7], M. ZADRO[6]

[1]*Dipartimento di Metodologie Fisiche e Chimiche per l'Ingegneria, Università di Catania, Catania, Italy*

[2]*Laboratori Nazionali del Sud-INFN, Catania, Italy*

[3]*Institut de Physique Nucleaire, Université Catholique de Louvain, Louvain-la-Neuve, Belgium*

[4]*Institut für Experimentalphysik III – Ruhr Universität Bochum, Germany*

[5]*Dipartimento di Fisica, Università di Catania, Catania, Italy*

[6]*Institut Ruđer Bošković, Zagreb, Croatia*

[7]*Ludwig Maximilians Universität München. Germany*

The ^{6}Li(^{6}Li,$\alpha\alpha$)^{4}He reaction has been studied in a kinematically complete experiment at E_{6Li} = 6 MeV, from which indirect evaluation of the ^{6}Li(d,α)^{4}He bare nucleus cross section was extracted by means of the Trojan Horse Method (THM). A new theoretical description, based on the DWBA approach, allows to extract information on the bare astrophysical S-factor, and therefore on the electron screening U_e. Both projectile and target break-up were analyzed and used to extract the averaged S(E)-factor. The results are compared with direct experimental data leading to a model independent value of the electron screening potential energy, U_e = 320 ± 50 eV, much higher than the adiabatic limit U_{ad}= 175 eV.

1 Introduction

The experimental measurement of cross sections of astrophysical interests has allowed a better understanding of many astrophysical processes. The energies which characterize the nuclear processes in the astrophysical context, are so low that the presence of atomic electrons must be taken into account. This effect is usually referred to as the electron screening [1]. But the different physical conditions make electron shielding in the stellar plasma and in the laboratory completely different. Nevertheless a complete understanding of electron screening in the laboratory can help to describe the effect of electron shielding in stellar plasma. In both cases it leads to an exponential increase of $\sigma(E)$ (or equivalently of S(E)) with decreasing energy with respect to the case of bare nuclides. This can be described, for the laboratory case, by an enhancement factor

$$f_{lab}(E)=\sigma(E)\backslash\sigma_b(E)=\exp(\pi\eta U_e\backslash E) \qquad (1)$$

where $\sigma(E)$ and $\sigma_b(E)$ refer to the cross sections of electron-shielded and bare nuclei respectively, and U_e is the electron screening potential energy.

In order to give useful inputs to astrophysics, a precise knowledge of the bare nucleus cross section must be available. Given the impossibility to directly measure the bare nucleus factor, extrapolations are generally used in order to obtain the bare cross section. We suggested the THM as a complementary and independent method to evaluate the bare nucleus cross section and therefore the electron screening potential. This method has been applied to the $^6Li(d,\alpha)^4He$ which is present in the reaction network taking place in the primordial nucleosynthesis in the framework of the Inhomogeneous Big Bang model.

2 Description of the Method

The basic assumption of the THM have already been discussed extensively elsewhere [2,3]; the method is based on a quasi-free reaction mechanism which allows to extract the cross section of an astrophysically relevant two-body reaction

$$A + x \rightarrow C + c$$

(here: $^6Li + d \rightarrow \alpha + \alpha$) at low energies from a suitably chosen three-body reaction

$$A + a \rightarrow C + c + b$$

(here: $^6Li + ^6Li \rightarrow \alpha + \alpha + \alpha$). This is done by assuming that the Trojan Horse ^6Li (either in the projectile and in the target) is predominantly composed of the clusters d and α. In the three-body process the wave function of the Trojan Horse $^6Li = \alpha \oplus d$ has a momentum distribution of finite width which must be independently known. In the framework of the distorted-wave impulse approximation (DWIA) the three-body cross section can be written following [3,4] in the form

$$\frac{d^3\sigma}{dE_{\alpha 1}d\Omega_{\alpha 1}d\Omega_{\alpha 2}} \propto KF\left|W\left(Q_{(\alpha\alpha)\alpha}\right)\right|^2 \frac{v_{12}}{v_{Li\text{-}d}} \frac{\exp(-2\pi\eta)}{E\,P_0(k_{Li-d}R)}\left(\frac{d\sigma}{d\Omega}\right)_{^6Li+d} \qquad (2)$$

where KF is a kinematic factor, $|W(Q_{(\alpha\alpha)n})|^2$ is the relative momentum distribution, P_0 is the penetration factor for l=0, v_{12} and $v_{Li\text{-}d}$ the relative velocities and $d\sigma/d\Omega$ is the cross section for the virtual reaction $^6Li(d,\alpha)^4He$. As regards the momentum distribution we adopted the function

$$|\Phi_{Li}(q)|^2 = \frac{e^{-x}}{1+x}$$

with $x=q^2/3555\ fm^{-2}$ and q the inter-cluster α-d momentum, as reported in [5]. In this way it is possible to extract the two-body cross section by inverting formula (2) thus obtaining the bare nucleus cross section for the two body reaction.

3 Experimental setup

The experiment [3] was performed using the EN Tandem Van de Graaff accelerator at the Institut Rudjer Boskovic, Zagreb. A ^{6}Li beam (energy E=6 MeV) was used to bombard a isotopically enriched lithium oxide target, (125 μg/cm^2 thick). The outgoing α particles were detected in coincidence by three position sensitive detectors (PSD) and mounted in a co-planar geometry. They were centered at the quasi-free angles (θ_1=60°, θ_2=-73° θ_3=-103° with respect to the beam axis) and covered an angular range $\Delta\theta$=14°. Each telescope had a geometrical acceptance of about 50x10 mm and offered an angular resolution of 0.1°. During all measurements, the target thickness was continuously monitored by observing the elastic scattering peak with a Si-detector, positioned at θ=40°. Detector signal were processed by standard acquisition electronic chain and sent to the acquisition system which allowed on-line monitoring of the experiment and the data storage on magnetic tape for off-line analysis.

4 Data analysis and results

Measurements of the α-α coincidence events were carried out at each incident energy. Using kinematical cuts, as explained in [3] spectra in the relative energy between the two α's were analyzed. After separating the quasi-free contribution from the sequential one by means of angular correlation and relative energy spectra as shown in [3] we have extracted the triple differential cross section for events with p_s<35 MeV/c). By applying equation (2) we have extracted the two body reaction cross section and thence the astrophysical factor S(E) according to the formula:

$$S(E)=\sigma(E)E\,\exp(2\pi\eta) \qquad (3)$$

where η is the Sommerfeld parameter. In picture 1 the comparison between the normalized S(E)-factors extracted from the target and projectile break-up is made.

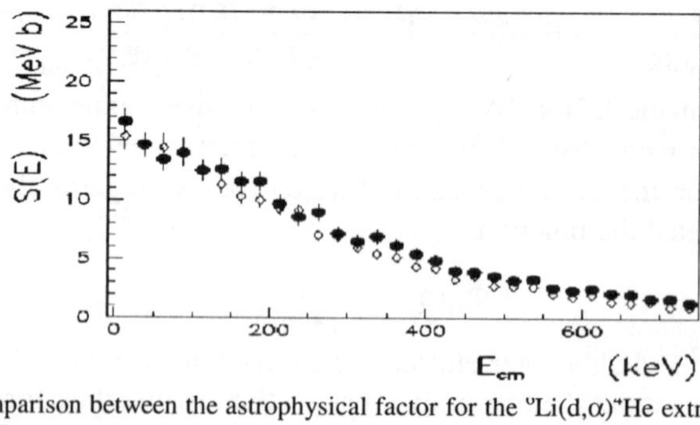

Figure 1. Comparison between the astrophysical factor for the ^{6}Li(d,α)^{4}He extracted from the projectile and target break-up.

We can notice how the two data-set are in agreement and how it represents a positive test for the validity of the method. Moreover it has also allowed an improved confidence in the final results for the S(E)-factor which has been obtained after averaging over the two data-sets. This is shown in picture 2 where the full dots represent the THM data and the open dots the direct measurement [6]. A second order polynomial fit is also shown as a dashed line. The normalization procedure, which is required by the THM was performed at the highest possible energies, i.e. for $E_{c.m.}$=600-700 keV. The zero order parameter, S(0) is in this case 16.9 ± 0.5 MeV b, which has to be compared to the direct measured 17.4 MeV b.

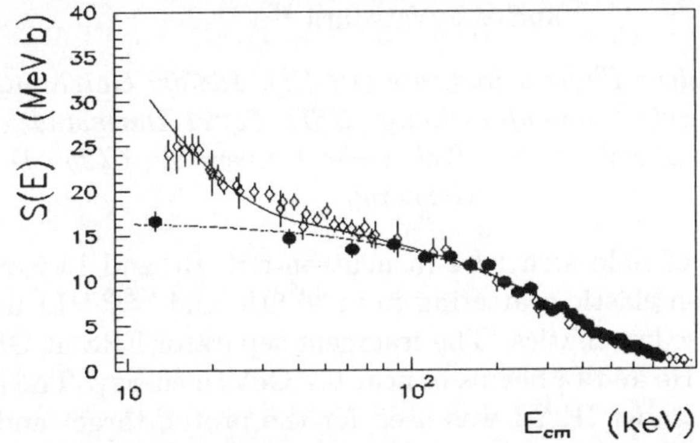

Figure 2. The average astrophysical S(E)-factor (full dots) compared with direct data (diamonds) from [9]. The dashed line represents a second order polynomial fit while the solid line represents the fit used for determining Ue.

The discrepancy at $E_{c.m.}$<100 keV is connected to the electron screening effect in the direct data. From the comparison of the THM and direct data, and applying equation (1) assuming the THM S(E)-factor as the bare one obtains the screening potential U_e=320 eV. This agrees with the evaluation from direct experiment [6] but is sensibly higher than the calculations for the adiabatic case (U_e=175 eV). The same systematic discrepancy was seen for other reactions of astrophysical interest such as the ^{7}Li(p,α)^{4}He. A better and synergic study, using both direct and indirect methods, is thus required.

References

1. Assembaum et al., Z. Phys. A, **327** (1987) 461.
2. G. Baur, Phys. Lett. B**178** (1986) 135.
3. C. Spitaleri, et al., Phys. Rev. C, **63** (2001) 055801.
4. M. Lattuada et al., submitted to Ap. J. (2001).
5. C. Spitaleri et al., Eur. Phys. Jour. A, **7** (2000) 181.
6. S. Engstler et al., Z. Phys. A, **342** (1992) 471.

STUDY OF HALO-STRUCTURE IN RADIOACTIVE He AND Li NUCLEI WITH PROTON ELASTIC SCATTERING

A.V. Dobrovolsky [1,2], G.D. Alkhazov [1], M.N. Andronenko [1], A. Bouchet [2], P. Egelhof [2], S. Fritz [2], G.E. Gavrilov [1], H. Geissel [2], C. Gross [2], H. Irnich [2], A.V. Khanzadeev [1], G.A. Korolev [1], G. Kraus [2], A.A. Lobodenko [1], G. Münzenberg [2], M. MUTTERER [3], S.R. Neumaier [2,3], T. Schäfer [2], C. Scheidenberger [2], W. Schwab [2], D.M. Seliverstov [1], T. Suzuki [2], N.A. Timofeev [1], A.A. Vorobyov [1], and V.I. Yatsoura [1].

[1] *Petersburg Nuclear Physics Institute (PNPI), 188300 Gatchina, Russia*
[2] *Gesellschaft für Schwerionenforschung (GSI), 64291 Darmstadt, Germany*
[3] *Institut für Kernphysik (IKP), Technische Universität, 64289 Darmstadt, Germany*

For the study of halo structure in neutron-rich He and Li isotopes, experiments on proton elastic scattering from ^{4,6,8}He and ^{6,8,9,11}Li nuclei were performed in inverse kinematics. The fragment separator FRS at GSI, Darmstadt provided the He and Li beams of near 0.7 GeV/u energy. The hydrogen-filled ionization chamber IKAR was used for the proton target and detector for recoil protons. Accurate data on absolute differential cross sections for proton elastic scattering at near-forward angles have been procured. Analysis of the cross section data with the aid of Glauber multiple scattering theory has allowed the nuclear matter radii of the He and Li isotopes to be determined, and the corresponding radial matter distributions to be deduced.

Details about experimental method and data analysis are given. The more recently measured elastic-scattering cross sections for the ^{6,8,9,11}Li isotopes are presented, and the nuclear matter radii and radial matter distributions deduced from the data at the present stage of the analysis are discussed.

PACS: 21.45.+v; 27.20.+n; 21.10.Gv; 25.40.Cm; 25.10.+s; 24.10.-i
Keywords: halo nuclei; proton elastic scattering in inverse kinematics; nuclear matter radii and matter density distributions in neutron-rich He and Li nuclei.

1 Introduction

Nuclear sizes and the spatial distributions of nuclear matter and charge are fundamental properties of nuclei, important for the understanding of their internal structure. Accurate experimental data are of particular interest for probing the predictions from microscopic theory. Over the years a large variety of experimental methods were applied, using leptonic probes (as electrons, muons ect.) for investigating nuclear charge distributions, and hadronic probes (as protons, α-particles, pions ect.) for the distributions of nuclear matter (see Ref. [1] for an overview). Whereas most former work was related to the stable nuclei, the recent experimental studies on short-lived radioactive nuclei performed at the radioactive ion beam (RIB) facilities have opened a rapidly expanding and exciting new field of nuclear structure physics.

One of the outstanding results was the discovery that nuclear matter in neutron-rich light nuclei may appear with a qualitative new structure, so-called "halo"-structure. Compared to the stable nuclei, in which all the protons and neutrons are essentially distributed uniformly over the nuclear volume and the nuclear r.m.s. radius R_m scales approximately with mass number A (by $R_m(A) = r_0\ A^{1/3}$), some light neutron-rich nuclei located at or near the neutron drip-line exhibit a far extended distribution of loosely bound valence neutrons (the "halo") surrounding a compact distribution of the majority of nucleons (the "core"). The phenomenon was discovered in the pioneering work of I. Tanihata et al. [2] where the total cross section σ_I from the interaction with various target nuclei was found to rise steeply when He and Li isotopes near the drip-line were used as the projectiles. In order to get a deeper insight into the structure of neutron-rich nuclei, the halo phenomenon has been subject of numerous studies during the last decade. Experiments on halo nuclei have also been a major activity at GSI, Darmstadt where projectile fragmentation at relativistic energies has proven to be very competitive in producing clean and intense secondary beams of radioactive nuclei with the aid of the fragment separator (FRS) [3]. Various experimental methods were applied since in many laboratories [4,5,6,7,8,9,10], such as beta-decay measurements following in-beam polarization by optical pumping, and the investigation of momentum distributions of the reaction products after fragmentation of the halo nuclei impinging on a nuclear target. The picture of nuclear halo structure was qualitatively confirmed. The halo nuclei are imagined as being composed of nucleons from the core and a number (one, two or four) of valence neutrons [11,12,13]. The halo structure manifests itself, besides the large interaction cross sections, by similar electric and magnetic moments of the halo nuclei and the nuclei corresponding to their cores, and by narrow momentum distributions of the reaction products following fragmentation.

In the present work, elastic proton scattering at intermediate energy was chosen as further experimental approach which, due to the availability of a well established theory, provides direct access to the size of the halo nuclei and their radial structure of nuclear matter. For stable isotopes, probing nuclear matter distributions by exploiting the differential cross section for elastic

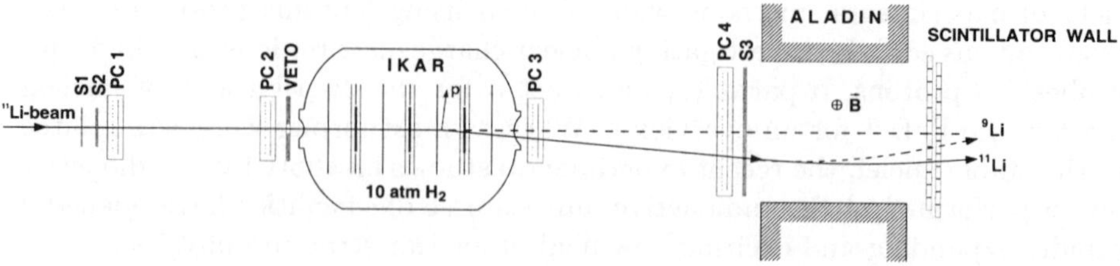

Figure 1. Experimental setup for the inverse-kinematics measurement of elastic $p^{6,8,9,11}$Li scattering at GSI, Darmstadt (see text).

proton scattering at intermediate energies ($\simeq$ 1 GeV) is a well proved method [14]. In this range of energy, elastic proton scattering from composite nuclei is described accurately by the diffractive multiple-scattering theory [15] and, related to that, the measured differential cross sections reflect the nuclear matter distributions of the target nuclei rather unambiguously. In an analogous manner, the structure of the short-lived radioactive nuclei was investigated in the present study via proton elastic scattering in inverse kinematics, using projectile nuclei of $\simeq$ 0.7 GeV/u energy from the secondary FRS beams at GSI and hydrogen as the proton target. The absolute differential cross sections for proton elastic scattering at near-forward angles were accurately measured.

In the present article we are presenting mainly the results obtained more recently on the ^{6,8,9,11}Li isotopes, and compare them with the previous data on $p^{4,6,8}$He scattering [16,17,18]. The measurement of the stable nucleus ^{6}Li served, akin to ^{4}He, for a test of the applied method in cases of stable nuclei.

2 Experiment

For the experiments, primary ^{18}O beams were extracted from the heavy-ion synchrotron SIS and focused on an 8 g/cm^2 beryllium production target at the entrance of the FRS. The helium and lithium ions produced by fragmentation of ^{18}O enter the FRS and are separated according to their magnetic rigidity. In part of the cases the separation power of the FRS was enhanced by inserting an achromatic (20 g/cm^2 aluminium) energy degrader at the dispersive central focal plane for achieving isotopically pure beams. Beam intensities were generally 10^3 s^{-1}, with the duty factor between 25% and 50%.

The arrangement of detectors for the scattering experiment is displayed in Fig. 1. The main constituent is the multiple ionization chamber IKAR [19,20] (filled with hydrogen gas at a pressure of 10 bar) which consists of six in-line axial ionization chambers. Beam particles interact with the protons inside IKAR which thus serves simultaneously as the target and recoil proton detector. The recoil protons registered in IKAR were measured in coincidence with the scattered He and Li particles. For the purpose of determining the scattering angle Θ_S of the projectiles a system of tracking detectors was applied. It consists of two pairs of two-dimensional position-sensitive multi-wire proportional chambers (PC1-PC2 and PC3-PC4), arranged up stream and down stream with respect to IKAR. A set of thin plastic scintillators (S1, S2) was used to provide a fast trigger signal of the events, and a circular-aperture scintillator (Veto) was used to select projectiles that enter IKAR within an area of 2 cm in diameter around the central axis. A thick scintillator (S3) behind PC4 provided for particle atomic-number identification by a ΔE measurement. In the experiment with the secondary beams from the lithium isotopes ^{6,8,9,11}Li, the experimental setup was supplemented with a magnetic-rigidity analysis of the scattered particles (by the ALADIN magnet and a scintillator wall behind) for a safe separation from the neutron break-up channels involved.

The resulting cross-sections $d\sigma/dt$ for the lithium isotopes ^{6,8,9,11}Li are displayed in Fig. 2, in the accessible range below 0.05 $(GeV/c)^2$ for the four-momentum transfer squared $|t|$. The values of $|t|$ were alternatively deduced from the proton recoil energies T_R and/or the projectile scattering angles Θ_S. We note that all the data presented here are still preliminary.

3 Data Analysis and Results

3.1 Nuclear Structure of $^{6,8,9,11}Li$.

For the analysis, the p^{6,8,9,11}Li elastic scattering cross sections were calculated from Glauber theory [15], using four different phenomenologic model densities for the description of the distributions of nuclear matter density. The parametrical model functions, denoted as SF (Symmetrized Fermi), GH (Gaussian-Halo), GG (Gaussian-Gaussian) and GO (Gaussian-Oscillator), have two free parameters each, allowing for reproducing an extended matter density as expected for the halo nuclei (for details see Refs. [16,18]). Conventionally, all distributions refer to the point-nucleon density. While parametrizations SF and GH assume the neutrons and protons to be uniformly distributed, GG and GO presume the nuclei to consist of core nucleons and halo neutrons with different spatial distributions. For calculating the cross sections with the required accuracy, exact numerical values (at the experimental energies) for the elementary pp and pn scattering amplitudes were deduced by interpolation

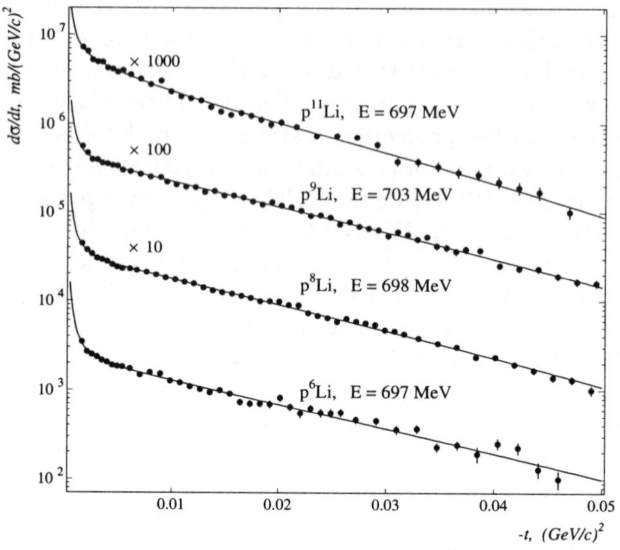

Figure 2. Cross sections $d\sigma/dt$ for elastic $p^{6,8,9,11}$Li scattering (preliminary data), as a function of the four-momentum transfer squared t. Energies are the eqivalent proton energies for normal kinematics. Lines are calculated cross sections using the GG parametrization (see text).

from literature data [18]. In turn, the free parameters in the model density distributions were deduced by least-square fit of the calculated cross sections to the experimental ones, in the t-range covered by the experiment.

For the isotopes ^{6,8,9}Li the cross sections calculated using all four parametrizations for the density distribution fit equally well (with a reduced χ^2/N value around unity) to the experimental data on $d\sigma/dt$. This was also the case for the former data on ^{6,8}He. The corresponding four values of the nuclear radii R_m also closely resemble each other, within rather small errors. Table 1 summarizes the mean valued for each isotope, and compares them with the ^{4,6,8}He results. In the case of ^{11}Li, the parametrizations GH, GG and GO have permitted good description of the data. The present results are thus rather independent from the parametrizations selected for the analysis. Calculated cross sections ensuing from the GG parametrization are shown as solid lines in Fig. 2, as examples. The χ^2/N values listed in Tab. 1 refer to

Table 1. Nuclear matter radii R_m, core radii R_c, and halo radii R_h, deduced from the measured elastic scattering cross sections $d\sigma/dt$. Energies E_p denote the eqivalent proton energies for normal kinematics. The listed χ^2/N values refer to the analysis with the GG parametrization, as example.

Nucleus	E_p (MeV)	R_m (fm)	R_c (fm)	R_h (fm)	χ^2/N
^{6}Li	697	2.44(7)	–	–	44.6/41
^{8}Li	698	2.45(6)	–	–	43.5/47
^{9}Li	703	2.43(7)	–	–	49.0/50
^{11}Li	697	3.62(19)	2.55(12)	6.54(38)	43.2/37
^{4}He	702	1.49(3)	–	–	69.7/76
^{6}He	721	2.30(7)	1.88(12)	2.97(26)	27.3/36
^{8}He	678	2.45(7)	1.55(15)	3.08(10)	43.5/40

the fits with the parametrization GG.

The results obtained on the nuclear matter density distributions represent a clear signature of a pronounced neutron halo in ^{11}Li, and less extended ones in ^{6}He and ^{8}He [16,18]. For attaining a deeper insight into the nuclear structure, i.e. on the extension of the cores and the halos inside the nuclei, assumptions on the partition of the nucleus into halo neutrons and core nucleons have to be made. This was done with the GG and GO parametrizations, assuming a ^{9}Li-like core in ^{11}Li and α-particle like cores in the neutron-rich He isotopes. The resulting core and halo radii obtained by taking the average values deduced from the GG and GO models are included in Tab. 1. The deduced halo radius of ^{11}Li is about twice the halo radii of ^{6}He and ^{8}He. The core radii are different for the ^{6}He and ^{8}He nuclei, and the ^{6}He core radius is significantly larger than the point matter radius $R_m = 1.49$ fm of the free ^{4}He. The core radius in ^{11}Li also slightly exceeds the measured radius of ^{9}Li considered to be the core in ^{11}Li. Larger core sizes in halo nuclei originate from the common motion of the core nucleons around the mass center of the nucleus, which increases the effective core size. The difference between the deduced R_c values and the R_m values for the nuclei thought to form the cores may thus be understood in terms of a different contribution of the valence neutrons to the internal core motion. Figure 3 depicts the distributions for the nuclear matter and the core nucleons in ^{11}Li (both for the GG and GO analysis) in comparison with the matter distribution of ^{9}Li. The ^{11}Li nucleus exhibits, as expected from previous reaction cross-section studies with nuclear targets, the by far most extended halo component from all the nuclei at hand.

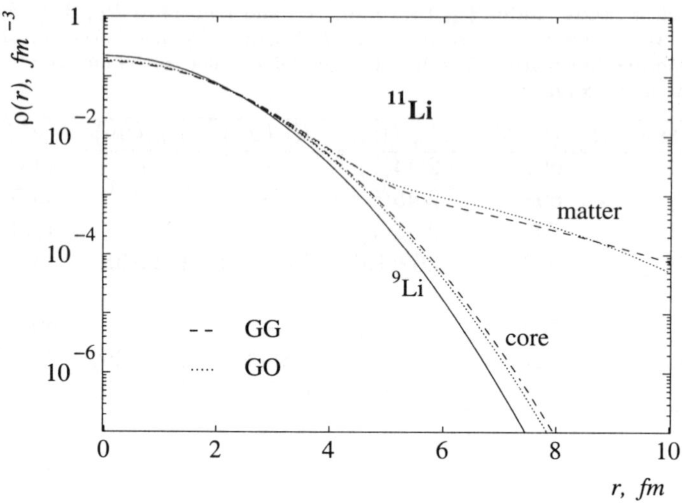

Figure 3. Matter density $\rho(r)$ for ^{11}Li, and matter density $\rho_c(r)$ for the core nucleons, deduced from the data on $d\sigma/dt$, using the parametrizations GG and GO. The solid line is the matter density $\rho(r)$ for ^{9}Li.

3.2 Comparison of Nuclear Structure in the He and Li Isobares of Mass Numbers A = 6 and A = 8.

It is striking from the data in Tab. 1 that matter radii R_m of the ^{6,8,9}Li nuclei are all the same within rather small experimental errors. They coincide with the R_m value for ^{8}He. It is interesting to show whether the method of proton elastic scattering applied in the present work is able to give an explanation for this behaviour in terms of nuclear structure. The present data, indeed, allow for a direct intercomparison of the matter densities in the Li and He isobares of either the mass number A = 6 or A = 8.

As shown in Fig. 4, the matter distribution in ^{6}Li is a little more extended than the one for the halo nucleus ^{6}He. The measured ^{6}Li cross section exhibits a significantly steeper slope than the one for ^{6}He, what clearly indicates a larger radius. On the other hand, the ^{6}Li matter distribution from the present work turns out to be in remarkably good agreement with the ^{6}Li nuclear charge distribution deduced from electron scattering [21]. It is thus conjectured that the ^{6}Li nucleus may feature also a halo structure with an α-like core and, consequently, a deuterium cluster forming kind of a halo.

The situation is quite different for the intercomparison of the ^{8}Li and

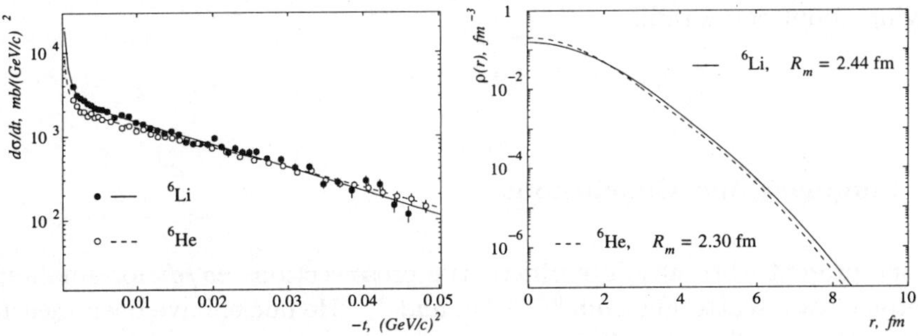

Figure 4. Measured proton elastic cross sections $d\sigma/dt$ (*left*) and deduced matter densities $\rho(r)$ (*right*) for ^{6}He and ^6Li.

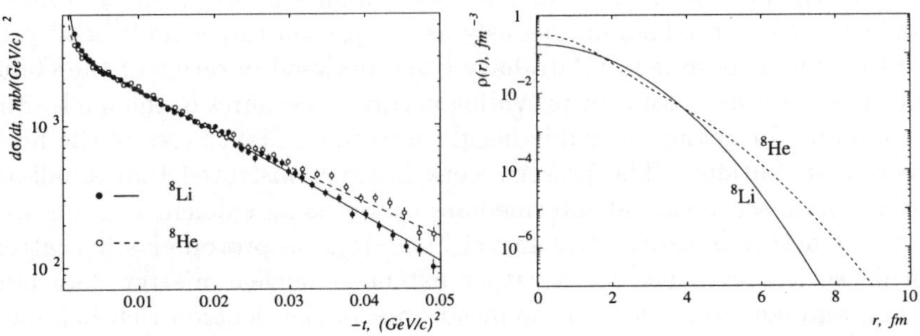

Figure 5. Measured proton elastic cross sections $d\sigma/dt$ (*left*) and deduced matter densities $\rho(r)$ (*right*) for ^{8}He and ^8Li.

^{8}He nuclei, shown in Fig. 5. While both the nuclei have essentially the same radius, their radial matter distributions are significantly different. Compared to the clear halo structure in ^{8}He, the matter distribution in ^{8}Li is significantly less extended, and can be approximated by a Gaussian shape. Accordingly, the t-dependence of the measured ^{8}Li cross section is close to an exponential function, whereas the ^{8}He cross section exceeds an exponential shape at the higher momentum transfers. The latter behaviour is a characteristic token for a halo structure [18]. The reason is that the contribution from the halo neutrons to the cross section decreases faster with increasing t than that from the core nucleons. As for the ^{9}Li nucleus, the present data indicate a Gaussian-like distribution of matter density, similar to that in ^8Li. Hence, there is clear evidence for the ^{6,8,9}Li nuclei to exhibit different internal structures, while having about equal radii.

4 Summary and Conclusions

In the present work absolute differential cross sections $d\sigma/dt$ for small-angle proton elastic scattering from ^{6,8,9,11}Li and ^{4,6,8}He nuclei have been measured with rather small errors. It is the first series of measurements of proton elastic scattering from short-lived radioactive nuclei at intermediate energy. The cross sections have been determined by an inverse-kinematics measurement using secondary beams from the FRS at GSI, Darmstadt, and gaseous hydrogen as the proton target. The high-performance hydrogen-filled ionization chamber IKAR served simultaneously as the proton target and recoil-proton detector. The experimental data have been analysed in terms of the Glauber multiple scattering theory for providing accurate measures of the nuclear matter radii and for getting valuable insight into the radial shapes of the nuclear density distributions. The present work has demonstrated that small-angle proton elastic scattering at intermediate energy is an efficient tool for studying the structure of exotic light nuclei. The data on proton elastic scattering provide a clear evidence for a rather extended nuclear matter distribution in ^{11}Li, and somewhat less pronounced ones in the neutron-rich helium isotopes ^{6,8}He. The experimental method was shown to be even sensitive enough for exhibiting different internal structures in the ^{6,8,9}Li nuclei, which feature about equal radii R_m. In this contexture, it is worthwhile to note, that the present data have already served as a test ground for the predictions from microscopic theories [18,22], e.g. the theoretical approaches of Refs. [23,24,25]. Two examples were also presented as contributions to this conference [26,27].

For attaining deeper insight into the structure of the studied halo nuclei, the measurement of proton elastic scattering has to be extended to the higher momentum transfer region. Then, in particular, the structure of the cores [28] and space correlations between the halo neutrons [29] could be addressed. Very recently, such a measurement on p^{4,6,8}He elastic scattering has been successfully completed, using a different setup furnished with liquid hydrogen as the proton target [30].

The present work has been supported by the German Minister for Education and Research (BMBF) under contract no. 06DA461, by the Scientific and Technological Cooperation Program between Germany and Russia, and by the Russian Fundamental Nuclear Physics Grant No. 121.04.

References

1. J.W. Negele *et al.*, *Adv. Nucl. Phys.* **19**, 1 (1989).
2. I. Tanihata *et al.*, *Phys. Rev. Lett.* **55**, 2676 (1985); *Phys. Lett.* B **160**, 380 (1985); *Nucl. Phys.* A **488**, 113c (1988).
3. H. Geissel *et al.*, *Nucl. Instr. Meth.* B **70**, 286 (1992).
4. K. Riisager, *Rev. Mod. Phys.* **66**, 1105 (1994).
5. P.G. Hansen *et al.*, *Ann. Rev. Nucl. Part. Sci.* **45**, 591 (1995).
6. I. Tanihata, *Prog. Part. Nucl. Phys.* **35**, 505 (1995).
7. I. Tanihata, *J. Phys.* G **22**, 157 (1996).
8. E. Arnold *et al.*, *Phys. Lett.* B **197**, 311 (1987).
9. E. Arnold *et al.*, *Phys. Lett.* B **281**, 16 (1992).
10. H. Emling, *J. Phys.* G **24**, 1561 (1998).
11. I. Tanihata *et al.*, *Phys. Lett.* B **289**, 261 (1992).
12. P.G. Hansen, *Nucl. Phys.* A **553**, 89c (1993).
13. I. Tanihata, *Nucl. Phys.* A **520**, 411c (1990); *Nucl. Phys.* A **522**, 275c (1991).
14. G.D. Alkhazov, S.L. Belostotski, A.A. Vorobyov, *Phys. Rep.* **42c**, 89 (1978).
15. R.J. Glauber, *Lectures in Theoretical Physics*, W.E. Britten and L.G. Dunham (eds.), Interscience, New York, Vol. 1, (1959) p.315.
16. G.D. Alkhazov *et al.*, *Phys. Rev. Lett.* **78**, 2313 (1997).
17. S.R. Neumaier *et al.*, *Nucl. Phys. A*, submitted.
18. G.D. Alkhazov *et al.*, *Nucl. Phys. A*, submitted.
19. A.A. Vorobyov *et al.*, *Nucl. Instr. Meth.* **119**, 509 (1974).
20. J.P. Burq *et al.*, *Nucl. Phys.* B **217**, 285 (1983).
21. G.C. Li *et al.*, *Nucl. Phys.* A **162**, 583 (1971).
22. P. Egelhof *et al.*, *Prog. Part. Nucl. Phys.* **46**, 307 (2001).
23. I.J. Thompson *et al.*, *Phys. Rev.* C **49**, 1904 (1994).
24. J. Wurzer, and H.M. Hofmann, *Phys. Rev.* C **55**, 688 (1997).
25. M. Tomaselli *et al.*, *Phys. Rev.* C **62**, 067305 (2000).
26. H. Lenske *et al.*, this proceedings.
27. M. Tomaselli *et al.*, this proceedings.
28. L.V. Chulkov *et al.*, *Nucl. Phys.* A **587**, 291 (1995).
29. G.D. Alkhazov, *Yad. Fiz.* **62**, 765 (1999); *Phys. Atom. Nuclei* **62**, 715 (1999).
30. F. Aksouh, O. Kisselev *et al.*, to be published.

EFFECTS OF SHELL STRUCTURE IN REACTIONS LEADING TO THE SAME COMPOUND NUCLEUS OR DIFFERENT ISOTOPES

A.K. NASIROV

Bogoliubov Laboratory of Theoretical Physics, JINR, 141980 Dubna, Russia
E-mail: nasirov@thsun1.jinr.ru
Institute of Nuclear Physics 702132 Ulugbek, Tashkent, Uzbekistan
E-mail: ion@suninp.tashkent.su

G. FAZIO, G. GIARDINA, A. LAMBERTO, R. RUGGERI, A. TACCONE

Istituto Nazionale di Fisica Nucleare, Sezione di Catania, and
Dipartimento di Fisica dell'Università di Messina, 98166 Messina, Italy
E-mail: giardina@nucleo.unime.it

F. HANAPPE

Université Libre de Bruxelles, CP 229, B-1050 Bruxelles, Belgium
E-mail: fhanappe@ulb.ac.be

R. PALAMARA

Facoltà di Architettura dell'Università di Reggio Calabria, Italy
E-mail: rpalamara@unirc.it

L. STUTTGÉ

Institut de Recherches Subatomiques, F-67037 Strasbourg Cedex 2, France
E-mail: stuttge@in2p3.fr

The role of the entrance channel in fusion-fission reactions was studied by the theoretical analysis of the experimental evaporation residue excitation functions for reactions leading to the same compound nucleus. The evaporation residues cross sections for xn-channels were calculated in the frame of the combined dinuclear system concept (DNS) and advanced statistical model (ASM). The revealed differences between experimental data on the evaporation residues in the ^{40}Ar+^{176}Hf, ^{86}Kr + ^{130}Xe and ^{124}Sn + ^{92}Zr reactions leading to the ^{216}Th* compound nucleus (CN) as well as that between the ^{48}Ca + ^{174}Yb and ^{86}Kr + ^{136}Xe reactions leading to the ^{222}Th* CN are explained by the different spin distributions of compound nuclei which are formed. It is shown that the intrinsic fusion barrier B^*_{fus} and size of potential well are different for every entrance channel.

1 Introduction

An appropriate way to understand the mechanism of the fusion-fission process in heavy ion collisions is the analysis and comparison of the measured excitation functions of evaporation residues for reactions leading to the same compound nucleus. Theoretical calculations performed in the framework of the model including the nuclear shell effect and shape of colliding nuclei allow us to come to useful conclusions about the mechanism of fusion-fission process. Often excitation functions of evaporation residues measured in various reactions leading to the same compound nucleus are different not only by the position of the maximum but also by the value of their maximums. The question on what characteristics of nuclei or what relevant degrees of freedom of the fusion-fission process are responsible for such difference in evaporation residue cross sections is a well actual problem to reveal optimal conditions for the synthesis of new super-heavy elements.

2 Evaporation residue production in the DNS concept

In the dinuclear system concept[1] the evaporation residue cross section is factorized as follows:

$$\sigma_{er}(E) = \sum_{\ell=0}^{\infty}(2\ell+1)\sigma_{\ell}^{fus}(E,\ell)W_{sur}(E,\ell). \tag{1}$$

Here the effects connected with the entrance channel are included by the partial fusion cross section $\sigma_{\ell}^{fus}(E)$ which is defined by the expressions:

$$\sigma_{\ell}^{fus}(E) = \sigma_{\ell}^{capture}(E)P_{CN}(E,\ell), \tag{2}$$

$$\sigma_{l}^{capture}(E) = \frac{\lambda^2}{4\pi}\mathcal{P}_{\ell}^{capture}(E), \tag{3}$$

where λ is the de Broglie wavelength of the entrance channel, $P_{CN}(E,\ell)$ is a factor taking into account the decrease of the fusion probability due to break up of the dinuclear system before fusion; $\mathcal{P}_{\ell}^{capture}(E)$ is the capture probability which depends on the collision dynamics and determines the amount of partial waves leading to capture.

According to dinuclear system concept[1], the fusion of two colliding nuclei is possible if the following two conditions are satisfied. The first one is overcoming Coulomb barrier along axis connecting nuclear centers by nuclei at incoming stage of collision and formation of the nuclear composite system (molecular alike so called dinuclear system). This process is called a capture. The second condition is the transformation of the dinuclear system into the more compact compound nucleus overcoming the intrinsic barrier (B^*_{fus}) which must be overcome by system during the motion on the mass (charge) asymmetry axis. So this stage is a fusion and it is realized by nucleon transfer from the light nucleus on the unoccupied states of heavy one at the limited size of overlap volume of nucleon density. For light and mediate nuclear system or for heavy nuclear system with largest mass asymmetry this B^*_{fus} barrier is very small or equal to zero and capture leads immediately to fusion. Therefore for those cases fusion cross section is calculated in frame of well known models. The barrier B^*_{fus} is determined by the difference between the maximum value of a driving potential and its value at the point corresponding to the initial charge asymmetry of the considered reaction (Fig.1). A driving potential holds significance in consideration of the fusion process as a motion of system along the charge (mass) asymmetry degree of freedom. Therefore, here more attention should be paid to that. It is obtained from the potential energy surface $U(A, Z; R, \ell)$ as a function of masses (charges) A_1, A_2 ($A_2 = A_{tot} - A_1$) of fragments forming the dinuclear system at the R_m value of internuclear distance corresponding to the minimum of their nucleus-nucleus potential $V(R)$. The driving potential $U(A, Z; \ell)$ calculated in this way is presented in Fig.1. The distribution of neutrons between two fragments by the given proton numbers Z_1 and Z_2 (or ratios A_1/Z_1 and A_2/Z_2 for the both of fragments) was determined by minimizing of potential $U(A_1, Z_1; R)$ as a function of A_1 for each Z_1.

$$U(A, Z; R, \ell) = B_1 + B_2 + V_0(Z, \ell, \beta_1, \alpha_1; \beta_2, \alpha_2; R) - (B_0 + V_{comp}(\ell)), \quad (4)$$

where, B_1, B_2 and B_0 are the binding energies of the nuclei forming a dinuclear system and of the compound nucleus, respectively. The values of $B_i (i = 0, 1, 2)$ were obtained from[2]; $\beta_i (i = 1, 2)$ are the fragment deformation parameters and α_i are the orientations of nuclei symmetry axis relative to the beam direction; $V_{comp}(\ell)$ is the rotational energy of the compound nucleus. The R_m is the position of the minimum of the nucleus-nucleus potential (bottom of the pocket) on the R axis for a given mass asymmetry A_1. The smallest value of excitation energy of the compound nucleus ($E^{*(min)}_{CN}$) is determined by the absolute maximum value of the driving potential laying on the way to fusion ($Z = 0$) from the point corresponding to the initial charge asymmetry (Fig.1). The shapes of the potential energy surface and driving potential depend on the orientations of nuclei relatively to axis connecting the centers of interacting nuclei. The presented results are obtained by averaging over contributions of different orientations. The quasi-fission is a decay of dinuclear system after capture moving along R in the $V(R)$ nucleus-nucleus interaction potential without reaching compact shape. Thus, for quasi-fission, it is necessary to overcome the Coulomb barrier on the R axis from internal side. The DNS concept allows one to reveal a reason of the strong decrease in the fusion cross section for a massive system or for a symmetric entrance channel due to the increase of B^*_{fus}.

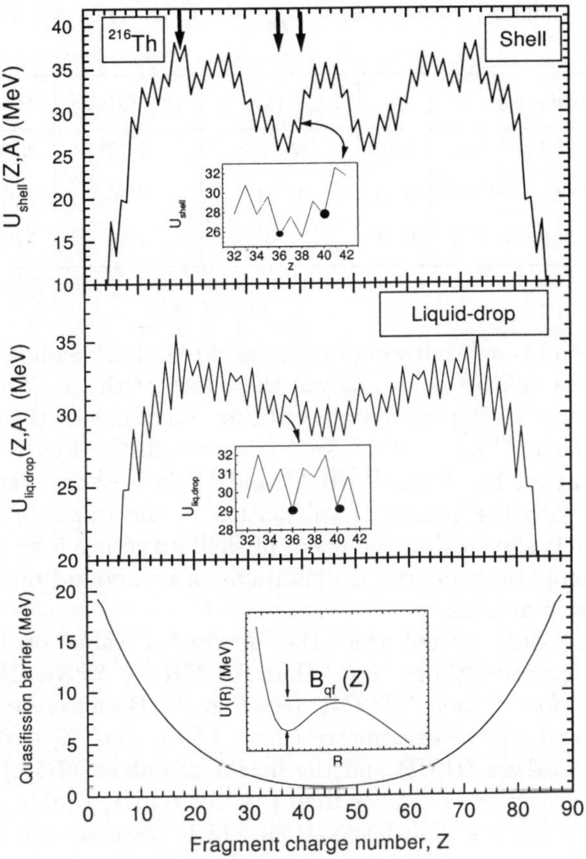

Figure 1. Driving potential $U(Z, A, R_m; l = 0)$ as function of the charge number Z of a fragmont of dinuclear system calculated by using of binding energies from the mass table[2] (top panel) and liquid-drop model (middle-panel). The dependence of quasi-fission barrier on the charge number (bottom panel).

3 Comparison of DNS model results and experimental data on the reactions leading to ^{216}Th

The role of the entrance channel on the formation of the compound nucleus and evaporation residues is a main point of our interest. The qualitative

Table 1. Charge asymmetry, intrinsic fusion B^*_{fus} and quasi-fission B^*_{qf} barriers for the reactions leading to ^{216}Th.

Reactions	η_Z	B^*_{fus} (MeV)	B_{qf} (MeV)	P_{CN}
(I) ^{40}Ar+^{176}Hf	0.60	2.31	7.12	0.580
(II) ^{86}Kr + ^{130}Xe	0.20	12.31	2.35	0.014
(III) ^{124}Sn + ^{92}Zr	0.15	9.87	1.35	0.081

difference between fusion excitation functions of reactions leading to the same compound nucleus allows us to analyze the effect of the shell structure on the fusion mechanism. Experimental excitation functions of the evaporation residues measured in ^{86}Kr + ^{130,136}Xe [3] reactions in the Flerov Laboratory (Dubna) and that for the ^{40}Ar+^{176}Hf [4,5] and ^{124}Sn + ^{92}Zr [6] reactions have been compared with the results of calculation in the frame of the method presented in[7]. It is shown that the effect of shell structure is revealed in the differences at comparison of spin distributions of compound nucleus formed by using different reactions.

Experimental data reveal that the maximum value of the ER for ^{40}Ar+^{176}Hf (I) is twelve times larger than for ^{86}Kr + ^{130}Xe (II) and three times larger than for ^{124}Sn + ^{92}Zr (III) (see Fig. 2). Because the ^{40}Ar+^{176}Hf reaction has largest charge asymmetry ($\eta_Z = (Z_2 - Z_1)/(Z_1 + Z_2)$) in comparison with two others (II,III) and the height of barrier (B^*_{fus}) in the way to fusion for this reaction is smaller than that for (II,III) (Table 1). The way to fusion is longer for the dinuclear system which has more mass symmetric configuration. The excitation functions of the ER calculated in the frame of advanced statistical model[8] are in good agreement with the experimental data. At these calculations, spin distributions for CN were found by the method presented in[7].

But an unusual phenomenon is that the maximum value of the ER for ^{124}Sn + ^{92}Zr is four times larger than for ^{86}Kr + ^{130}Xe nearly at the same value of E^*_{CN}. These reactions lead to the same (^{216}Th*) CN. The mass asymmetry of (III) ($(Z_2 - Z_1)/(Z_1 + Z_2)$=0.148) is smaller than one of (II) (0.203). It is not clear from the driving potential calculated using the binding energies B_1, B_2 and B_{CN} determined by the liquid-drop model (see middle panel of Fig.1), since the value of the fusion barrier B^*_{fus} for the ^{124}Sn + ^{92}Zr reaction is not lower than that for the ^{86}Kr + ^{130}Xe reaction. The reason of this phenomenon may be established from the analysis of the driving

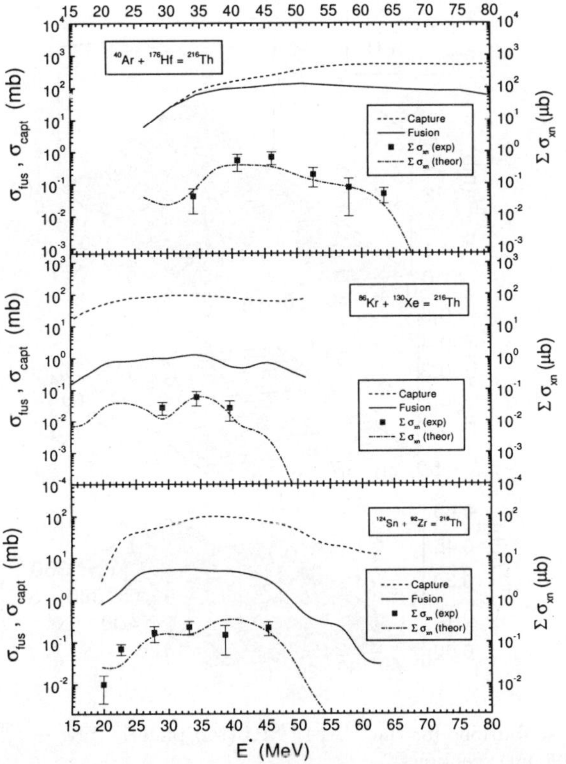

Figure 2. Comparison of the calculated capture and fusion excitation function (left axis) as well as calculated total excitation functions of evaporation residue (right axis) with the experimental data for $^{40}Ar+^{176}Hf$, $^{124}Sn + ^{92}Zr$, and $^{86}Kr + ^{130}Xe$ combinations leading to the ^{216}Th CN.

potential which includes the effect of the shell structure. The driving potential calculated by using binding energies of nuclei from Table in[2] discloses even qualitatively difference between values of B^*_{fus} for reactions (II) and (III). As seen from the Table 1 for the former reaction, B^*_{fus} is larger than for the latter one. Another reason is seen from the analysis of spin distributions of compound nuclei formed in these three reactions.

As seen from Fig.3, spin distribution of CN formed in reaction (I) (top panel) has largest volume in comparison for the reactions (II) (middle panel) and (III) (bottom panel). It is an expected result. The matter is that why the volume of in spin distributions of CN corresponding to reaction (III) is larger

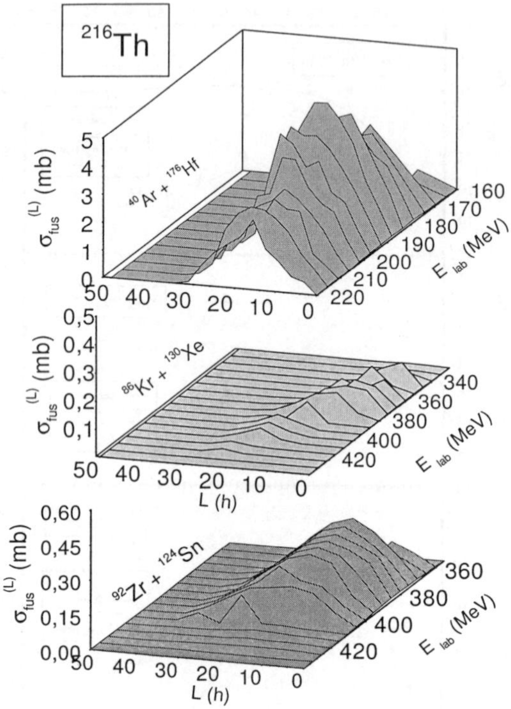

Figure 3. Spin distributions for the ^{40}Ar+^{176}Hf (top panel), ^{86}Kr + ^{130}Xe (middle) and ^{124}Sn + ^{92}Zr (bottom) reactions.

than that for reaction (II). This could be connected with the $\sigma_l^{capture}(E)$ and P_{CN} which determine partial fusion cross sections (2). The last is calculated from the equation of motion. The number of partial waves ℓ which contribute to $\sigma_l^{capture}(E)$ is determined by the size of the potential well in the entrance channel (the high of quasi-fission barrier B_{qf}) and the friction coefficient of the radial motion. The use of the calculated friction coefficients leads to a gradual dissipation of kinetic and rotational energies[9]. It was shown that by collisions of massive nuclei, despite of continuous dissipation, the capture becomes impossible at beam energy values larger than the Coulomb barrier because of the small size of the well in the nucleus-nucleus potential which is a function of R. The dissipation is not enough to trap colliding nuclei in the potential well to create a necessary condition for fusion at low values of angular momentum which allow to the dinuclear system to fuse. At largest values of beam energy, capture is possible only for high angular momentum. In this case the formed dinuclear system can exist as a molecular state forming a super-deformed shape or undergoes quasi-fission because B^*_{fus} increases by angular momentum of dinuclear system. Therefore the maximum of the calculated spin distributions has tendency to move to larger values of angular momentum at beam energies well above the Coulomb barrier.

The competition between fusion and quasi-fission is calculated on the basis of the suggestion that at the given excitation energy of the dinuclear system, $E^*_{DNS} = E_{cm} + Q_{gg}$, the probability of realizing the complete fusion in competition with the quasi-fission, starting from an entrance channel where DNS is formed, is related to the ratio of level densities on the intrinsic fusion and quasi-fission barriers

$$P_{CN} = \frac{\rho(E^*_{DNS} - B^*_{fus})}{\rho(E^*_{DNS} - B^*_{fus}) + \rho(E^*_{DNS} - B_{qf})}, \tag{5}$$

where $\rho(E^*_{DNS} - B^*_K)$ is the level density of the dinuclear system on the intrinsic fusion barrier (B^*_{fus}) and quasi-fission barrier (B_{qf})[7]. This is seen in the spin distributions for the ^{124}Sn + ^{92}Zr and ^{86}Kr + ^{130}Xe reactions (Fig.3). The driving potential is the same for all the reactions leading to the same compound nucleus. Therefore the intrinsic fusion barriers for these reactions under discussion can be compared. It is seen in Fig. 1 (top panel) from the curve of the driving potential that B^*_{fus} is smaller for the reaction (III) than for (II). This is connected with the increase of the potential energy at $Z > 37$ due to the shell effects.

The calculation of the driving potential using binding energies determined by liquid drop model shows that in this case the shape of the curve is different than one calculated using values of binding energies from the mass table in [2]. Therefore, an observed smallness of the excitation function of evaporation residues for the ^{86}Kr + ^{130}Xe reaction than that for ^{124}Sn +^{92}Zr is concluded to be connected with the peculiarity of the shell structure of nuclear fragments forming the dinuclear system. Due to large difference between Q-values of these three reactions leading to the ^{216}Th* CN the centers of their excitation functions are placed at different values of the excitation energy.

3.1 Comparison of reactions induced by ^{86}Kr on the ^{130}Xe and ^{136}Xe targets

Another observed interesting phenomenon was revealed at the comparison of the experimental data on reactions induced by the ^{86}Kr projectile on the ^{130}Xe and ^{136}Xe targets. The ER in ^{86}Kr + ^{136}Xe (IV) was about 50 times

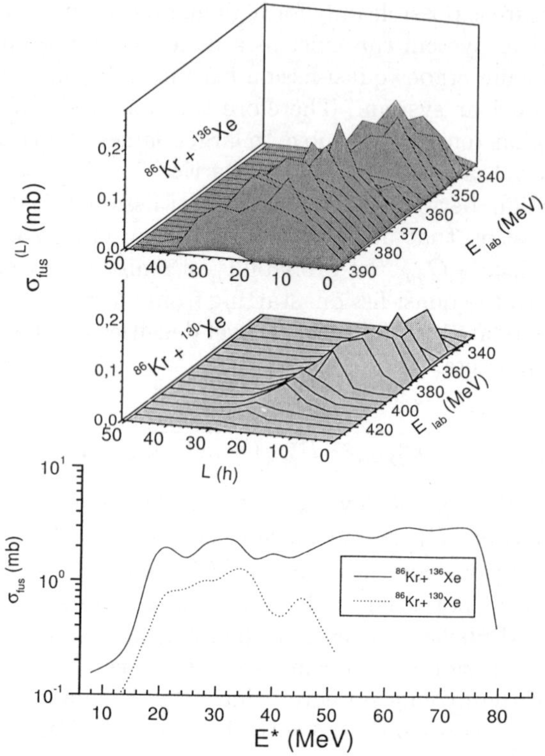

Figure 4. Spin distributions for the ^{86}Kr+^{136}Xe (top panel) and ^{86}Kr + ^{130}Xe (middle) reactions. Fusion excitation functions are compared in the bottom figure.

larger than in ^{86}Kr + ^{130}Xe (II). This result is related to two characteristics of the fusion-fission mechanism. The first is that the surviving probability of the ^{222}Th* CN is larger than one of ^{216}Th* because the fission barrier of the compound nucleus is larger for the former isotope than for the last one and the neutron separation energy in ^{222}Th* is smaller than in ^{216}Th*. Therefore the ratio Γ_n/Γ_f is larger for ^{222}Th* than ^{216}Th*. The second reason is that the fusion cross section calculated by (2) for the reaction (IV) is larger than for (II). It is seen from the difference in the spin distributions presented in Fig.4 for these two reactions.

This difference could be explained by two facts: the size of the well in the nucleus-nucleus potential for the ^{86}Kr + ^{130}Xe reaction is smaller than that for ^{86}Kr + ^{136}Xe and the intrinsic fusion barrier B^*_{fus} is larger for the former reaction than for the last one. Because the excess of neutron number in the ^{136}Xe target, the nuclear interaction with the beam is more attractive and the potential well is wider in comparison with the nucleus-nucleus potential between ^{86}Kr and ^{130}Xe. The calculated excitation functions for the different evaporation residue channels in the ^{86}Kr + ^{136}Xe and ^{86}Kr + ^{130}Xe reactions are shown in Fig. 4.

Acknowledgments

We are grateful to Prof. R.V. Jolos, Prof. V.V. Volkov, Prof. Dr. W. Scheid, Drs. G.G. Adamian, and N.V. Antonenko for helpful discussions. One of the authors (A.K.N.) thanks the Russian Fund of Basic Research (Grant No. 01-02-16033) for financial support, and he is grateful to the UNST Uzb-45, Uzbekistan State Scientific-Technical Committee (Grant No. 7/2000) for partial support. A.K.N. would also like to express his gratitude to the Università of Messina (Italy) for the warm hospitality during his stay, and to the Bonino-Pulejo Foundation (FBP) of Messina for the help given in developing the collaboration with the group of Prof. G. Giardina.

References

1. V.V. Volkov, N.A. Antonenko, E.A. Cherepanov, A.K. Nasirov, V.P. Permjakov, *Phys. Lett.* B **319**, 425 (1993); *Phys. Rev.* C **51**, 2635 (1995).
2. G. Audi, A.H. Wapstra, *Nucl. Phys.* A **595**, 509 (1995).
3. Yu.Ts. Oganessian, A.Yu. Lavrentev, A.G. Popeko, R.N. Sagaidak, A.V. Yeremin, S. Hofmann, F.P. Heßberger, V.Ninov, Ch. Stodel, *JINR FLNR Scientific Report 1995-1996. Heavy Ion Physics*, B.I. Pustylnik ed., p. 62 (JINR, E7-97-206, Dubna), 1997.
4. D. Vermeulen, H.-G. Clerc, C.-C. Sahm, K.-H. Schmidt, J.G. Keller, G. Münzenberg, W. Reisdorf, *Z. Phys.* A **318**, 157 (1984).
5. H.-G. Clerc, J.G. Keller, C.-C. Sahm, K.-H. Schmidt, H. Schulte, D. Vermeulen, *Nucl. Phys.* A **419**, 571 (1984).
6. C.-C. Sahm, H.-G. Clerc, K.-H. Schmidt, W. Reisdorf, P. Armbruster, F.P. Heßberger, J.G. Keller, G. Münzenberg, D. Vermeulen, *Nucl. Phys.* A **441**, 316 (1985).
7. G. Giardina, S. Hofmann, A.I. Muminov, A.K. Nasirov, *Eur. Phys. J.* A **8**, 205 (2000).
8. A. D'Arrigo, G. Giardina, M. Herman, A.V. Ignatyuk, A. Taccone, *J. Phys.* G **20**, 365 (1994).
9. G.G. Adamian, R.V. Jolos, A.I. Muminov, A.K. Nasirov, *Phys. Rev.* C **56**, 373 (1997).

INFLUENCE OF EXCITED RADIOACTIVE NUCLEI FOR RESULTS IN LARGE-SCALE NUCLEAR CHRONOMETRY

V.S.OLKHOVSKY
Institute for Nuclear Research, Kiev-252028, Ukraine
E-mail: olkhovsk@kinr.kiev.ua

It is shown that the known methods of the nuclear chronometry of the astrophysical processes have to be revised and corrected, taking into account distinctions between lifetimes of the ground and excited states for radioactive nuclei and also the multiple processes of γ-emissions from excited states with consequent γ-absorptions by ground in large masses of radioactive matter. In the result of this, *(1)* the corrected "nuclear clocks" indicate to *smaller* values of the durations of real decay processes, *(2)* the usual (non-corrected) "nuclear clocks" indicate to the *upper limits* of the durations of real decay processes.

1. Up to 2000 the lifetimes of *only ground states* of decaying nuclides have been taken into account in all known methods of nuclear chronometry (see, for instance, [1-4]). However during s - and r - processes of nucleosynthesis not only ground states but also all possible excited states of synthesized nuclei are being formed. It is known that the α- and β- decays of excited nuclei go on much more rapidly than the decays from the ground states. In particular, it follows directly from the Geiger and Nutall law for the α-decay (exposed in every university manual on nuclear physics) that lifetime τ_{ex} of a α-decaying nucleus very strongly depends on the α-particle kinetic energy (and hence on the excitation energy of a decaying nucleus). In many cases lifetime τ_{ex} is diminished by several (4-5) orders with the increasing of the α-particle energy by 1 - 2 MeV ! As to the β-radioactive nuclei, one can expect also a relatively rapid decreasing of lifetime τ_{ex} with the increasing of the nucleus energy excitation up to the release of free beta-radioactive neutrons with $T_{1/2} \approx 13$ min. But up to now no systematic experimental study of the dependence of lifetimes of the excited radioactive nuclei relative to the α- and β-decay on their excitation energy had been undertaken because of the difficulties connected with much more rapid (within $10^{-13} - 10^{-9}$ sec) and strong γ-decays of excited nuclei. For the same reason it had always (up to our last papers [5,6]) assumed that there is no practical reason to take into account the noticeably more slow processes of the α- and β-decays from these states.

In the last published author's results [5,6], it had been shown that inside large ($\rightarrow$ infinite) masses of stellar, planetary and meteorite substance a part of radioactive nuclei can be long supported in excited states for the extremely wide temperature interval (from $0\ ^{o}C$ till very large values) due to chains of subsequent emissions and quasi-resonance absorptions of γ-quanta by nuclei-chronometers when energy losses, caused by recoils during emitting and absorbing, are compensated by nucleus thermal kinetic energy.

2. Let the α-decay widths of the ground and first excited states be Γ_α^0 and Γ_α^1 respectively, and the γ-decay width of the first excited state be Γ_γ and we note: $\Gamma_\alpha^0 = \Gamma_0$ and $\Gamma_\alpha^1 + \Gamma_\gamma = \Gamma_1$.
In standard nucleo-chronometry techniques together with the probabilities $L(t)$ and $W(t)$ one uses the abundances P and D of parent and daughter nuclei connected with L and W by the following evident relations:

$$L(t-t_o)=P(t)/P(t_o), \quad W(t-t_o)=[P(t)-P(t_o)]/P(t_o) \tag{1}$$

and

$$P(t)+D(t) = P(t_o)+D(t_o), \tag{2}$$

or

$$p(t)[\exp(\Gamma_0(t-t_o)/\hbar)-1] - d(t) + d(t_o) = 0 \tag{2a}$$

where $p=P/D_x$ and $d=D/D_x$, D_x being the abundance of *another* stable isotope. Measuring p and $d=D$ in different samples, we obtain the plot (2a) at the plane (p, d) which has the form of a straight line, the slope of which with respect to axis p permits simply to define age $t-t_o$ (see, for instance,[1]).

Taking into account the *single* intermediate γ- absorptions by radioactive (for definiteness, α-radioactive) nuclides, we generalize the identity (2) in [8,9] in a following way:

$$D(t)=D(t_0)+ \\ + [P_0(t_o)+P_1(t_o)][1-\exp(-\Gamma_0(t-t_o)/\hbar)]+P_1(t_o)Q_\lambda \exp(-\Gamma_0(t-t_o)/\hbar) \tag{3a}$$

with $\quad Q_\lambda=(\Gamma_\alpha^1/\Gamma_1)qM_o[P_0(t_o)+P_1(t_o)(2\Gamma_1+c\hbar/\lambda)/2(\Gamma_1+c\hbar/\lambda)],$

or

$$D(t)=D(t_0)+[P_0(t_o)+(1-Q_\lambda)P_1(t_o)][1-\exp(-\Gamma_0(t-t_o)/\hbar)]+Q_\lambda P_1(t_o), \tag{3b}$$

where we used the expression $M(t) = qM_o[1-\exp(-ct/\lambda)]$ for the γ- absorption probability during γ- quantum propagation inside the matter sample (c is the light velocity and λ is the inversed γ--quantum absorption coefficient inside the matter, q is the dimensionless quantity, which depends on the small loss of the γ-quantum energy due to the γ-scattering by nuclei and by electrons. The orientative estimations of q gave values which are approximately within $(1/2, 1)$.

Taking into account not only single (one-step) γ- absorptions but all possible *multiple* γ- absorptions, one can evaluate an every step of γ- absorptions by an equal contribution with $M \cong M_o(2\Gamma_1+c\hbar/\lambda)/2(\Gamma_1+c\hbar/\lambda)$ and easily sum up them in (3a) or (3b). And then, if Γ_α^1/Γ_1 is not very small (let us say, ≥ 0.1), one can also roughly take $N\Gamma_\alpha^1/\Gamma_1 \cong 1$ in (3b) (N being an effective number of the considered γ- absorption steps).

We can confirm and illustrate the realisticity of such valuations with the help of the following simple evident reasonings. For lifetimes τ_γ relative γ-decay larger than 10^{-13} *sec* and the Doppler width $D= 2[\varepsilon_{recoil}\, kT]^{1/2}$, T being the sample temperature (k is the Boltzman constant) at the approximation $\varepsilon_{recoil} \rightarrow \varepsilon_{recoil}^{(o)} = (E_1 - E_0)^2/2Am_n c^2$ (A and m_n are the atomic number of parent nuclei and the nucleon mass respectively), the widths $\Gamma_\gamma = \hbar/\tau_\gamma$ satisfy the following relation $\Gamma_\gamma << D$ and, moreover, for $A \cong 250, \varepsilon_\gamma \cong 50\ keV$ and $N \cong 10$ the quantity $N\varepsilon_{recoil}$ satisfies the following expressions $N\varepsilon_{recoil} < D$ when $T > 300\ ^oK$ (for terrestrial lumps) and $N\varepsilon_{recoil} << D$ when $T \cong 10^9\ ^oK$ (inside stars). Although up to now the partial α-decay widths for excited states experimentally are not studied we can expect that the condition $N < 10$ is rather realistic on account of the Geiger and Nutall law, at least for high-energy excited states or for isomeric states. - So, if the values of Γ_γ and $N\varepsilon_{recoil}$ get into such spreads D we can generalize (3b) for multiple γ- absorptions and write (with $Q_\lambda = q$)

$$D(t) = D(t_0)+[P_0(t_o)+P_1(t_0)\ (1-q)][1-\exp(-\Gamma_0(t-t_o)/\hbar)] + qP_1(t_o)\,. \qquad (3c)$$

We underline that the result (3c) is valid at the approximations of (a) infinitely large medium volumes, (b) sufficiently large times $t-t_o$ (which are much larger than mean lifetimes of excited states and mean times of the free flight of γ-quanta inside the medium and also than times of quantum oscillations caused by different interference processes described by applying the Krylov-Fock theorem[7],generalized in [8,9]) and (c) the condition $(\Gamma_0/\Gamma_\alpha^1)N \geq 1$.

And now we shall pass to the analysis of the new relation (3a) or (3c) in comparison with the previously known equation (2a) both of which are basic for the determination of the age of a pattern, *with and without* taking into account the intermediate γ- absorptions inside the sample matter. For the convenience we re-write (2a), taken for the same conditions as (3c), using the same terms as in (3c) :

$$D(t)=D(t_o)+[P_0(t_o)+P_1(t_o)][1-\exp(-\Gamma_0(t-t_o)/\hbar)]. \qquad (2b)$$

From the simple comparison of (2b) and (3c) one can see that for the same $D(t_o),P_0(t_o),P_1(t_o),t_o$ and Γ_0 *(i)* at any moment t the value of $D(t)$ in (3c) is ***larger*** than in (2b) by the quantity $P_1(t_o)\exp(-\Gamma_0(t-t_o)/\hbar)$, *(ii)* the same value $D(t)$ is obtained in (3c) at an ***earlier*** moment t than in (2b) and *(iii)* the larger is the contribution of P_1 into the sum $P_0 + P_1$, the ***earlier*** is the moment t in (3c) in comparison with (2b) at which the same value of $D(t)$ is obtained.

Now we illustrate the inference *(ii)* by the following instance : if $P_0(t_o)=P_1(t_o)= (1/2)P(t_o)$, then the same value of $D(t)$ is obtained for the values of $t=t_{usual}$ in (2c) and $t=t_{real}$ in (3c) which are connected by the following striking relation

$$t_{real} = t_{usual} - (\Gamma_0 / \hbar)\ln 2 \tag{4}$$

(of course we imply that $t_{usual} - t_o > (\Gamma_0 / \hbar)\ln 2$).
- So, one can see that final results of evaluations of all time intervals, characteristic for evolution of chains of decays inside large masses, can be essentially diminished *at practically any temperature above 0° C (273° K)*. It is necessary to underline the following : the importance of considering the thermal motion of decaying nuclei is in no way connected with the possibilities of nucleus excitations but does consist only in the reality of quasi-resonance γ-quantum absorptions when the recoil energy losses do not bring the γ-quantum energy out of the limits of the Doppler width (for γ-decay) caused by the nucleus thermal motion. Therefore, any changes of the temperature during the history of a star or a planet etc do not noticeably influence on the decay velocity if the Doppler width remains to be sufficiently large.

In particular, for example, if the instant $(t-t_o)_{usual} = \tau(1+a)\ln 2$ with a being any small number ($a<<1$), corresponds to a certain abundance of the daughter nuclei $D(t-t_o)$ in equation (2a), then to the same abundance of the daughter nuclei $D(t-t_o)$ in equation (3c) with $q=1$ the instant $(t-t_o)_{real} = \tau a$ (i.e. $<< \tau$) corresponds! In other words, billions years obtained in the usual nuclear-chronometry method can correspond to much smaller number of years (for instance, several thousands years or even any possible small time interval) at the scale of one mean lifetime of the nuclei-chronometers.

References

1. J. Audouze and S. Vauclair, **An Introduction to Nuclear Astrophysics.** ***The Formation and the Evolution of Matter in the Universe***, Reidel Publish.Comp.,1980, Ch.VIII.
2.R.Cayrel, V.Hill, T.Beers, *Nature,* **409**, 661 (2001, 8 Febr.).
3.C.Sneden, *Nature,* **409**, 673 (2001, 8 Febr.).
4.G. B. Dabrymple, ***The Age of the Earth*** (Stanford Univ., Stanford), 1991.
5.V.S.Olkhovsky, V.I.Grantsev, Ukrainsky fizichesky zhurnal, **45**, 1413(2000).
6.V.S.Olkhovsky, Atti dell'Accademia Peloritana dei Pericolanti, Sci.Fis., Mat. e Nat.(Messina), **LXXVII** (1999).
7.N. S. Krylov and V. A. Fock, Zh. Eksp. Teor. Fiz., **17**, 93 (1947).
8.V. S. Olkhovsky, Izv. AN SSSR, ser. fiz., **49**, 938 (1985).
9.V.S.Olkhovsky, N. L. Doroshko and G. A. Prokopets, Izv.AN SSSR,ser.fiz., **54**, 988 (1990).

NEW MAGIC NUMBER, $N = 16$, NEAR THE NEUTRON DRIP-LINE

AKIRA OZAWA
The Institute of Physical and Chemical Research (RIKEN),
Hirosawa 2-1, Wako-shi, Saitama 351-0198, Japan
E-mail: ozawa@rarfaxp.riken.go.jp

The neutron separation energies (S_n) and the interaction cross-sections (σ_I) for the neutron-rich *p-sd* and *sd* shell region have been surveyed in order to search for a new magic number. Very recently, both measurements have reached up to the neutron drip-line, or close to the drip-line, for nuclei of Z • 8. The neutron-number (N) dependence of S_n shows clear breaks at N=16 near to the neutron drip-line, which shows the creation of a new magic number. A neutron number dependence of σ_I shows a large increase of σ_I for N=15, which supports the new magic number. The origin and influence of the new magic number are also discussed.

1 Introduction

The shell structure is one of the very important quantities concerning nuclear structure. Nuclei show a closed shell structure for the following neutron (N) and proton (Z) numbers: 2, 8, 20, 28, 50, 82 and 126. They are called nuclear magic numbers. The appearance of magic numbers has been explained by Mayer and Jensen at 1949, introducing a spin-orbit coupling.

Recent improvements of radioactive beam (RI beam) technology allow measurements of the mass, half-life, size and other properties of unstable nuclei. Furthermore, nuclear reactions using RI beams have been studied very much [1]. Thus, based on the results, some interesting characteristics of unstable nuclei, such as a halo and a skin, have been revealed. In the points of nuclear magic numbers, the disappearance of some traditional magic numbers has been studied both experimentally and theoretically. For example, the disappearance of the N = 20 closed shell was shown in ^{32}Mg experimentally in terms of a low-lying 2^+ level [2] and a large $B(E2;0^+_1\rightarrow 2^+_1)$ value [3]. The low excitation energy of 1^- state and the deduced large $B(E1;0^+_1\rightarrow 1^-_1)$ value in ^{12}Be provide a consistent picture of the N = 8 shell melting in the nucleus [4]. Although the disappearance of magic numbers has been discussed, as shown above, no appearance of a magic number has been shown experimentally so far.

From theoretical points of view, the appearance of new magic numbers in a neutron-rich region has been predicted within the framework of the energy-density formalism [5]. The authors in Ref. [5] predicted the appearance of N = 16, 34 and

58 as neutron magic numbers for unstable nuclei, together with a weakening of the shell closure at $N = 20$ and 28.

2 Search for new magic numbers

Recent improvements of secondary-beam techniques allow us to now measure the mass and the interaction cross-sections (σ_I) for nuclei on the neutron drip-line or close to the drip-line. One of the powerful mass measurements is a direct time-of-flight mass determination. The principle of this technique requires only a determination of the magnetic rigidity and velocity of an ion. That is, for a particle of mass m and charge q traversing through an achromatic system, the magnetic rigidity ($B\rho$) is related to the velocity (v) as $B\rho = mv/q$. Thus, a precise measurement of the rigidity and velocity would allow the mass of an ion to be deduced. Direct mass measurements have reached to the drip-line and near to the drip-line for Z • 8 in a recent mass evaluation [6]. On the other hand, measurements of σ_I can be performed by the transmission method, which have been extensively measured at the projectile fragment separator facility (FRS) at GSI. FRS is the only available RI beam facility for relativistic energies (~1 A GeV). Recently, σ_I measurements have reached to the drip-line for Z • 8, except for ^{22}C [7].

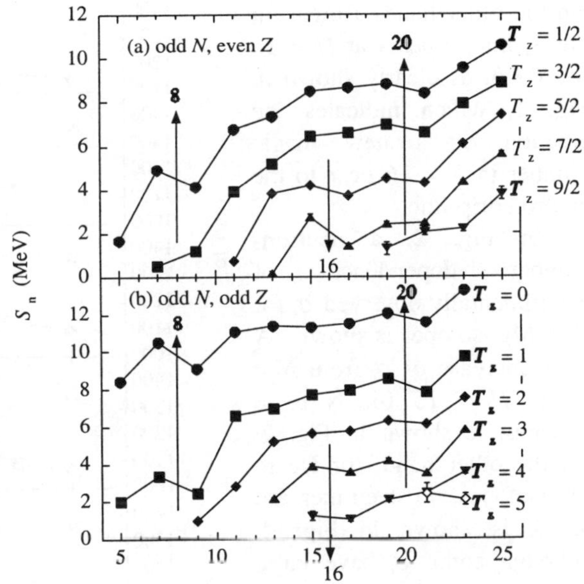

Figure 1. Neutron-number (N) dependence for experimentally observed neutron separation energies (S_n).

The neutron-number dependence of experimentally observed S_n for nuclei with odd N and even Z (odd N and odd Z) is shown in Fig. 1 (a) (Fig. 1 (b)), respectively. S_n is defined by the difference of the binding energies (BE) ($S_n(N,Z) = BE(N,Z) - BE(N\text{-}1,Z)$), where BE are determined by recent mass measurements. In Fig. 1, a magic number appears as a decrease of S_n along with an increase of N [8]. The traditional magic numbers (N=8, 20) are clearly seen close to stable nuclei as breaks in the small isospin numbers (T_z) lines. However, the break at $N = 8$ ($N = 20$) disappears at neutron-rich $T_z = 3/2$ ($T_z = 4$), which is also known concerning other experimental quantities, as discussed above. On the other hand, a break in the S_n line appears at $N = 16$ for T_z• 3, as clearly shown in Fig. 1, which indicates the creation of a new magic number in $N = 16$ near to the neutron drip-line.

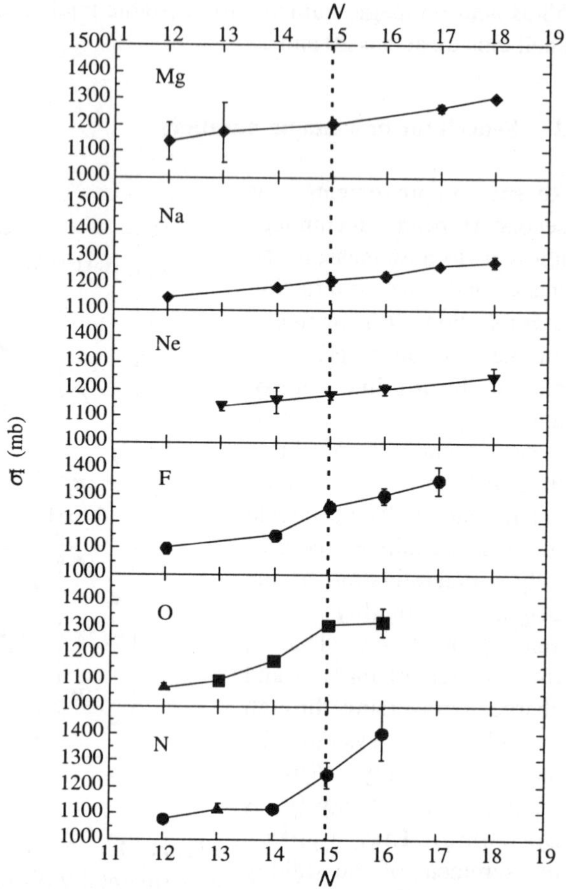

Figure 2. Neutron-number (N) dependence for experimentally observed interaction cross-sections (σ_I) for N to Mg isotopes on C targets.

In Fig. 2, a neutron-number dependence of experimentally observed σ_I for N to Mg isotopes is shown. A steep increase of σ_I from $N = 14$ to $N = 15$ for N to F isotopes, is shown in Fig. 2. On the other hand, for Ne to Mg isotopes, no steep increase of σ_I is shown in Fig. 2, although some σ_I have large error bars. It is noted that a clear difference occurs at $T_z = 3$, which suggests some correlation for the new magic number $N = 16$, as shown in S_n. The recent development of a Glauber model analysis for few-body systems allows one to distinguish the single-particle wave-functions for the valence nucleon ($2s_{1/2}$ or $1d_{5/2}$ orbitals) [7]. The results of an analysis show a dominance of the $2s_{1/2}$ orbital for the valence neutron in ^{22}N, ^{23}O and ^{24}F. On the other hands, in ^{25}Ne, ^{26}Na and ^{27}Mg, the nature of a mixing of $2s_{1/2}$ and $1d_{5/2}$ orbitals is shown. Thus, a valence neutron in the $N = 15$ chain shows the following tendency: the purity of the $2s_{1/2}$ orbital is larger when T_z is larger. This conclusion supports the creation of a new magic number at $N = 16$ near to the neutron drip-line, since a clear single-particle structure is suggested for $N = 15$ nuclei near to the neutron drip-line.

3 The origin and influence of the new magic number

The origin of this new magic number may be due to neutron halo formation. Fig. 3 shows the single-particle orbitals of a neutron in the normal spherical Woods-Saxon potential. The single-particle orbitals were calculated for $A/Z = 3$ nuclei just to show the effect of neutron excess (A is the mass number). Normal shell gaps N=8, 20, 28 are seen in the region for a binding energy of about 6 to 8 MeV, as expected.

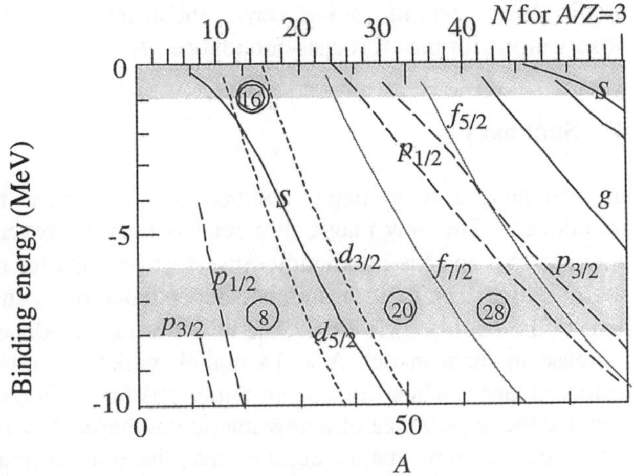

Figure 3. Spectrum of single-neutron orbitals, obtained by the Woods-Saxon potential, for $A/Z = 3$ nuclei.

However, for a weakly bound system in $A/Z = 3$ nuclei, the spacing, and even the ordering, of the orbitals changes. The most pronounced is the s orbital for $N = 9$, 10. In the region of a binding energy below 1 MeV, the $2s_{1/2}$ orbital is below the $1d_{5/2}$ orbital. This fact is clearly observed as an abnormal ground-state spin-parity of ^{15}C (J^{π}=1/2$^+$). It is also observed as a strong contribution of the $2s_{1/2}$ orbital to the formation of a neutron halo in ^{11}Li, ^{11}Be and ^{14}Be (the mixing of $(2s_{1/2})^2$ and $(1p_{1/2})^2$ in the halo wave function). The lowering of the s orbitals is due to halo formation. A neutron halo is formed since the orbital with a low angular momentum gains energy by extending the wave function. This effect is largest in the s orbital and next in the p orbital. The effect for the p orbitals is also seen in Fig. 3.

When the neutron number increases for a weakly bound system, strong mixing between $2s_{1/2}$ and $1d_{5/2}$ orbitals appears. Thus, the energy gap between these two orbitals and $1d_{3/2}$ become much larger. At weakly bound $N = 16$, the $2s_{1/2}$ and $1d_{5/2}$ orbitals are filled by neutrons leaving a large gap to the $1d_{3/2}$ orbital, as shown in Fig. 3. Thus, the mechanism that forms a neutron halo and makes the lowering of low-angular momentum orbitals, for weakly bound neutrons, is also essential for the appearance of the magic number N=16 near to the neutron drip-line. This mechanism may be common in nuclei near to the neutron drip-line. It therefore may destroy or produce other magic numbers in heavier elements, too.

The new magic number may give some influence on the r-process path. One of characteristics of the path is that the r-process goes through neutron magic numbers. Thus, if new magic numbers appear in the heavier neutron-rich region, such as near to N=82 and 126, the r-process path may be modified. Such a modification would solve any inconsistencies between calculations and observations for the r-process abundance [9].

4 Summary

In summary, a new magic number, N = 16, near to the neutron drip-line is introduced. The new magic number has been surveyed by the neutron separation energies (S_n) and the interaction cross-sections (σ_I) for the neutron-rich *p-sd* and *sd* shell region. The neutron-number dependence of S_n shows clear breaks at N = 16 near to the neutron drip-line. The neutron-number dependence of σ_I shows a large increase in neutron-rich N = 15 nuclei, which shows that the purity of the $2s_{1/2}$ orbital is larger when the isospin number is larger in our analysis. These two facts indicate the appearance of a new magic number at N = 16 near to the drip-line. The origin of this new magic number may be due to neutron halo formation. The mechanism to form a new magic number may be common in nuclei near to the neutron drip-line. Other magic numbers in heavier elements may be produced. The new magic number may change the possible r-process path.

References

1. I.Tanihata, Experimental Techniques in Nuclear Physics, p.343 (Walter de Gruyter, 1997).
2. D.Guillemaud-Mueller et al., Nucl. Phys. A 426, 37 (1984).
3. T.Motobayashi et al., Phys. Lett. B 346, 9 (1995).
4. H.Iwasaki et al., Phys. Lett. B 491, 8 (2000).
5. M.Beiner et al., Nucl. Phys. A 249, 1 (1975).
6. G.Audi and A.H.Wapstra, Nucl. Phys. A 595, 409 (1995).
7. A.Ozawa et al., Nucl. Phys. A in press.
8. A.Bohr and B.R.Mottelson, Nuclear Structure vol. 1, p.192 (W.A.Benjaming, 1969).
9. S.E.Woosley et al., Astrophys. J. 433, 229 (1994).

CONSTRAINED MOLECULAR DYNAMICS APPROACH TO FERMIONIC SYSTEMS

M.PAPA

Istituto Nazionale Fisica Nucleare-Sezione di Catania del Sud, Via Santa Sofia 44, 95123 Catania, Italy
e-mail: papa@lns.infn.it

T. MARUYAMA

Advanced Science Research Center, JAERI, Tokai, Ibaraki, 319-1195, Japan
e-mail: maru@hadron02.tokai.jaeri.go.jp

A. BONASERA

Istituto Nazionale Fisica Nucleare-Sezione di Catania, Corso Italia 57, 95129 Italy
e-mail: bonasera@lns.infn.it

We propose a Constraint Molecular Dynamics model for Fermionic system. In this approach the equations of the nuclear many-body problem are solved by imposing that the one-body occupation number $\overline{f}_i$, evaluated for each particle, can assume only values less or equal to 1. Some comparison to data is given.

1 Introduction

Heavy ion collisions in the medium energy region have been described in a large variety of semiclassical approaches to the many-body problem. As it is well known the correlations treated in the one-body approach are not able to describe the large fluctuations which develops in a multifragmentation process. This difficulty can be solved by adopting more suitable treatments of the N-body problem like molecular dynamics. One of the first molecular dynamics model is the so called quantum molecular dynamics (QMD) In this model there are no condition reflecting the Fermionic nature of the nuclear problem. The Fermionic nature of the nuclear problem has been considered in the Fermionic molecular dynamics and more recently in the antisymmetrized molecular dynamics (AMD) For these cases the wave function of the system is expressed as a single Slater determinant of N wave packets preserving the Fermionic nature of the system, but the CPU time necessary to work out calculations for systems with total mass around 200 is very large for practical studies. In the present work we propose a new molecular dynamics model: the Constraint Molecular Dynamics (CoMD), which aims to overcome the above mentioned limitations.

2 The model

In the present approach [1] each nucleon state is represented by a Gaussian wave-function ϕ_i of width σ_r. The total wave-function is assumed to be a direct product of these wave-functions. The N-body distribution function is the direct product of the single particle distribution functions f_i. The f_i are obtained by the Wigner transform of the wave-functions ϕ_i. In this paper we take the dispersion in momentum σ_p as a parameter as well as that in coordinate space. Therefore the distribution functions of each nucleon in phase-space is expressed through the product of two gaussian in r and p space with centers $\langle \mathbf{r}_i \rangle$ and $\langle \mathbf{p}_i \rangle$ and widths $\sigma_{r,p}$ The equations of motion of $\langle \mathbf{r}_i \rangle$ and $\langle \mathbf{p}_i \rangle$ are derived using the time-dependent variational principle which gives the usual Hamiltonian equations. The total energy H for A particles with mass m consists of the kinetic energy, the two-body and three- body effective interactions. The term $T_0 = \frac{3\sigma_p^2 A}{m}$ arising from the gaussian width in p-space will be omitted in our future considerations on the binding energies and on the nucleus-nucleus collision. This is a crucial assumption. In fact, if one keep such a term in trying to reproduce the right binding energies part or all of the Fermi motion will come from it. Concerning the effective interaction we use a simple local Skyrme(Sk0) parameterization with a 200 MeV compressibility as described in many review articles [2]. Moreover a Weizsacker symmetry term and a surface one are included [1].

The equations of motion for the nucleons have been solved using a fourth-order Runge-Kutta method coupled with two numerical algorithms which aim to take into account the effects of the residual interaction and the Fermionic nature of the many-body problem under study. One consists of the usual two-body nucleon-nucleon scattering. The other algorithm constraints at each time step for all the nucleons i the following quantities:

$$\overline{f}_i \equiv \sum_j \delta_{\tau_i,\tau_j}\delta_{s_i,s_j} \int_{h^3} f_j(\mathbf{r},\mathbf{p};\langle \mathbf{r}_j \rangle, \langle \mathbf{p}_j \rangle)\, d^3r\, d^3p \leq 1 \qquad (1)$$

The ground state configuration of the nuclei is obtained using a coupling between a cooling/warming procedure with the constraint. In such a way it has been possible, at variance with the others molecular dynamical models, to obtain stable states (up to 2000 fm/c) in which the effective kinetic energy of the nucleons is near the value predicted by the Thomas Fermi Model. The free parameters affecting the effective average potential σ_r, σ_p and the surface term have been chosen to reproduce in a wide range of masses ($A \simeq 30 - 208$) the nuclear radii and the binding energy [1].

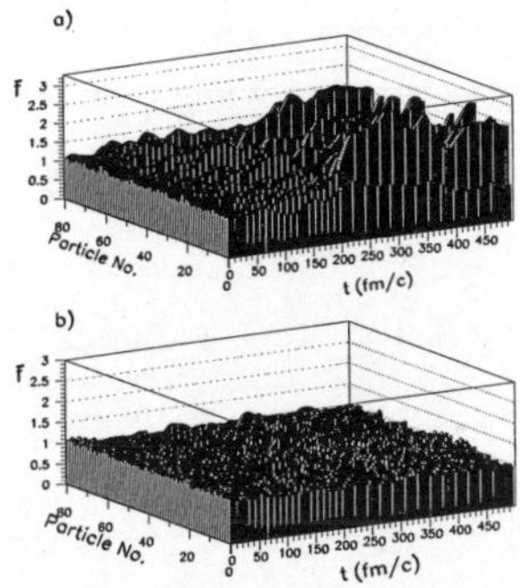

Figure 1: Occupation probability as function of the particle index for $^{40}Ca +^{40} Ca$ at 35 MeV/A central collision evaluated with the QMD a) am CoMD b) approaches respectively.

2.1 The Nucleus-Nucleus Collision

In Fig. 1 we display the time dependence of $\overline{f}_i$ for all particle index i in a central collision $^{40}Ca +^{40} Ca$ at 35 MeV/A. The two panel a) and b) show the results of QMD and CoMD models respectively, obtained with exactly the same initial condition. The difference between the two cases is quite clear. In the QMD case after a short time (about 50 fm/c) the values of the $\overline{f}_i$ for most of the particles start to be greater than 1. In the CoMD case, due to the constraint, the phase space occupation $\overline{f}_i$ is on the average less than 1 at each time step. We note a good uniformity of $\overline{f}_i$ as function of the particle index i and time t. These results obviously will also produce an increase of the collision rate with respect the QMD case.

With this ingredient the model has been successfully tested in to reproduce the charge distributions for ^{40}Ca+^{40}Ca and ^{197}Au+^{197}Au at 35 MeV/A[1]. The observed agreement is good. In particular we observe that QMD calculations fails in to reproduce the multifragmentation of the system producing on the average a binary process and an α particle rate which is a factor ten smaller than the CoMD case. In the CoMD model in fact, as due to the Pauli blocking correlations, α particles manifests as preferred clustered configuration. Finally in Fig.2 we show the comparison between the average isotope distribution measured by the REVERSE apparatus for $^{124}Sn +^{64} Ni$ and $^{112}Sn +^{58} Ni$[3] and the CoMD calculations including the experimental filter. The agreement is quite good showing slight memory effect in the average ratio $\frac{N}{Z}$.

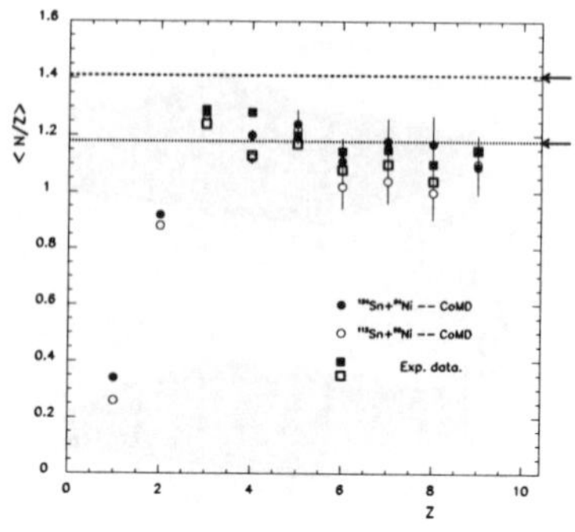

Figure 2: Comparison between experimental and calculated average isotope distributions for the systems $^{124}Sn +^{64} Ni$ and $^{112}Sn +^{58} Ni$[3] at 35 MeV/A. The filtering effect of the experimental apparatus has been taken into account.

3 Conclusion

The CoMD model proposed in the present work is able to reproduce with the same set of parameters both the main characteristic of stable nuclei in a wide region of mass ($A = 30 \sim 208$) and the charge and average isotope experimental distributions produced in the collision of different systems: ^{40}Ca$+^{40}$ Ca, $Au + Au$ and $Sn + Ni$ at $E_{\rm lab} = 35$ MeV/nucleon. Calculations on more exclusive data are in progress. The constraint, introduced to describe the nuclear many-body problem, affects the dynamics of the nucleus-nucleus collision for two main reasons:
a) the nucleon-nucleon collision rate is higher with respect to the QMD case;
b) the constraint for low momentum particles produces on the average a non local repulsion effect.
Both these effects determine the disassembly of the highly excited intermediate systems formed at the beginning of a nuclear collision. Finally we note that the typical CPU time needed to follow the time evolution of systems of mass number around 80 for 300 fm/c is quite short: about 10 sec in a 600 MHz Unix machine.

1. M. Papa, T. Maruyama, A. Bonasera, Phys,ReV.C 64 (2001) and references there in.
2. A. Bonasera et al,
3. see S.Pirrone et al, contribution to this conference.

EXPERIMENTS ON SUPER-HEAVY NUCLEI AT GANIL

J. PÉTER [1] AND THE FULIS COLLABORATION: N. ALAMANOS [3], N. AMAR [1], J.C. ANGÉLIQUE [1], R. ANNE [2], G. AUGER [2], F. BECKER [2], R. DAYRAS [3], A. DROUART [3], J.M. FONTBONNE [1], A. GILLIBERT [3], S. GRÉVY [1], D. GUERREAU [2], F. HANAPPE [1], R. HUE [2], A.S. LALLEMAN [2], N. LECESNE [2], T. LEGOU [1], M. LEWITOWICZ [2], R. LICHTENTHÄLER [4], E. LIÉNARD [1], W. MITTIG [2], F. DE OLIVEIRA [2], N. ORR [1], G. POLITI [5], Z. SOSIN [6], M.G. SAINT-LAURENT [2], J.C. STECKMEYER 1, C. STODEL [2], J. TILLIER [1], R. DE TOURREIL [2], A.C.C. VILLARI [2], J.P. WIELECZKO [2], A. WIELOCH [6]

1 : LPC Caen, France, 2 : GANIL, Caen France, 3 : DAPNIA/SPhN, C.E.N. Saclay, France 4 :I.F.U.S.P., São Paulo, Brazil 5 : Univ. di Catania, Italy 6 : Ins. Fyziki Uniw., Krakow, Poland

1 Introduction

GANIL has several assets which lead us to start a programme on the study of super-heavy nuclei. First, a high beam intensity. Since the cross sections are very low, it is necessary to use beam intensities as large as can stand the target, at least in the microAmpere-particle range (6.27 10^{12} projectiles/s), which are produced by ECR sources. Second, a powerful velocity filter: the Wien filter LISE3.

The experimental set-up is described and the performances reached on known isotopes are given in the second and third parts. Short and long range plans for future experiments are described in the following parts: use of $^{76}Ge_{32}$ beams and, mostly, use of Pb beams to benefit from the assets of inverse kinematics.

2 Experimental set-up.

A complete set-up has been built and tested (see figure). The main tool is the Wien filter with crossed magnetic and electric fields. It is divided into two identical halves and followed by a dipole magnet. It had been designed for the purification of exotic secondary beams produced by fragmentation in the LISE magnetic device and several modifications were found to be necessary for the study of complete fusion evaporation residues (ER's) and the rejection of the primary beam : - a reaction chamber was added just before LISE3, - the upper plate in the first half of the filter was moved up, the beam is thus deflected after this first half without hitting the plate and stopped on a water-cooled plate - two pairs of independently movable slits

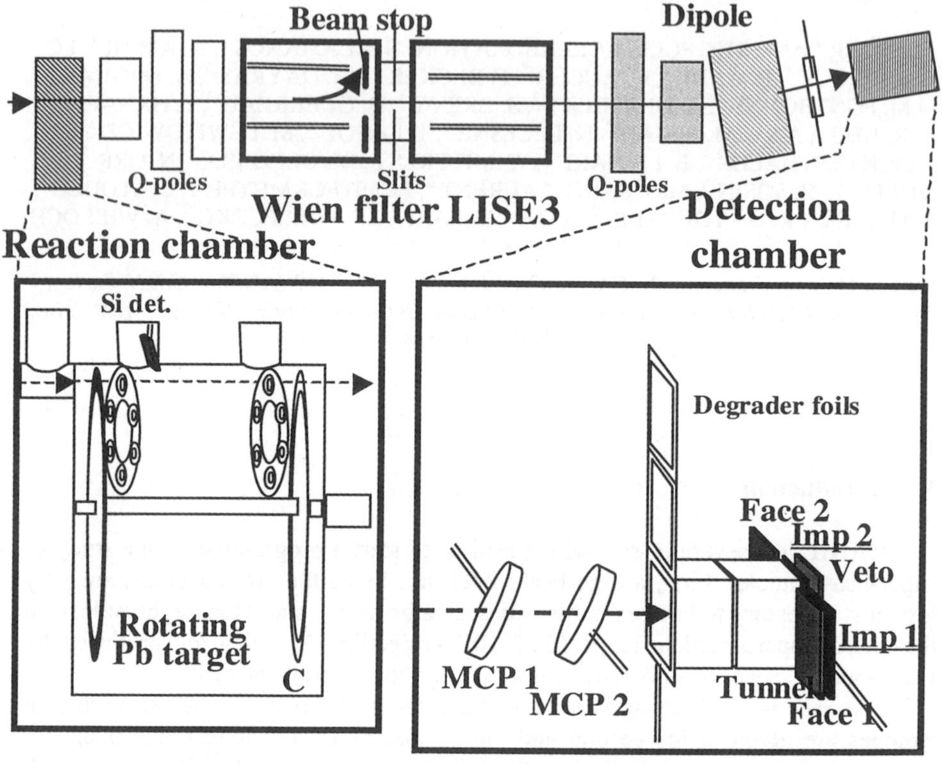

Figure 1 –Experimental set-up

and a beam profiler were installed at mid-filter. The suppression of unwanted products is improved by a dipole magnet located after the velocity filter.

The reaction chamber contains a wheel bearing 35 targets on a diameter of 670 mm and rotating at 2000 RPM, allowing targets with a low melting temperature (Pb, Bi) to sustain intense beams (1 particleμA= 6.25 10^{12} particles/s, or more). The time structure of the beam is synchronised with the rotation. A Si detector continuously monitors the status of each target. Smaller rotating targets are used for targets having a higher melting temperature.

The velocity of each product is obtained with 2 aluminized mylar foils and micro-channel plate detectors; its kinetic energy and localisation are given by a X-Y Si implantation detector (Imp). The energy of α–particles and fission fragments escaping from Imp is measured with a "tunnel" of 8 Si detectors. A Si veto detector is installed behind Imp to reject light particles which might punch through it. When emission of α–particles or fission fragment in the Imp is identified via electronics, the beam is immediately stopped for a short while, during which Imp1 is moved out and Imp2 comes in; then the beam is sent again. α–particles escaping from Imp1 are then detected by the Si detector Face1. It is thus possible to study α–emission or spontaneous fission on long durations without any background. Specific electronics and data acquisition systems were developed. A fast analysis program allows us to identify α–decay chains or spontaneous fission on line.

The transmission of fusion nuclei from the target position through the quadrupoles, filter and final dipole was studied via a simulation code. The optics was checked and the Wien filter was calibrated using low energy (0.35 MeV/u) ions. The response of the whole set-up was checked via fusion reactions with known cross sections and α decay chains: $^{204\text{-}206}$Fr nuclei formed via ^{86}Kr + ^{nat}Sb fusion reactions with cross sections of 10 to 300 μb [Sah85], $^{260,261}Sg_{106}$ nuclei formed via ^{54}Cr + ^{208}Pb with cross sections of 300-500 pb [HofMün].

3 Experiments.

3.1 Studies with $^{260,261}Sg_{106}$

With the ^{208}Pb targets and a ^{54}Cr beam of 40nAp, the silicon implantation detectors measured a counting rate of 5-10 Hz. 2 different energies were used : 4.698 and 4.756 MeV/u. The excitation energies correspond mostly to the 1 neutron evaporation channel ($^{261}Sg_{105}$, cross section peaking at 500 pb) and to the tail of the 2 n excitation function [HofMün].

With the on-line analysis, we indeed observed 2 decay chains of ^{261}Sg at 4.698 MeV/u whereas at 4.756 MeV/u we observed 9 chains with the characteristics (lifetimes and α energies) attributed to ^{261}Sg and one attributed to ^{260}Sg. The decay chains were identified by position correlation in the Si detector between the implantation of the evaporation residue and the decay by α emission or fission A transmission efficiency above 60 % and a suppression factor of the primary beam of $2*10^{10}$ were achieved.

3.2 Search for element 118.

In 1999 ^{86}Kr on ^{208}Pb studied at Berkeley produced 3 events with a chain of 6 α's which were attributed to element 118 and its descendants, with a cross section ~2.2 pb [Nin99]. The same system studied with the velocity filter SHIP in Darmstadt did not lead to the observation of any event attributable to element 118, with an upper limit of 1 pb [Hof99]. Since Ganil is able to deliver a high intensity beam of ^{86}Kr (15 μA with charge 10+ from CSS1, i.e. nearly 10^{13} projectiles/s) it was decided to try to obtain additional information: i) possible α decay before the minimum time of 120 μs in the Berkeley experiment. This required a faster acquisition system. ii) detection of α emitted more than 1 mn after implantation of the ER: the movable implantation detector described in the previous section was installed. iii) ER's may be created in an isomeric state which decays via electron capture followed by an electron cascade with an unknown half-life, modifying the ionic charge and strongly reducing the transmission. The carbon stripper foil which re-establishes the ionic charge distribution was located 3 times further, i.e. at larger times, than in the SHIP experiment.

With a total dose of 1.1×10^{18} ions at 5.27 MeV/u no event attributable to a super-heavy nucleus was observed. A similar result was obtained also at RIKEN [Mor00].

4 Plans with $^{76}Ge_{32}$ Beams.

Cold fusion reactions were used to produce elements 102 to 112 with ^{208}Pb or ^{209}Bi targets and the heaviest isotopes of projectiles ranging from ^{48}Ca to ^{70}Zn [HofMün]. ^{76}Ge projectiles fusing with the heaviest isotopes of targets ranging from Os_{76} to Tl_{81} offer the advantage of forming compound nuclei ranging from Z= 108 to 113 heavier by 2 neutrons than the compound nuclei formed with ^{208}Pb or ^{209}Bi (except for 111). A similar possibility is offered by $^{82}Se_{34}$ projectiles [Sto00]. A practical advantage is that only one beam development is needed.

The difficult point is to estimate which cross sections will be obtained. Compared to ^{208}Pb - ^{209}Bi induced reactions, the entrance channel is more symmetric and there is no closed shell in the partners. Both points are known to reduce the fusion cross section. An estimate of the reduction factors can be made using the ER cross sections observed for the compound nucleus ^{220}Th formed via different entrance channels having similar Bass barriers, as shown in figure 2 [Sah84, Sah85, Ver84, Sto98]. At the same excitation energy, 45 MeV, the effect of the deexcitation step is expected to be the same and the difference in the xn cross sections is attributed to entrance channel characteristics.

Ar_{18}+$Hf_{72}$and Zn_{30}+Nd_{60} systems have no closed shell in any partner. Even though it has a smaller Bass barrier, the more symmetric system Zn+Nd has a cross section smaller by a factor 14 than Ar+Hf. Assuming a steady variation of the cross section with the charge asymmetry, the expected cross sections for Ca_{20}+Yt_{70} and Zr_{40}+Sn_{50} are estimated. The actual cross sections are much higher, by factors 6 and 23 respectively, due to the role of the closed shells [Mol97]: 20n and 28p in $^{48}Ca_{20}$, 50p in Sn_{50}.

These factors may now be used to estimate the reduction factor which apply when the same compound nucleus (and ER) is formed via ^{76}Ge+...instead of ^{208}Pb-

^{209}Bi+...since the excitation energies at the Bass barrier are very close. The loss of shell effects is estimated to reduce the cross section by an order of magnitude. At Z=106 the decrease of charge (and mass) asymmetry reduces the cross section by a factor about 12. This asymmetry disappears for ^{76}Ge+^{208}Pb reactions (element 114).

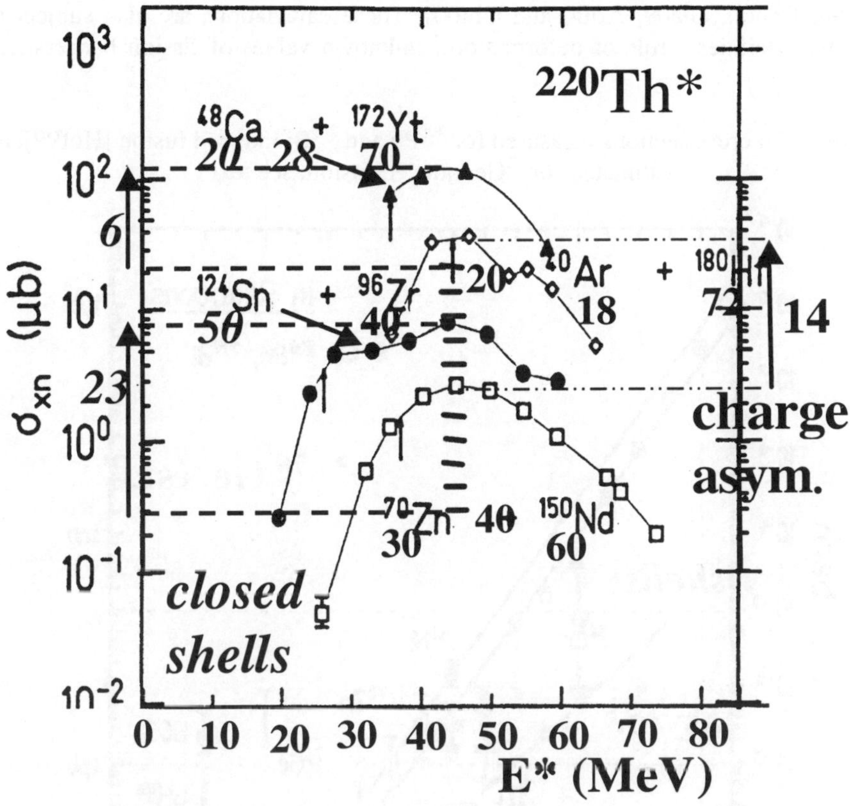

Figure 2 – Excitation functions of evaporation residues (xn) of ^{220}Th compound nuclei formed via different entrance channels [Sah84,Sah85,Ver84,Sto98].

Therefore, the dashed line in figure 3 represents the cross sections which would be obtained for Ge induced reactions leading to the same 1n ER's as ^{208}Pb-^{209}Bi induced reactions. For Z=106, 107 and 111, this is the case. For Z=108,109,110,112,113, compound nuclei may be formed with 2 additional neutrons. Detailed measurements made for Th isotopes show that 2 more neutrons in the compound nucleus increase the residue cross section by a factor 9 [Sto98]. As seen in fig. 3, for Z=110 the use of ^{64}Ni instead of ^{62}Ni on ^{208}Pb results in an increase by a factor ~5. An average factor 7 is used in figure 3.

The cross sections thus estimated are smaller than ^{208}Pb-^{209}Bi induced reactions. ^{84}Se induced reactions make more symmetric entrance channels and therefore smaller cross sections. Of course these empirical estimates are subject to large uncertainties: the reduction factors may be larger or smaller, the role of target deformation is not taken into account… Theoretical predictions were asked to authors who reproduced the cross sections of ^{208}Pb and ^{209}Bi induced reactions: [Ada98, Den00, Gia00, Ari00 and Oht00]. These calculations are also subject to large uncertainties (role of deformation, unknown values of fission barriers,…), but

Figure 3 : 1n cross sections measured for ^{208}Pb and ^{209}Bi induced fusion [Hof99]and estimated for ^{76}Ge induced fusion: see text

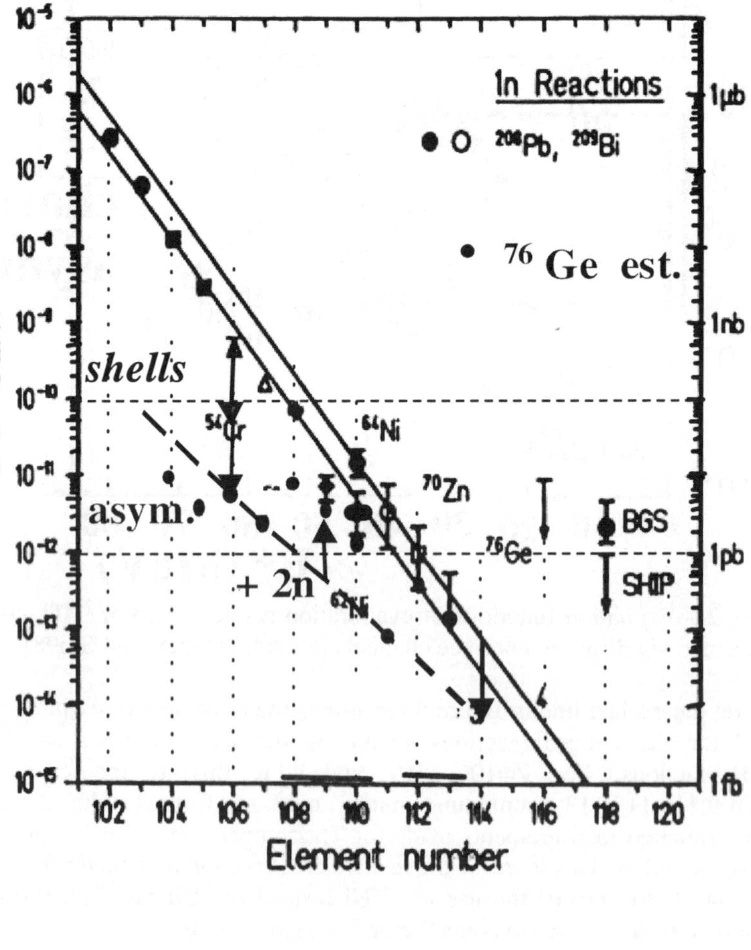

they confirm that ^{82}Se projectile is less favorable than ^{76}Ge and that, even with ^{76}Ge, the cross sections are expected to be well below those of ^{208}Pb-^{209}Bi induced reactions [private communications].

A specific interest of these reactions [Sto00] is the direct cold fusion production of ^{273}110 (^{76}Ge+^{198}Pt =^{273}110 + 1n) in order to establish the production of element 112. Indeed different α–decay energies and lifetimes had been observed .for the few events of ^{273}110 produced in the decay of ^{277}112 obtained via cold fusion (^{70}Zn+^{208}Pb) [HofMun.] and in direct hot fusion production of this isotope: ^{34}S+^{244}Pu=^{273}110+5n [Laz95]. At this conference, S. Hofmann showed that the decay chains of ^{277}112 are now understood. The α spectroscopy of ^{273}110 is of interest since the population of these different levels (especially isomeric states) may be due to overreaching the deformed shell N=162, but it would require a "large" number of events, which would need a very long beam time with cross sections in the pb range.

5 Assets of inverse kinematics.

The use of inverse kinematics has an obvious drawback: the velocity difference between the beam and the evaporation residues is much smaller and larger magnetic and electric fields are necessary in a velocity filter. LISE3 is powerful enough to obtain the same separation at mid-filter (fig. 1). Another difficulty is the larger energy deposit in the target and the stripper foil. The targets have higher melting temperatures than Pb or Bi., but mechanical effects (micro-holes) are more important.

There are however several assets relative to usual kinematics. The first gains concern the beam time:

- the target thickness is no longer limited by multiple scattering of the ER's in the target. Therefore, instead of covering only 3-4 MeV in excitation energy, a target thick enough to cover the whole excitation function of interest may be used: 8 to 12 MeV, thereby maximising the counting rate. The most important application is the search for an unknown isotope. Indeed different mass defects in different mass tables lead to incident energies differing by several MeV whereas the excitation function is rather narrow (a few MeV). In usual kinematics, lengthy measurements at several incident energies must be made, but only one in inverse kinematics.
- the strong forward focussing of ER's ensures a better transmission, even after α–decay in flight.
- the ionic charge distribution of ER's is better estimated and its relative width is smaller, improving the transmission..

The other gains concern the quality of the data:

- the energy deposited by an implanted ER is large and measured with a much better accuracy so that it is able to provide, combined with the time-of-flight, an estimate of the mass. This is valuable additional information

- due to the deeper penetration of ER's in the implantation detector, both spontaneous fission fragments are stopped in this detector and the fission total kinetic energy is easily measured. α-particles emitted backwards leave at least 2 MeV in this detector and cannot be missed.

6 : Plans with Pb beams.

This method is suitable for several aspects of super-heavy nuclei studies and can be applied to all systems. We will first use it to the reactions which were found to have the largest cross sections up to Z=112: cold fusion of ^{208}Pb with neutron-rich isotopes [HofMün]. In order to get new information on super-heavy nuclei, several systems, listed in the table below, can be studied for different purposes:

		208Pb +...			207Pb +...		
	Z	Tget	ACN	E*Bass	Tget	ACN	E*Bass
Rf	104	50Ti	258	23	50Ti	257	23
Db	105	**51V**	**259**	**25**	**51V**	**258**	**25**
Sg	106	54Cr	262	22	54Cr	261	22
Bh	107	**55Mn**	**263**	**24**	**55Mn**	**262**	**24**
Hs	108	58Fe	266	21	58Fe	265	20
Mt	109	**59Co**	**267**	**23**	**59Co**	**266**	**22**
	110	64Ni	272	17	64Ni	271	16
	111	**65Cu**	**273**	**19**	**65Cu**	**272**	**18**
	112	70Zn	278	12	**70Zn**	**277**	**12**
	113	**71Ga**	**279**	**13**	**71Ga**	**278**	**13**
	114	**76Ge**	**284**	**9**	**76Ge**	**283**	**9**
	115	**75As**	**283**	**13**	**75As**	**282**	**13**
	116	**82Se**	**290**	**3**	**82Se**	**289**	**2**
	117	**81Br**	**289**	**7**	**81Br**	**288**	**6**

bold letters: new reactions

new paths — new isotopes

1 - **New paths to known isotopes with ^{208}Pb**. The same compound nuclei and ER's with odd Z formed at GSI with a ^{209}Bi target and an even Z projectile can be formed with a ^{208}Pb beam and a target with an odd Z. Two points are of interest: 1- to measure the cross section and to compare it to the known cross section with ^{209} Bi. That will give a clue to the respective roles of macroscopic and structure

effects on the ER cross section: charge asymmetry of the entrance channel and closed shells in the partners. Indeed fusion with ^{208}Pb is less favoured due to the larger Coulomb repulsion and excitation energy at the interaction barrier, but this may be partially compensated by its exactly filled shells.
2- to reach different levels in the ER, especially isomeric states indicated by an unusually large value of alpha-particle energy.

2 - **Studies of Spontaneous Fission and α energies**. For elements ranging from Fm_{100} to Rf_{104} the mass distributions are known to vary rapidly from symmetric to asymmetric shapes as a function of Z and A of the fissioning nucleus. The total kinetic energy (TKE) distributions also exhibit different shapes [DHof95] and can be measured. In direct kinematics most events leave a part of their energy in the entrance windows of the implantation detector and one of the tunnel detectors, and the TKE of each event is affected by a different loss. As noted above, in inverse kinematics the whole TKE is measured in the implantation detector. Such studies will be made first on nuclei with a large spontaneous fission probability and cross sections in the hundreds of pb range: $^{260}Sg_{106}$ (about 50 % of s. f.) and its α decay product $^{256}Rf_{104}$.

3 - **New isotopes with ^{207}Pb**. Isotopes lighter by 1 neutron than the known isotopes of several elements can be produced with ^{207}Pb instead of ^{208}Pb: with ^{55}Mn the 2n channel leads to ^{260}Bh107 - ^{59}Co, 1n : ^{265}Mt109 - ^{65}Cu, 1n : ^{271}111 - ^{70}Zn, 1n : ^{276}112. For these last two cases, a chain of 3 α 's will be observed before reaching a known isotope, i.e. 3 new isotopes will be observed in one event.

^{276}112 offers the special interest of being an even-even nucleus. Its α decay and spontaneous fission are unhindered and can be more easily compared with theoretical predictions than nuclei with unpaired nucleon(s). In addition its level scheme is expected to be simpler.

4 - **New elements.**Above Z=109, the counting rate is very low. For ^{277}112 measurements were made at one beam energy only, which is not necessarily at the maximum of the excitation function. Indeed the 1n excitation function seems to have its maximum shifted to smaller values with Z (due to the decrease in the fission barrier height) and also to become narrower [HofMün]. The broad energy coverage provided by inverse kinematics may reveal a larger cross section value.

The 1n channel cross section decreases quickly with Z. It may be that the 0n cross section (radiative capture, i.e. compound nuclei formed at excitation energies smaller than the neutron separation energy Sn, 7-8 MeV) becomes larger than the 1n cross section, even though the fusion probability drops very much at the corresponding incident energy. Added to the uncertainty on the compound nucleus mass defect, the exploration of the broad range of incident energies corresponding to 0n + 1n excitation functions leads to lengthy measurements at several incident energies in direct kinematics : 6 beam energies were tried for Sc+Pb in a search for element 116 [HofMün]. With a Pb beam, an excitation energy range of 12 MeV or more can be covered with a single incident energy.

7 References

Ada98: G. G. Adamian et al., Nuc. Phys A 633, 409 (1998)
Ari00: Y. Aritomo et al., Proc. Int. Workshop Fusion, Dubna, May 2000, World Sci.
Den00 : V. Yu. Denisov and S. Hofmann, Phys. Rev. C 61,034606 (2000)
Dhof95 : D.C. Hoffman and M. R. Lane, Rad. Acta **70/71**, 135 (1995).
Gia00: G. Giardina et al., Eur. Phys. J. A 8, 205 (2000)
Hof99 : S. Hofmann, Proc. Int. Conf. Sevilla (1999)
HofMün : S. Hofmann, G. Münzenberg Rev. Mod. Phys. 2000)
Möl97 : P. Möller et al, Z. Phys. A 359, 251 (1997)
Nin99 : V. Ninov et al., Phys. Rev. Lett. **83**, 104 (1999)
Oht00 : M. Ohta, Proc. Int. Workshop Fusion, Dubna, May 2000, World Sci.
Dhof95 : D.C. Hoffman and M. R. Lane, Rad. Acta **70/71**, 135 (1995)
Mor00: K. Morita, comm. at Int. Workshop Fusion, Dubna, May 2000
Mor91 : W. Morawek et al. Z. Phys. **A341** (1991) 75 and report GSI-91-26
Sah85 : C.C. Sahm et al., Nuc. Phys. A441,316 (1985)
Sto98 : C. Stodel, PhD thesis, LPC Caen (1998)
Sto00 : C. Stodel et al., Tours Symposium on Nuclear Physics, September 2000

ISOSPIN PHYSICS WITH THE REVERSE EXPERIMENT

S.PIRRONE FOR REVERSE COLLABORATION§

INFN-Sezione di Catania Italy - E-mail:pirrone@ct.infn.it

§M. ALDERIGHI(1), A. ANZALONE(2), R. BARNA'(3), V. BARAN (2), I. BERCEANU (4), J.BLICHARSKA(10), A. BONASERA(2), B. BORDERIE(5), R. BOUGAULT(6), M. BRUNO(7), G. CARDELLA(8), S.CAVALLARO(2), A. CHBIHI(17), M. COLONNA(2), M. D'AGOSTINO(7), R. DAYRAS(9), E. DE FILIPPO(8), D. DE PASQUALE(3), M. DI TORO(2), E. GERACI(2), F. GIUSTOLISI(2), A. GRZESZCZUK(10), P. GUAZZONI(11), D. GUINET(12), M. IACONO MANNO(2), A. ITALIANO(3), S. KOWALSKI(10), A.LANCHAIS(7), G. LANZANO'(8), G. LANZALONE(8,5), N. LE NEINDRE(6,7), S. LI(13), S. LO NIGRO(8), C. MAIOLINO(2), Z. MAJKA(18), T. PADUSZYNSKI(10), A. PAGANO(8), M. PAPA(8), M. PETROVICI(4), E. PIASECKI(14), S. PIRRONE(8), G. POLITI(8), A. POP(4), F. PORTO(2), M. F. RIVET(5), E. ROSATO(15), S. SAMBATARO(8,*), G. SECHI(1), V. SIMION(4), M.L. SPERDUTO(2), J. C. STECKMEYER(6), C. SUTERA(8), A. TRIFIRO'(3), M.TRIMARCHI(3), G. VANNINI(7), M. VIGILANTE(15), J.P.WIELECZKO(17), J. WILCZYNSKI(16), H. WU(13), Z. XIAO(13), L. ZETTA(11), W. ZIPPER(10)

*1)Istituto Fisica Cosmica, CNR and INFN Milano 2)LNS and Dipartimento di Fisica Universita' di Catania 3)INFN and Dipartimento di Fisica Universita' di Messina 4)Institute for Physics and Nuclear Engineering, Bucharest, Romania 5)IPN, IN2P3-CNRS and Universita' Paris-sud, Orsay, France 6)LPC,ISMRA University of Caen,France 7)INFN and Dipartimento di Fisica Universita' di Bologna 8)INFN and Dipartimento di Fisica Universita' di Catania 9)DAPNIA-SPhN,CEA Saclay, France 10)University of Silesia, Katowice, Poland 11)INFN and Dipartimento di Fisica Universita' di Milano 12)IPN, IN2P3-CNRS and Université Claude Bernard, Lyon, France 13)Institute of Modern Physics Lanzhou, China 14)Institute of Experimental Physics, University of Warsaw, Poland 15)INFN and Dipartimento di Fisica Università di Napoli1 6)Institute of Nuclear Studies Otwock-Swierk, Poland 17)GANIL, CEA, IN2P3-CNRS,Caen,France 18)Institute of Physics, Jagellonian University, Krakow, Poland. *)Deceased on January 27th,2001*

Heavy ion collisions at intermediate energies offer the possibility to investigate the role of the isospin degree of freedom in determining the evolution of the reaction mechanism in an energy domain dominated by the competition between one body and two body nuclear dynamics. The REVERSE experiment has been carried on the period September 1999 – May 2000, at the Laboratory Nazionali del Sud in Catania, with the forward part of the 4π multidetector CHIMERA. This part of the device consists 688 Si-CsI (Tl) telescopes arranged in nine ring and has been used to study the inverse reactions $^{112,124}Sn+^{58,64}Ni$, ^{27}Al, at E_{beam} 35 MeV/A. Much care was devoted in calibrating the detectors in energy, charge and mass for the proper identification of light heavy nuclei and charged particles was obtained. The first physical results are presented.

1 Introduction

The REVERSE experiment is the first experiment performed with the 4π multidetector CHIMERA at the Laboratori Nazionali del Sud (LNS) in Catania. It represents part of a large experimental campaign suggested for 2000-2003 and it is in the framework of heavy-ion collisions at intermediate energy, (10 MeV/A ≤ E/A ≤100 MeV/A). This energetic range has been subject of intense experimental and theoretical activities [1] during the last decades. In particular, many efforts have been devoted to the study of the properties of the hot and compressed nuclei produced in the multifragmentation reactions.

More recently, this research field was enriched by the observation that the isospin degree of freedom plays an important role in the dynamical evolution of the two interacting nuclei system[2]. To search experimental signals on phase transition, on isospin dependence of nuclear multifragmentation and on the symmetry term of the nuclear matter equation of state, we realised the reactions $^{112}Sn+^{58}Ni$ and $^{124}Sn+ ^{64}Ni$, using beams delivered at E_{beam}= 35 MeV/A by the cyclotron at LNS.

2 Experimental system and identification procedures

The CHIMERA detector, fig.1, is characterised by a large coverage (94%) of the total solid angle, by a very high granularity (1192 detection modules) and by a very low detection threshold for heavy ions (< 0.5 MeV/A) and light charged particles (≈ 1 MeV/A). The basic module is a telescope formed by a 300μm silicon detector and a CsI(Tl) read by a photodiode. The experiment has been performed

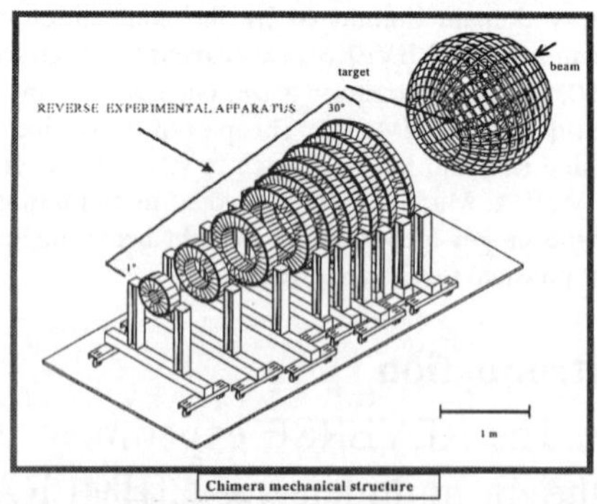

Figure 1 . CHIMERA detector

by using the nine rings of the CHIMERA array, namely 688 telescopes, covering an angular polar range from 1° to 30°, together with 40 INDRA silicon strip detectors in order to detect fragments emitted in a single plane in the angular range from 30° to 90°. The used identification methods are:

- ΔE−E for charge identification of the more energetic particles and for mass identification of the charge identified up to Z=9
- E-TOF method for mass identification and velocity measurements of the less energetic particles
- pulse shape discrimination method for isotopic light charged particles (LCP) identification, from H to Be element.

A typical matrix ΔE−E is plotted in fig.2 for the reaction ^{124}Sn+^{64}Ni at 35 MeV/A; charges up to Sn are clearly seen. Particle identification function allowed us to linearize the matrix in order to obtain one-dimensional charge spectra, used for charge calibration in the data analysis.

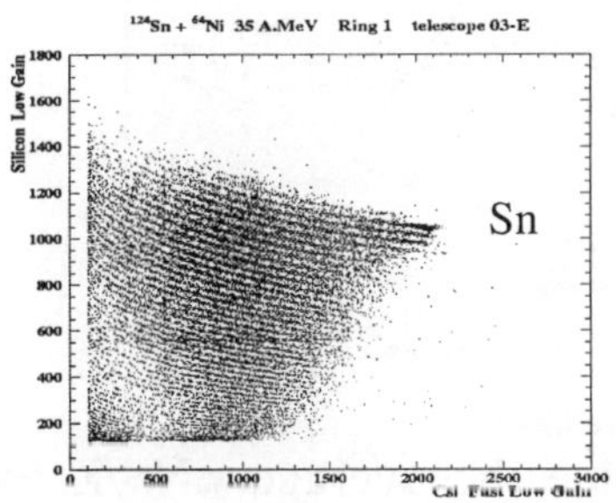

Figure 2

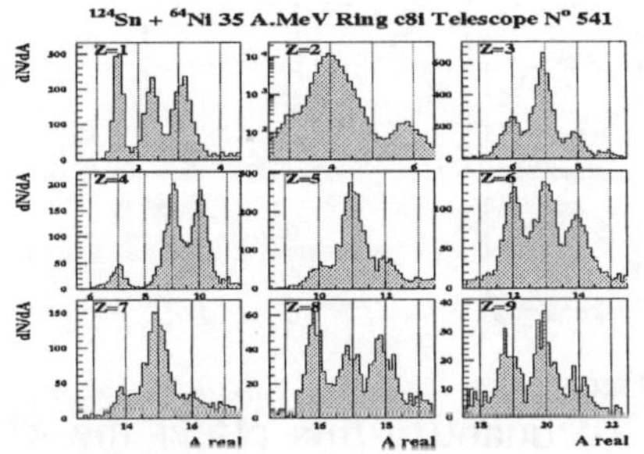

Figure 3

The ΔE−E method has been also extended also for isotopic identification of light ions with charge up to 9. This is shown in fig.3, in which the one-dimensional spectra for isotopes ranging from Z=1 to Z=9 are plotted. Mass identification (see fig.4) and velocity measurements of the reaction products stopped in the silicon detector has been done by using the TOF technique. The LCP identification is

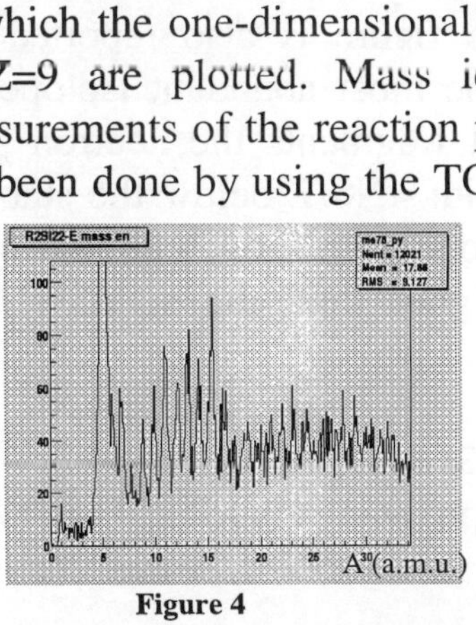

Figure 4

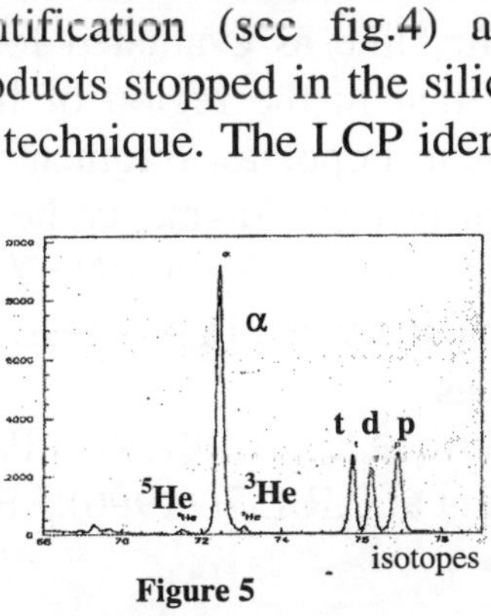

Figure 5

performed by pulse shape discrimination methods and linearization. In fig.5 a typical LCP identification spectrum is shown.

3 Results

An example of interesting observable for isospin investigation is in fig.6a), in which we compare the isotopic distributions for the two systems $^{112}Sn+^{58}Ni$ and $^{124}Sn+^{64}Ni$, at E_{beam} 35 MeV/A. In this figure are reported the relative abundances of isotopes for element from Z=3 to Z=9, as a function of the mass number. A memory effect of the isospin entrance channel is clearly seen. Neutron rich system favours the production of neutron rich isotopes.

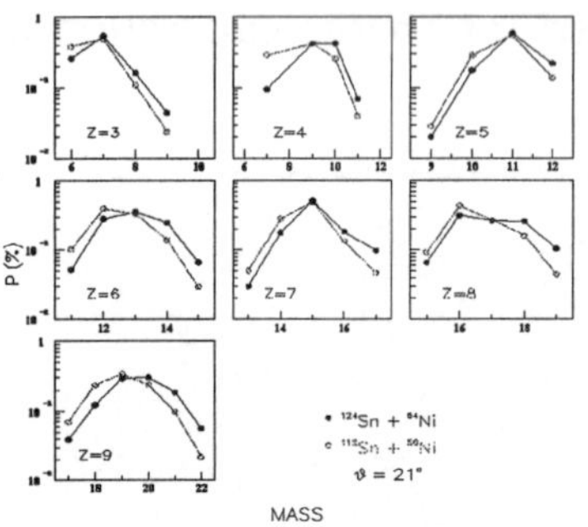

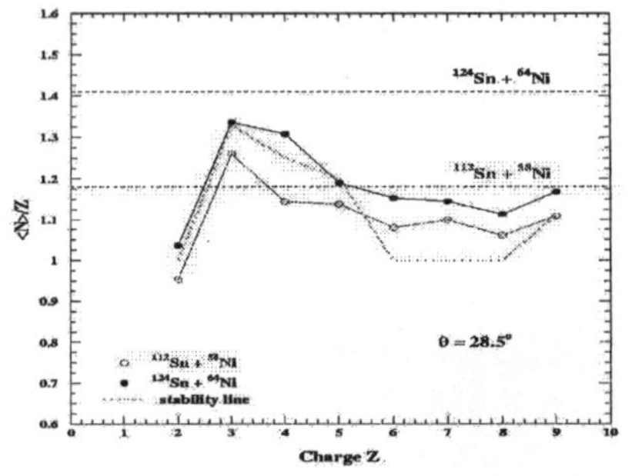

Figure 6a) **Figure 6b)**

To quantify this effect the <N>/Z value the average <N>/Z for each fragment is plotted for the same reactions in fig.6b). The <N>/Z value in the neutron rich system is systematically larger than in the neutron poor system. Indipendently of the system the same odd-even structure can be also observed. In the figure is also reported the <N>/Z stability line as evaluated for the most abundant isotopes in nature. Note that in the region of light fragments the neutron poor system trends to popolate fragments with <N>/Z below the stability line while, for heavier fragments the two system show a similar trend to populate fragments with <N>/Z higher then the stability line. Reaction simulations are in progress.

References

1. L.G.Moretto et al., A.Rev.Nucl.Part.Sc.43 (1993).
2. BaoAnLi et al,P.R.L.76(1996);J.Besprosvany et al,P.L.B45 (1989)

COLLISIONAL RELAXATION OF COLLECTIVE VIBRATION IN HOT NUCLEI

V. A. PLUJKO, M.O.KAVATSYUK AND O.M. GORBACHENKO
Taras Shevchenko National University, Kiev, Ukraine
E-mail: plujko@mail.univ.kiev.ua

The collisional relaxation times of collective vibrations are studied using both the transport approach and doorway state mechanism with hierarchy of particle-hole configurations in heated nuclei. It is shown that 1)the relaxation times of the kinetic method are rather slowly dependent on multipolarity of the Fermi surface distortion and mode of the collective motion; 2) variations of the in-medium two-body cross-sections with energy lead to non-quadratic dependence of the collisional relaxation time both on temperature and on collective motion frequency.

1 Introduction

The relaxation time method is widely used as the simplest and rather accurate approach for simulation of the collisional relaxation rate which determines the damping of the collective excitations as well as transport coefficients for viscosity and heat conductivity [1,2]. Relaxation time method can be applied to describe the decay rate of arbitrary mode of motion, but an explicit form of the relaxation time depends on specific features of the mode. In this contribution, the collisional relaxation times responsible for the width of the collective vibrations are studied.

2 Expressions for relaxation times of collective motion

The collisional relaxation times can be calculated using the collision integral of the transport equation [1,2,3]. The analytical expressions for collective relaxation times of the damping of the collective vibrations with frequency ω has the following general form in low-temperature and low-frequency limits $(T, \hbar\omega \ll \epsilon_F)$[2,3,4]

$$\frac{\hbar}{\tau_\ell^{(\pm)}} = \frac{2}{3\pi}\frac{m}{\hbar^2}\left[< \sigma'_{av}\Phi_\ell^{(+)} > + < \sigma'_{pn}\Phi_\ell^{(\pm)} >\right]\{(\hbar\omega)^2 + (2\pi T)^2\}, \qquad (1)$$

where $\sigma'_{av} = (\sigma'_{nn} + \sigma'_{pp})/2$; $\sigma'_{jj'} \equiv d\sigma_{jj'}/d\Omega$ is in-medium differential cross-section for scattering of the nucleons j and j' (here, $j = n$ or p, and similarly $j' = p$ or n). The symbol $< \ldots >$ in Eq.(1) denotes averaging over angles of the relative momenta of the colliding particles and the functions $\Phi_\ell^{(\pm)}$ define the angular constraints on nucleon scattering within the distorted layers of the Fermi surface with multipolarity ℓ. These times are proportional to the relaxation times defining the damping widths of the isoscalar (+) and the isovector (−) vibrations in regime of rare collisions in the Fermi liquid. The time $\tau^{(+)}_{\ell=2}$ at $\omega = 0$ is the thermal relaxation time determining the viscosity coefficient of the Fermi liquid.

It can be obtained in the case of the isotropic scattering with independent of energy the angle-integrated cross sections $\sigma_{jj'}$ that

$$\frac{\hbar}{\tau^{(\pm)}_{\ell}} = \frac{1}{\alpha^{(\pm)}_{\ell}}\left[(\hbar\omega/2\pi)^2 + T^2\right], \quad \frac{1}{\alpha^{(\pm)}_{\ell}} = \frac{8m}{3\hbar^2}\left[c_\ell\sigma_{av} + d^{(\pm)}_{\ell}\sigma_{np}\right], \qquad (2)$$

where $c_\ell = 1-((2-(-1)^\ell)/(2\ell+1))$, $d^{(-)}_\ell = ((1-(-1)^\ell/(2\ell+1))$, $d^{(+)}_\ell = 0$; $\sigma_{av} = [\sigma_{pp}+\sigma_{nn}+2\sigma_{np}]/4$ is the in-medium spin-isospin averaged nucleon-nucleon cross section. The magnitude of the in-medium cross section $\sigma_{jj'}$ is taken usually in the form $\sigma^{(free)}_{jj'}$ with a factor $F = \sigma_{jj'}/\sigma^{(free)}_{jj'}$ which reduces the in-medium cross-section in comparison with the free space cross section. The parameter $\alpha^{(\pm)}_\ell$ is equal to $\alpha^{(\pm)}_\ell = \widetilde{\alpha}^{(\pm)}_\ell/F$, $\widetilde{\alpha}^{(\pm)}_\ell = 4.18/\left[c_\ell + 1.3d^{(\pm)}_\ell\right]$, MeV, when the values $\sigma^{(free)}_{av} = 3.75\ fm^2$ and $\sigma^{(free)}_{np} = 5\ fm^2$ are adopted; these values correspond to the free space cross sections near Fermi energy.

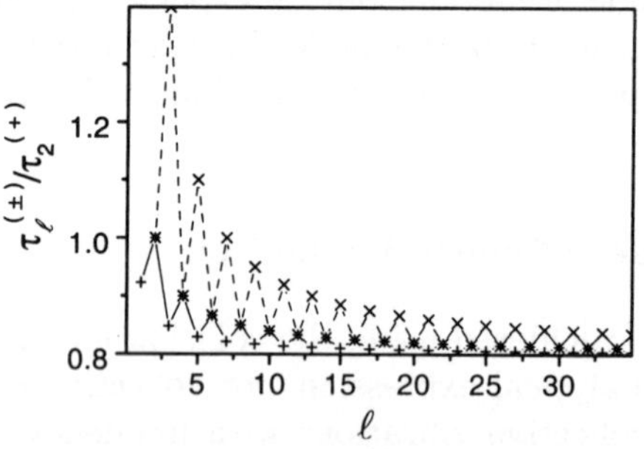

Fig.1: The relative relaxation times $\tau^{(\pm)}_\ell/\tau^{(+)}_2$ versus multipolarity ℓ. Solid and dashed lines connect the values which correspond to isoscalar and isovector modes of vibrations respectively.

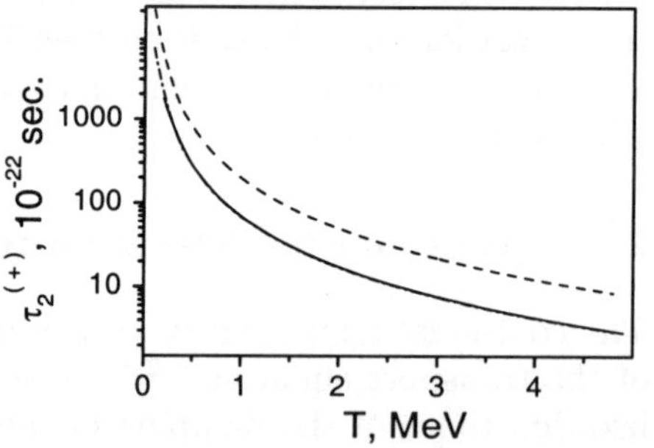

Fig.2: The relaxation time $\tau^{(+)}_2$ in dependence of T at $\hbar\omega = 0$. Solid line corresponds to results with isotropic energy-independent cross-sections at $F = 0.5$. Dot-dashed and dashed lines are calculations for the Skyrme and Cogny forces respectively.

The relative relaxation times $\tau^{(\pm)}_\ell/\tau^{(+)}_{\ell=2}$ at $F = const$ are shown on Fig.1 in relation to the multipolarity ℓ of the distorted layers of the Fermi surface which are accessible to particle collisions. Solid and dashed lines connect the values which correspond to isoscalar and isovector modes of vibrations respectively. The magnitudes of the relaxation times are different for isoscalar and isovector modes of vibrations and they are dependent on the multipolarity ℓ. As seen from the Fig.1, the collisional relaxation times rather slowly vary with multipolarity ℓ and with collective motion mode at isotropic scattering with energy independent free cross sections. The dependence of the viscosity relaxation time $\tau_{\ell=2}^{(+)}(\omega = 0, T)$ on the temperature is shown in Fig.2. Solid line corresponds to the relaxation times for isotropic energy-independent cross-sections with $F = 0.5$. Dot-dashed and dashed lines correspond to the relaxation times with in-medium cross-sections calculated for the Skyrme forces and for the Cogny forces respectively[5]. The calculations with isotropic energy-independent cross-sections at $F = 0.5$ and those ones with energy and angular dependent cross-sections for Skyrme forces are in close agreement. The collisional rates $1/\tau_{\ell}^{(\pm)}$ given by Eqs.(1) and (2) are quadratic both in temperature and in frequency with the same relationship between the components much as in the zero sound attenuation factor of heated Fermi liquid within the Landau prescription.

Let's consider a new approach for collision relaxation time calculation based on a formulae for the transition rates of the particle-particle transition between thermal state with collective vibrations and incoherent particle-hole configurations [4]. In this method the initial state is adopted as a mixture of a collective state and a thermal state which is approximated by uncorrelated superposition states of $mp - mh$ configuration with m excited particles and holes corresponding to the most probable number of excitons $\bar{n} \equiv 2m$ at given temperature T. Next it is accepted common feature that vibration states are formed by coherent superposition of many (predominantly correlated) one-particle one-hole configurations. As a result, the expression for relaxation time has the following form

$$\frac{\hbar}{\tau_c(\omega, T)} = \pi g^3 \mathcal{M}^2(n_< = \bar{n} + 2, E = \hbar\omega + U) \frac{(\hbar\omega + aT^2)^2}{3 + 0.843aT}, \tag{3}$$

where $\mathcal{M}^2(n_<, E)$ is the effective mean square matrix elements for transitions between incoherent exciton states with $n_<$ for minimal value of number of the excitons[7]; $U = aT^2$ is the thermal energy; $a = \pi^2 g/6$, g is the single nucleon state density at the Fermi surface.

According to the exciton model studies [6] the effective mean square matrix elements $\mathcal{M}^2(n_i, E)$ is energy-independent at low excitation energies and it is inversely proportional to energy at higher excitations. The behaviour of collisional relaxation, as given by Eq.(3), with energy-independent magnitude of the mean square matrix element agrees with estimation (2) based on kinetic equation approach at low temperatures $T \ll \hbar\omega$. There are different estimates for the mean square matrix element with dependence on energy and number of excitons. The expression for $\mathcal{M}^2(n, E)$ without assumption on a uniform sharing of the excitation energy E into n excitons was proposed in Ref. [8]: $\mathcal{M}^2(n_<, E) = ((n_< + 1)/4)\mathcal{K}_B/(A^3 E)$, where quantity $\mathcal{K}_B$ is not changed with E and n but can be dependent on numbers of protons and neutrons. If this value of $\mathcal{M}^2$ is employed as the squared matrix element, the collisional relaxation time τ_c is a linear function of the collective energy $\hbar\omega$ and thermal energy U, namely, $\hbar/\tau_c(\omega, T) = (\hbar\omega + U)/\alpha_e$, $\alpha_e = (\mathcal{K}_B\pi/4)(g/A)^3$. The corresponding expressions for the relaxation time have the same form as that one obtained within test particle approach, when collisions were simulated by modeling s-wave scattering between pseudoparticles [9]. The dependence of the matrix element $\mathcal{M}^2$ (i.e., the in-medium cross-section) on energy leads to non-quadratic dependence of the relaxation times on temperature and collective vibration frequency.

The approach (3) makes possible to take into account the energy dependence of the in-medium cross-sections in a simple phenomenological way by the use of the parametrization of the mean square matrix element $\mathcal{M}^2$ of particle-particle collisions from exciton model of nuclear reactions.

3 Acknowledgments

We are very thankful to Dr. Emil Betak for discussions and comments. This work is supported in part by the IAEA(Vienna) under contract 302-F4-UKR-11567.

References

1. M.M. Abu-Samreh, H.S. Koehler, *Nucl. Phys.* A **552**, 101 (1993).
2. V. M. Kolomietz, A.G. Magner, V.A. Plujko, *Z. Phys.* A **345**, 131 (1993).
3. V.A. Plujko, *Acta Phys. Pol.* B **30**, 1383 (1999).
4. V. A. Plujko, O.M. Gorbachenko and M.O.Kavatsyuk, *Acta Physica Slovaca*, in press, (2001).
5. O. Yilmaz, A. Gokalp, S. Yildirim, S. Ayik, *Phys. Lett.* C **472**,258 (2000); LANL e-print- http://xxx.lanl.gov/abs/nucl-th/9909057.
6. E. Gadioli, P.E. Hodgson, *Pre-Equilibrium Nuclear Reactions.* Clarendon Press, Oxford, 1992.
7. C.Kalbach, *Z.Phys.* A **287**, 319 (1978).
8. P. Oblozinsky, *Phys. Phys.* C **35**,407 (1987).
9. A. Bonasera, G.F. Burgio, M. Di Toro, *Phys. Lett.* B **221**, 233 (1989).

MASS-ENERGY DISTRIBUTIONS OF FISSION FRAGMENTS OF SUPERHEAVY NUCLEI PRODUCED IN THE REACTIONS WITH ^{48}CA IONS

E. V. PROKHOROVA, M. G. ITKIS, YU. TS. OGANESSIAN, E. M. KOZULIN, A. A. BOGATCHEV, I. M. ITKIS, M. JANDEL, J. KLIMAN, G. N. KNIAJEVA, N. A. KONDRATIEV, I. V. KORZYUKOV, L. KRUPA, I. V. POKROVSKI, V. A. PONOMARENKO, A. YA. RUSANOV AND V. M. VOSKRESENSKI

Flerov Laboratory of Nuclear Reactions, Joint Institute for Nuclear Research, Dubna, 141980, Russia,
E-mail: lena@cv.jinr.ru

F. HANAPPE, B. BENOIT AND T. MATERNA

Universite Libre de Bruxelles,1050 Bruxelles, Belgium

N. ROWLEY AND L. STUTTGE

Institut de Recherches Subatomiques, F-67037 Strasbourg Cedex, France

G. GIARDINA

Dipartimento di Fisica dell' Universita di Messina 98166 Messina, Italy

A series of experiments on the fission of superheavy nuclei formed in the reactions with ^{48}Ca ions, namely, the ^{48}Ca+^{208}Pb, ^{48}Ca+^{232}Th, ^{48}Ca+^{238}U, ^{48}Ca+^{244}Pu, ^{48}Ca+^{248}Cm reactions, was carried out during the past two years at the Flerov Laboratory of Nuclear Reactions (JINR, Dubna). Experimental data were obtained on the mass and energy distributions of fission fragments of compound nuclei 102-116 in an excitation energy range of 13-44 MeV. The capture cross section σ_{cap} and fusion-fission cross section σ_{FF} were also estimated for these nuclei. As a result of the experiments it was found that the mass distribution of fission fragments for compound nuclei ^{286}112, ^{292}114, ^{296}116 and ^{306}122 is asymmetric. Its nature, in contrast to the asymmetric fission of actinides, is determined by the shell structure of the light fragment with the average mass 132-134. It was established that the TKE for fission and that for quasi-fission of superheavy compound nuclei are significantly different. The analysis of the σ_{FF} data and the σ_{EV} data [1] obtained in the study of the same elements testifies to the fact that at excitation energies E*~30-35 MeV the ratio σ_{FF}/σ_{EV} is changing insignificantly up to the nucleus Z=116. This fact allows one to hope for a successful synthesis of even heavier nuclei in asymmetric reactions.

The experiment was carried out using the beams delivered by the U-400 accelerator of the FLNR (Dubna). For the registration of reaction products a two-arm time-of-flight spectrometer CORSET [2] and a neutron multidetertor DEMON [3] were used.

Figure 1 shows the results of measurements of mass-energy distributions (MED) of fission fragments produced in the reaction $^{48}Ca+^{208}Pb$ at a ^{48}Ca ion energy of 220-242 MeV. Two-dimensional Mass-TKE matrixes, mass distributions and the average TKE and its variance σ^2_{TKE} as a function of the fission fragment masses are shown in the figure (see from top to bottom). It is well seen that the mass distributions of fission fragments in the case of symmetric masses A/2±30 are nearly Guassian in shape, and the dependencies <TKE>(M) are parabolas. Such behavior of the above mentioned properties of fission fragments is typical for the fission of heated nuclei and is predicted by the liquid-drop [4] or diffusion [5] models.

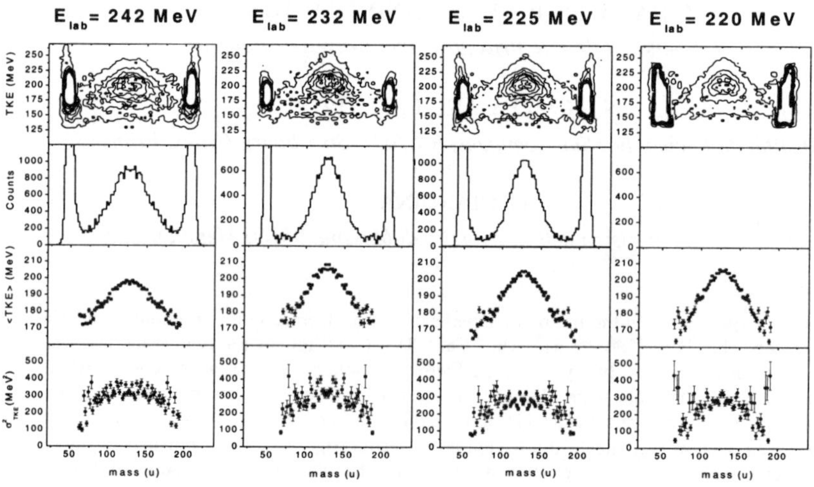

Figure 1. Two-dimensional TKE-M matrixes, mass yields, the average TKE and the varainces σ^2_{TKE} as a function of the masses for the reaction $^{48}Ca+^{208}Pb$ in the energy range E_{lab} = 220-242 MeV

At the same time, in the case of the light fission fragments with $65 \leq M_L \leq 100$ and complementary heavy ones, the so-called «shoulders» are observed in the mass distributions. The «shoulders» correspond to the enhanced yield of reaction products as compared with that expected in the case of classic fission. In this region of fission fragment masses an increase in the TKE and irregularities in the σ^2_{TKE} (M) distribution are observed at all the four ion energies (Fig.1) as compared with the classic fission. For finding the characteristics of the symmetric fission MED and ^{256}No quasifission "shoulders" (Fig. 2), the fission fragment mass distributions were fitted by the Gaussian (G). On the left-hand side of Fig. 2 the mass distribution of fission fragments are shown in the logarithmic scale as a function of the parameter $(M - A/2)^2$. In such a presentation the Gaussian is a linear function, and it is seen from Fig.2 that starting roughly from the value of the parameter $(M - 128)^2 \cong 800\text{-}900$ ($M_L \cong 100$) the experimental points are deviating upwards from G which describes well the initial part of the mass distributions. The mass yields and the G-fit of mass distributions are shown in the central part of Fig.2 in the linear scale. On the right-hand side of the figure the extracted quasifission «shoulders» are shown (for the case of light fragments only), obtained as a difference between the experimental fission fragment yield and the G-function: $Y_q(M) = Y_{exp}(M) - Y_G(M)$. In the mass distribution structures for quasi-fission peaks the manifestation of shell effects (Z=28, N=50) is cleary seen in the right-hand part of Fig. 2.

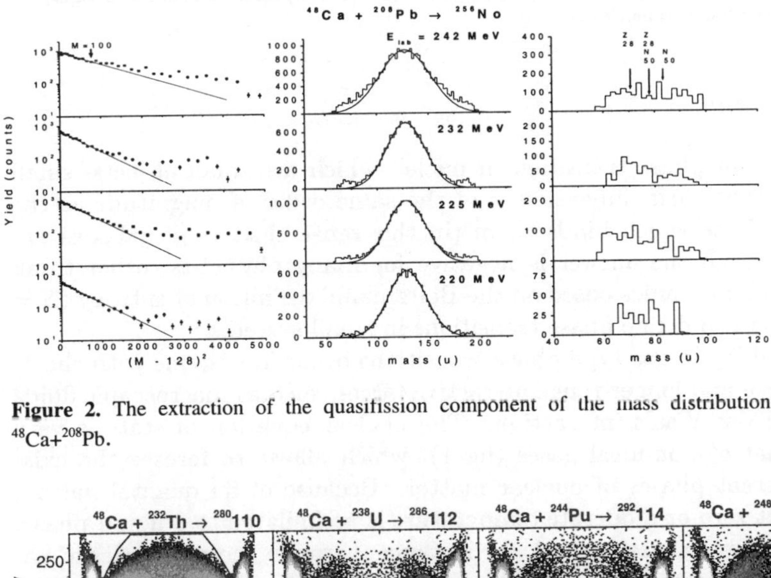

Figure 2. The extraction of the quasifission component of the mass distribution for the reaction $^{48}Ca+^{208}Pb$.

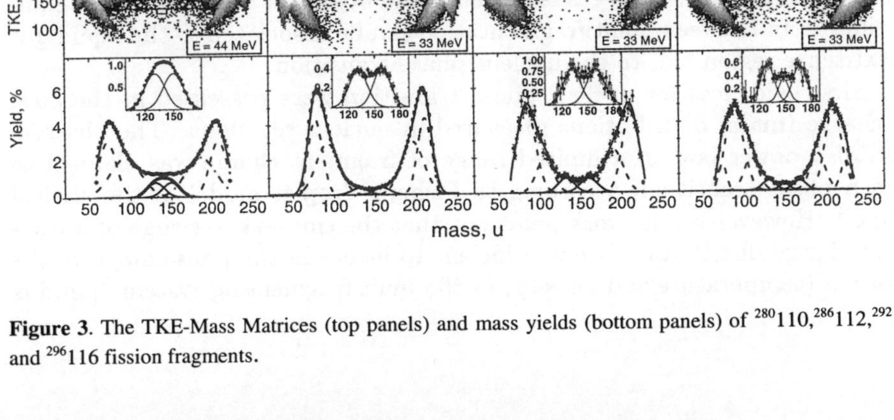

Figure 3. The TKE-Mass Matrices (top panels) and mass yields (bottom panels) of $^{280}110$, $^{286}112$, $^{292}114$ and $^{296}116$ fission fragments.

THERMODYNAMICS OF HOT NUCLEI: MULTIFRAGMENTATION AND PHASE TRANSITION

M.F. RIVET
Institut de Physique Nucléaire, IN2P3-CNRS, F-91406 Orsay Cedex, France.

The dynamics and thermodynamics of phase transition in nuclei are studied through experimental results on multifragmentation of systems with masses around 200 nucleons. Nuclear multifragmentation is indeed the phenomenon connected to the liquid-gas phase transition of nuclei. Negative heat capacities and spinodal decomposition are observed in a limited range of available energy, which sign the penetration of the systems in the coexistence zone of the phase diagram and thus a first order phase transition.

1 Introduction

Can we speak of phase transition in nuclei, which are small objects, small in the sense that their dimension is of the same order of magnitude as the range of the force which binds them (in this sense clusters of galaxies are small)? Although the answer is negative for many physicists, other think that statistical mechanics based on the Boltzmann' definition of entropy ($S = k.\log W$) allows to define phase transitions in small systems [1]

In general liquid-gas type phase transitions occur in systems with short-range repulsive and longer-range attractive forces, such as macroscopic fluids with the van der Waals interaction. The nuclear equation of state is very similar to that of non-ideal gases (fig 1), which allows to foresee the existence of different phases of nuclear matter. Because of its quantal nature, the nucleus at zero or moderate temperature is assimilated to a liquid phase. The gas phase was characterized, for instance, through the properties of vaporised quasi-projectiles from 95 AMeV Ar+ Ni reactions, fully predicted by modelling a van der Waals gas of fermions and bosons in thermodynamic equilibrium [3]. In between, multifragmentation can be associated to the liquid-gas coexistence region, i.e. to the nucleus phase transition.

First evidences for nucleus phase transition were presented in the 80's, in charge (mass) distributions measured in nuclear reactions. The observation of a power law on a limited range of fragment charge was claimed to sign a phase transition, according to Fisher's droplet model at the critical point [4]. However recent work point out that the simple knowledge of a fragment charge distribution is not sufficient to locate in the phase diagram the position (temperature and density) of the multifragmenting system [5], and is

Figure 3 shows mass-energy distributions and mass yields of fission fragments formed in reactions between ^{48}Ca ions and targets of ^{232}Th, ^{286}U, ^{244}Pu, ^{248}Cm. In this case the distributions fundamentally differ from the similar distributions obtained in the reaction ^{48}Ca+^{208}Pb. The main decay channel is quasifission in this case. The mass distributions have two humps, with a peak corresponding to the heavy fission fragment 208 (the doubly magic ^{208}Pb). Nevertheless, in the region of symmetric fission (masses A/2±20) the process of fusion-fission prevails, in our opinion. In the framings of the lower panel the mass yields for the fusion-fission process are presented. They are obtained as a difference between the experimental mass yields and the fit describing the quasifission peaks. It is seen that the mass distributions are asymmetric, and the mass of the light fission fragment is 132-134. Only in the case of the reaction ^{48}Ca+^{232}Th the distribution looks like a gaussian with a flat top. This situation can be explained by an increase in the excitation energy and thus a decrease in the influence of the shell effects. The capture cross section σ_{cap}, the fusion-fission cross sections σ_{FF} and quasifission cross section σ_{QF} were also estimated for these nuclei (Fig.4).

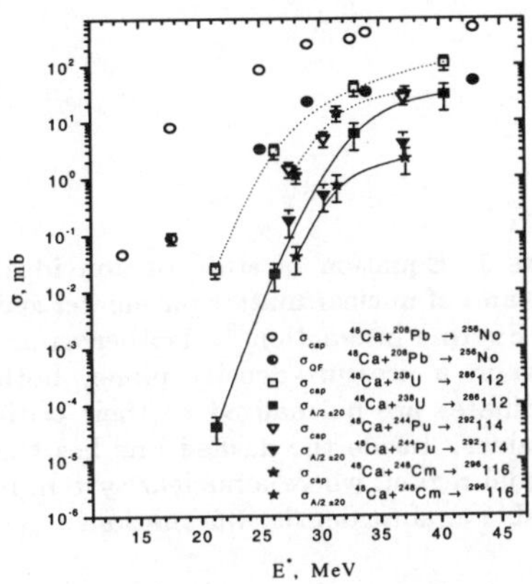

Figure 4. The capture cross section σ_{cap}, and the fusion-fission cross section σ_{ff} as a function of the excitation energy

Acknowledgement

We thank the INTAS foundation (Grant YSC 4258) and the Organizing Committee of the Conference for the financial support. The authors are also grateful to M.Morozova for her help in the preparation of this contribution.

References

1. Oganessian Yu.Ts. *et al*, *Eur. Phis. J*, **A 5** (1999) 63.
2. Kondratiev N. A *et al* in *Proc.4-th Int. Conf. on Dynamical aspects of nuclear fission (DANF'98)*, Casta-Papiernicka, Slovak Republic, October 1998 (World Scientific, Singapore, 2000) p.431
3. Mouatassim S. *et al*, *Nucl. Instrum. Methods*, **A 359** (1995) 330
4. Nix J. R. and. Swiatecki W. J, *Nucl. Phys*. **71** (1965) 1
5. Adeev G. D. *et al*, *Sov. J. Part. Nucl*., **19** (1988) 529

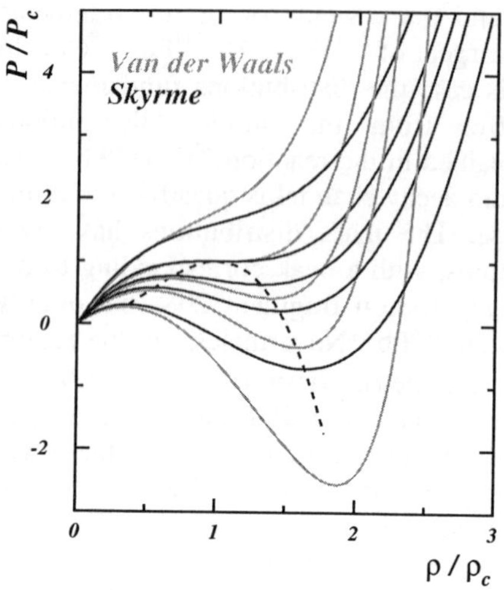

Figure 1. Equation of state of non ideal gases and of nuclear matter parameterized by a Skyrme interaction [2]. Isotherms are drawn in a pressure-density plane; both coordinates are normalised to their critical values. Inside the dashed line lies the spinodal region, where a nuclear system is unstable against density fluctuations.

thus not a good signal of critical behaviour. More recently, in the same line, critical exponents were reported to have measured values in agreement with those of a liquid-gas model [6,7]. Another approach was the determination of a caloric curve, linking the energy and the temperature of nuclei. The observation of a plateau, similar to what is observed for the ebullition of water, was claimed as a proof of the nuclear phase transition [8]. However as the only way nuclear physicists may use to modify temperature, energy, density of nuclei (and thus to eventually induce a phase transition) is a more or less violent nuclear collision, one is dealing with isolated systems and not with a system with external constraints (constant pressure in the case of boiling water for instance). As underlined in recent works on lattice-gas models, one should consider the small system evolution in a three-dimensional space, e.g. pressure, temperature and volume [9]. None of these being *a priori* constant, one cannot predict the shape of the caloric curve, which may or may not present a backbending at the phase transition, depending on the trajectory followed in this space. Conversely abnormally large fluctuations of part of the total energy in the region of the phase transition were demonstrated to be a robust state variable signing a phase transition. These large fluctuations correspond to negative values of the heat capacity of nuclei, and to the spinodal region of the phase diagram (see fig 1).

Some of the recent experimental studies of nuclear phase transition thus follow a new approach : firstly one must choose clean experimental conditions, where the multifragmenting sources can be precisely characterized, for instance a quasi-projectile, or a fused system merging the majority of the nucleons of the colliding nuclei. Providing that an ensemble of such sources explore a large fraction of phase space, forming a statistical sample representative of thermodynamic equilibrium, thermodynamic studies can be envisaged. Additional information on the nuclear EOS may be derived from the dynamics of the collision, and more specifically of the phase transition.

2 Thermodynamics of phase transition

Systems with about 200 nucleons have been studied through different conditions. The MULTICS collaboration isolated quasi-projectiles from 35 AMeV Au+Au reactions, with a mass remaining close to the Au mass[7,10]. Fused systems from central Xe+Sn collisions at different energies were scrutinized by the INDRA collaboration[11]. In both cases, the equilibrium state of the sample of sources was inferred from a comparison of data with the predictions of the Copenhagen Statistical Model for Multifragmentation [12]. An exemple is given in fig 2, where one can observe that several measured and calculated distributions of global variables are in good agreement [13].

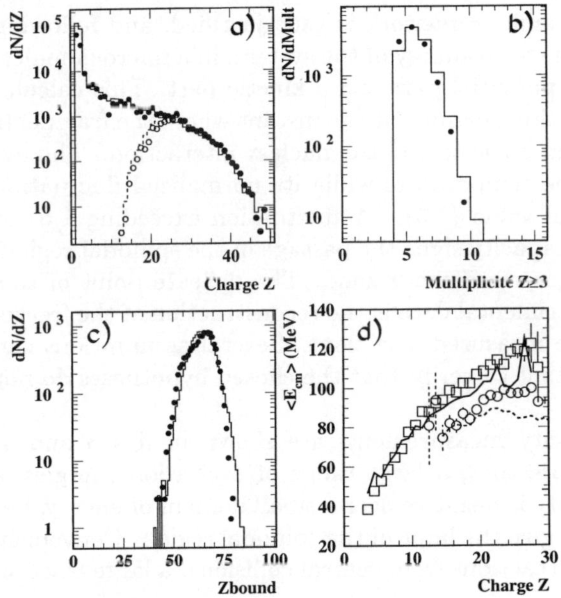

Figure 2. Comparison of experimental data from central 32 AMeV Xe+Sn collisions (lines) with SMM calculations (A_S=202, Z_S=85, ε= 5 MeV, V=3V_0) - symbols. The charge, multiplicity and Z_{bound} distributions are well reproduced, as well as the average fragment kinetic energy, provided the addition of a 0.6 MeV radial expansion energy.

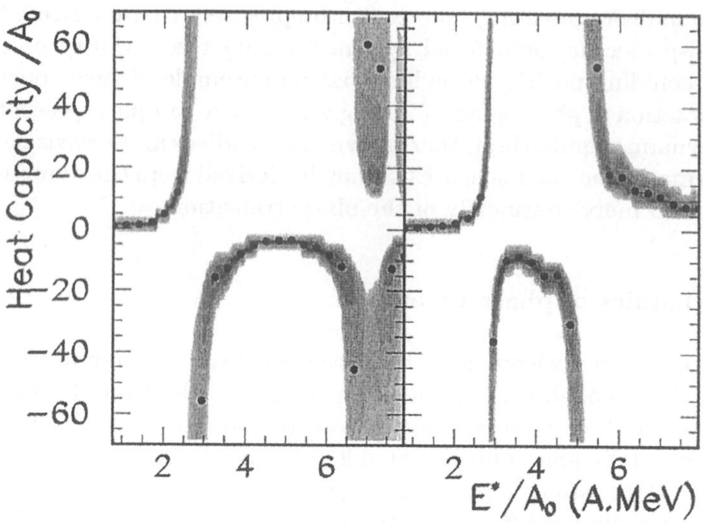

Figure 3. Heat capacity as a function of the energy of Au quasi-projectiles. The two parts of the figure correspond to different hypotheses for the freeze-out reconstruction. From [10]

A study in a thermodynamic framework is thus justified, and following the method proposed in [14], the total energy of the system, in a microcanonical framework, is divided into a potential part and a kinetic part. This calculation is valid at the freeze-out time, defined at the instant where the fragments born after multifragmentation cease to feel the nuclear interaction. The average kinetic energy gives the temperature while its normalized fluctuation is compared to the canonical value (1.5). A fluctuation exceeding 1.5, or equivalently a negative heat capacity signs the passage in the spinodal region, which lies inside the liquid-gas coexistence zone. The delicate point of this method when applied to experimental data is the reconstruction of the freeze-out configurations out of the measured partitions. Several assumptions can be envisaged and it is important to verify that the chosen hypotheses do not create spurious fluctuations [15].

Experimental heat capacity measurements are shown in figs 3 and 4. When studying Au quasi-projectiles, a large range of excitation energies is explored, and the heat capacity is negative over a small domain of energy, between two divergencies which give the limit of the spinodal region. Conversely in fusion-multifragmentation reactions from central collisions, a large fraction

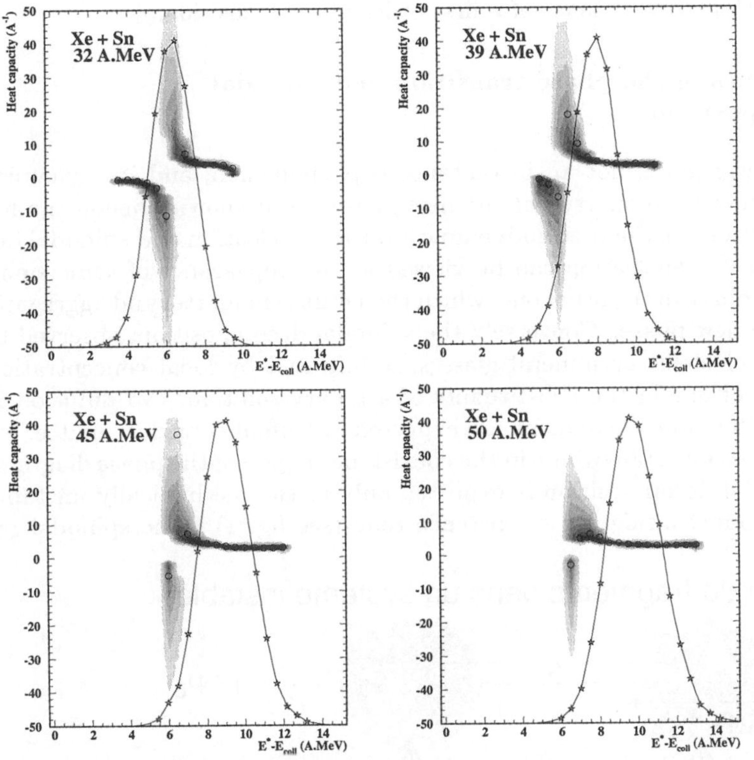

Figure 4. Heat capacity as a function of the energy of the fused systems formed in central Xe+Sn collisions between 32 and 50 AMeV. The lines show the energy distributions, and the shaded zone the heat capacities and their errors. From the INDRA collaboration [11].

of the incident energy is deposited into the system, leading to narrow distributions of excitation energy. For 32 and 39 AMeV Xe+Sn, the energy deposited is close to that of the second divergency. The values of energy corresponding to the divergencies should not yet be taken as definitive, as they depend on the way freeze-out configurations are reconstructed. One should just note for the moment that the observed signal persists when these hypotheses are varied within reasonable limits. At higher bombarding energy, 45 and 50 AMeV, the heat capacity is positive again. This will be further discussed in the next section, but one can already state that the observation of negative heat capacities directly signs a liquid-gas coexistence zone in the phase diagram of nuclei, and thus the existence of a first order phase transition.

3 Dynamics of the phase transition and spinodal decomposition

A phase transition is not an instantaneous phenomenon, and its dynamics strongly influences properties of the new phase. Well known phenomena for phase transitions at the thermodynamic limit are nucleation and spinodal decomposition [16]. Nucleation can be viewed as the appearance of some small seeds of one phase in the other one, which then diffuse in matter and aggregate, creating the new phase. Conversely the spinodal decomposition, observed in the cooling of steels or mineral glasses, is initiated by local concentration fluctuations of one of the constituants of a binary solution. No diffusion is necessary. The same phenomena are expected in "infinite" nuclear matter [17]. Nucleation can occur anywhere in the coexistence region of the phase diagram, while spinodal decomposition is expected only in the mechanically unstable region, contained inside the coexistence zone (see fig. 1). The spinodal re-

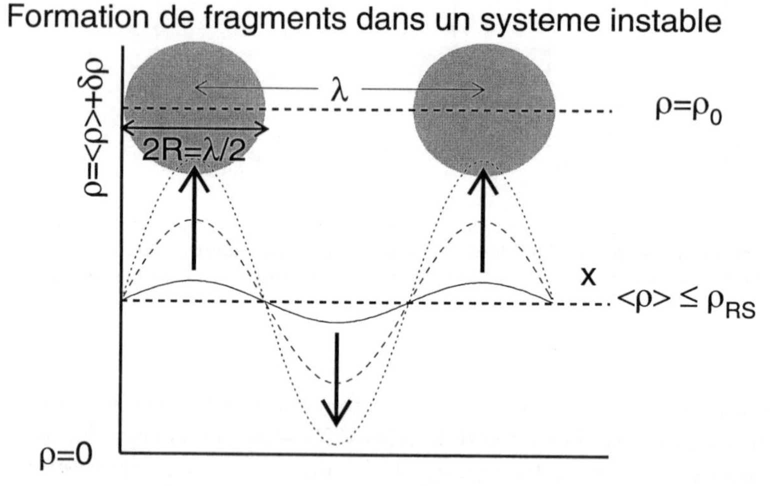

Figure 5. Schematic view of the groth of density fluctuations leading to spinodal decomposition. From [18].

gion corresponds to low densities of matter and the spinodal decomposition is initiated by the amplification of local density inhomogeneities; when the amplitude of the fluctuation approaches the average density value, the system breaks into small fragments whose size is directly related to the wavelength of the fluctuations (fig 5). In this picture, one expects that spinodal de-

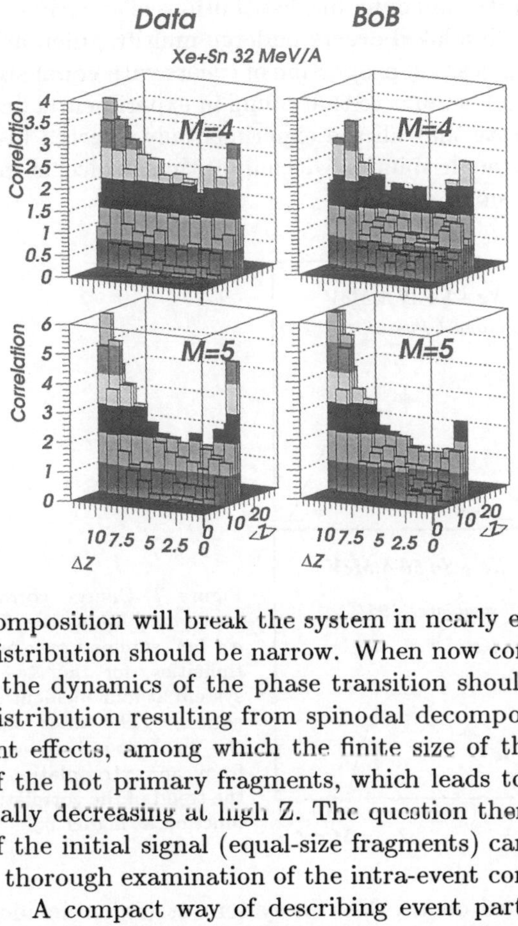

Figure 6. Charge correlations for events with fragment multiplicities 4 (top) and 5 (bottom). The left column refers to experimental data, and the right column to results of a dynamic simulation (BOB). From the INDRA collaboration [20]

composition will break the system in nearly equal-size fragments: the charge distribution should be narrow. When now considering nuclei - small systems - the dynamics of the phase transition should be the same, but the charge distribution resulting from spinodal decomposition will be blurred by different effects, among which the finite size of the system and the deexcitation of the hot primary fragments, which leads to charge distributions exponentially decreasing at high Z. The question then is whether some reminiscence of the initial signal (equal-size fragments) can be found in the data through a thorough examination of the intra-event correlations.

A compact way of describing event partitions is to build higher order charge correlation functions, defined as the ratio of yields Y(<Z>,ΔZ) for correlated and uncorrelated fragments; <Z> is the average fragment charge and ΔZ the standard deviation *per event* [21]. In such a representation, any privileged partition will peak above the neighbour bins at the corresponding <Z>,ΔZ bin . Charge correlation functions built for experimental events are shown in fig. 6 for the Xe+Sn system at 32 AMeV [22]. For all fragment multiplicities the experimental charge correlation decreases with decreasing ΔZ, but a sudden enhancement appears for ΔZ = 0-1. This peak for narrow event charge distributions corresponds to a small proportion of events (0.1%) which may appear as meaningless. The same analysis was performed on events generated in semi-classical simulations of collisions (BOB) [19]; These simulations account in great detail for the static and dynamic distributions of experimental observables [23,24]. Although all simulated events undergo multifragmentation through spinodal decomposition, a small proportion of events with equal-sized fragments is observed (fig 6 right), similar to that found in experimental data. A natural step forward is to state that all the experimentally selected fused systems also multifragment through spinodal decomposition, which is here observed for the first time in a small system.

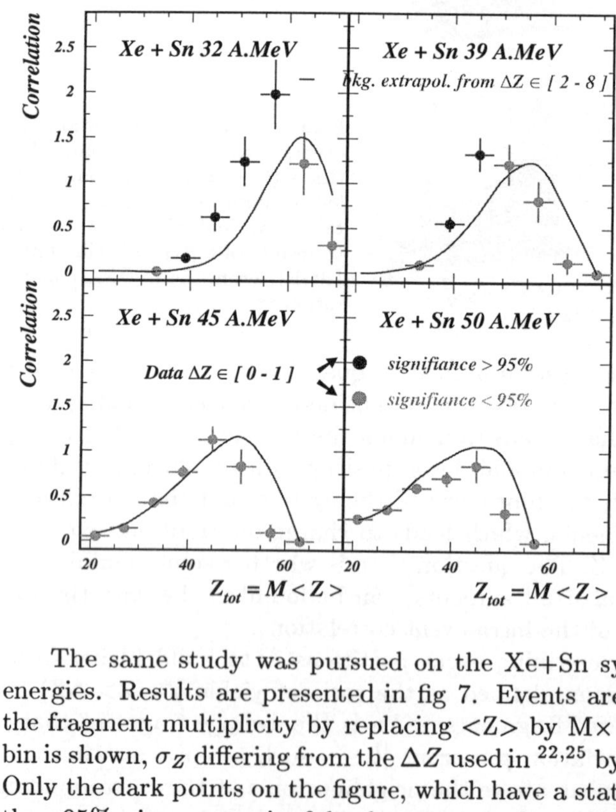

Figure 7. Charge correlation functions in the first σ_Z bin, summed over multiplicities, for the Xe+Sn system at four incident energies. The lines show the background obtained from an extrapolation of the trend of the correlation function at higher σ_Z.

The same study was pursued on the Xe+Sn system at higher incident energies. Results are presented in fig 7. Events are now summed whatever the fragment multiplicity by replacing <Z> by M× <Z>. Only the first σ_Z bin is shown, σ_Z differing from the ΔZ used in [22,25] by the ratio $\sqrt{(M-1)/M}$. Only the dark points on the figure, which have a statistical significance larger than 95%, sign a meaningful enhancement of equal-sized fragment partitions. The signal observed at 32 AMeV is still present, independently of the detailed calculation of σ_Z (and it concerns twice as more events), and also exist at

39 AMeV, but disappears from 45 AMeV up.

Both the negative heat capacities and the spinodal decomposition are observed in a narrow range of available energy, between 5 and 10 AMeV, for central Xe+Sn collisions. Dynamic and thermodynamic studies converge to indicate that the system goes and stays long enough in the spinodal region, which shows the existence of a first order phase transition for nuclei with masses around 200.

At higher energies both signals disappear. Several explanations can be envisaged: i) Multifragmentation still occurs through spinodal decomposition, but the higher excitation energy of the nascent fragments and their subsequent deexcitation completely wash out the partitions with equal-sized fragments. ii) the system passes rapidly across the spinodal region so that the density fluctuations have no time to develop. Multifragmentation is then caused by another mechanism, nucleation, or even vaporization if the system goes very rapidly toward the gas region of the phase diagram.

4 Conclusions

Signature of the liquid-gas phase transition of nuclei has been found in the negative branch of the heat capacity, and in the enhancement of equal-sized fragment partitions, reminiscence of a spinodal decomposition of nuclei. From a theoretical point of view, other signatures have been proposed and are being confronted to experimental data. Let us mention as exemples the distribution of complex zeros of the canonical partition function [26], the universal scaling laws of the fluctuations of an order parameter [27], or some topology anomalies (bimodality) of an order parameter distributions [28,29]. Although there are still debates on the type of phase transition which is observed, all these signals indicate without doubt the passage from an ordered to a disordered phase of nuclear matter.

References

1. D.H.E. Gross, preprint arXiv:cond-matt/0105313.
2. H. Jaqaman, A.Z. Mekjian and L. Zamick, *Phys. Rev.* C **27**, 2782 (1983).
3. B. Borderie et al. (INDRA collaboration), *Eur. Phys. J.* A **6**, 197 (1999).
4. M.E. Fisher, *Physics* **3**, 255 (1967).
5. X. Campi, H. Krivine and N. Sator, *Nucl. Phys.* A **681**, 485c (2001).
6. J.B. Elliot et al. (EOS collaboration), *Phys. Lett.* B **418**, 34 (1998)
7. M. D'Agostino et al., *Nucl. Phys.* A **650**, 329 (1999).

8. J. Pochodzalla et al. (ALADIN collaboration), *Phys. Rev. Lett.* **75**, 1040 (1995)
9. P. Chomaz, V. Duflot and F. Gulminelli, *Phys. Rev. Lett.* **85**, 3587 (2000).
10. M. D'Agostino et al., *Phys. Lett.* B **473**, 219 (2000).
11. R. Bougault et al., (INDRA coll.) Proc. of the XXXVIII Int. Winter Meeting on Nuclear Physics, Bormio, Italy (2000), ed I. Iori, Ricerca scientifica ed educazione permanente, p 404.
12. J. Bondorf et al., *Phys. Reports* **257**, 133 (1995)
13. N. Le Neindre, PhD thesis, Université de Caen, LPCC T 99 02.
14. P. Chomaz and F. Gulminelli, *Nucl. Phys.* A **647**, 153 (1999).
15. M. D'Agostino et al., preprint arXiv:nucl-ex/0104024.
16. P. Papon, J. Leblond and P.H.E. Meijer, Physique des transitions de phase, edited by Dunod, Paris (1999)
17. D. Idier, M. Farine, B. Remaud and F, Sébille, *Ann. Phys. Fr.* **19**, 159 (1994).
18. J. D. Frankland, PhD thesis, Université Paris XI Orsay, IPNO-T-98-06.
19. A. Guarnera et al., *Phys. Lett.* B **373**, 267 (1996).
20. G. Tăbăcaru et al. (INDRA collaboration), Proc. of the XXXVIII Int. Winter Meeting on Nuclear Physics, Bormio, Italy (2000), ed I. Iori, Ricerca scientifica ed educazione permanente, p 433.
21. L.G. Moretto et al., *Phys. Rev. Lett.* **77**, 2634 (1996).
22. B. Borderie et al. (INDRA collaboration), *Phys. Rev. Lett.* **86**, 3252 (2001).
23. J.D. Frankland et al. (INDRA collaboration), *Nucl. Phys.* A **689**, 940 (2001).
24. G. Tăbăcaru, Ph.D thesis, Université Paris XI Orsay, IPNO-T-00-13.
25. B. Borderie et al. (INDRA collaboration), Proc. of the XXXIX Int. Winter Meeting on Nuclear Physics, Bormio, Italy (2001), ed I. Iori, Ricerca scientifica ed educazione permanente, p 73 and preprint arXiv:nucl-ex/0106007.
26. O. Mülken and P. Borrmann, *Phys. Rev.* C **63**, 024306 (2001).
27. R. Botet et al., *Phys. Rev. Lett.* **86**, 3514 (2001).
28. Ph. Chomaz, F. Gulminelli and V. Duflot, preprint arXiv:cond-mat/0010365.
29. O. Lopez et al. (INDRA collaboration), Proc. of the XXXIX Int. Winter Meeting on Nuclear Physics, Bormio, Italy (2001), ed I. Iori, Ricerca scientifica ed educazione permanente, p 61.

ENTRANCE AND EXIT CHANNELS FOR VERY HEAVY AND SUPERHEAVY ELEMENTS

G. ROYER AND K. ZBIRI

Laboratoire Subatech, 4 rue A. Kastler, 44072 Nantes Cedex 03, France
E-mail: royer@subatech.in2p3.fr

R. A. GHERGHESCU

National Institute for Physics and Nuclear Engineering, P.O. Box MG-6, RO-76900 Bucharest, Romania

The fusion barriers experimentally investigated as well as the α and fission exit channels for superheavy elements have been determined within a Generalized Liquid Drop Model including the proximity effects, the asymmetry, an accurate nuclear radius and an adjustment to reproduce the experimental Q value. Double-hump fusion barriers appear in cold fusion reactions and, then, possible fast-fission events in the external well while warm fusion reactions lead to one peak potential barriers and to excited nuclei. The moment of emission of the excess neutrons or even an α particle is crucial to decide between complete fusion and fast fission.

1 Fusion barriers in the entrance channel

The synthesis of very heavy elements has apparently strongly advanced recently using both cold and warm fusion reactions. The only observed decay mode is often the α emission.

In the entrance channel the fusion barrier has been determined within a generalized liquid drop model including the proximity effects for the macroscopic part and the asymmetric two center shell model for the microscopic contributions [1,2,3]. The barrier has been adjusted to reproduce the experimental Q-value. The cold fusion reactions Fe, Ni, Zn, Ge, Se and Kr on Pb and warm fusion reactions Ca on U, Pu and Cm have been investigated. In such an approach the saddle-point corresponds always to two separated sphere configurations; the attractive nuclear forces compensating for the repulsive Coulomb forces. The often used approximation starting the fusion process from the contact point seems very rough; the pure Coulomb barrier is too high while if the proximity energy is included only at this point then the energy is too low.

In the cold fusion reactions a macroscopic potential pocket appears at large deformations and the energy is almost constant till the spherical compound nucleus. The microscopic corrections lead to double humped fusion barriers (see figure). In the deep minimum between the two peaks fast fission

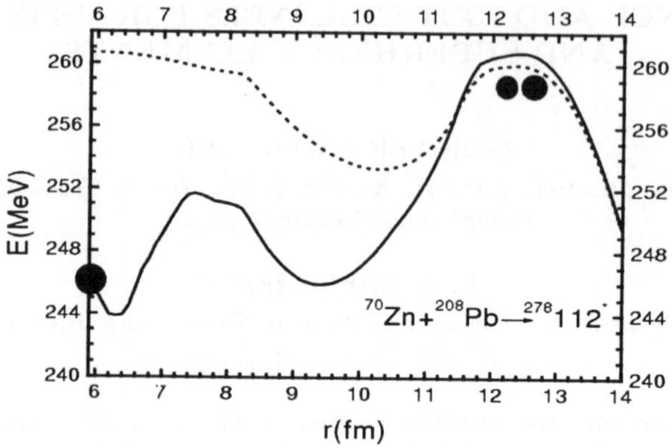

or incomplete fusion processes might appear since the neck between the two nuclei is formed and exchanges of nucleons occur. An open question is to know whether at these large deformations the nucleon shells can take form to stabilize the nuclear system before investigating a peculiar exit channel. The moment of emission of the excess neutron which evacuates a part of the excitation energy is also crucial to decide between complete fusion and fast fission processes. For the heaviest systems, the inner barrier is the highest.

In the warm fusion reactions, there is no double-humped barriers and the system descends automatically toward a quasi-spherical shape but with an excitation energy which is much higher. The emission of several neutrons or even an α particle is energetically possible.

The fusion barrier heights and positions given by the GLDM reproduce accurately the experimental data. They can be reproduced by the following formulas:

$$E_{0,fus} = -19.38 + \frac{2.1388 Z_1 Z_2 + 59.427(A_1^{1/3} + A_2^{1/3}) - 27.07 \ln(\frac{Z_1 Z_2}{A_1^{1/3}+A_2^{1/3}})}{(A_1^{1/3} + A_2^{1/3})(2.97 - 0.12 \ln(Z_1 Z_2))}, \tag{1}$$

$$R_{0,fus}(fm) = (A_1^{1/3} + A_2^{1/3}) \left[1.908 - 0.0857 ln(Z_1 Z_2) + \frac{3.94}{Z_1 Z_2}\right]. \tag{2}$$

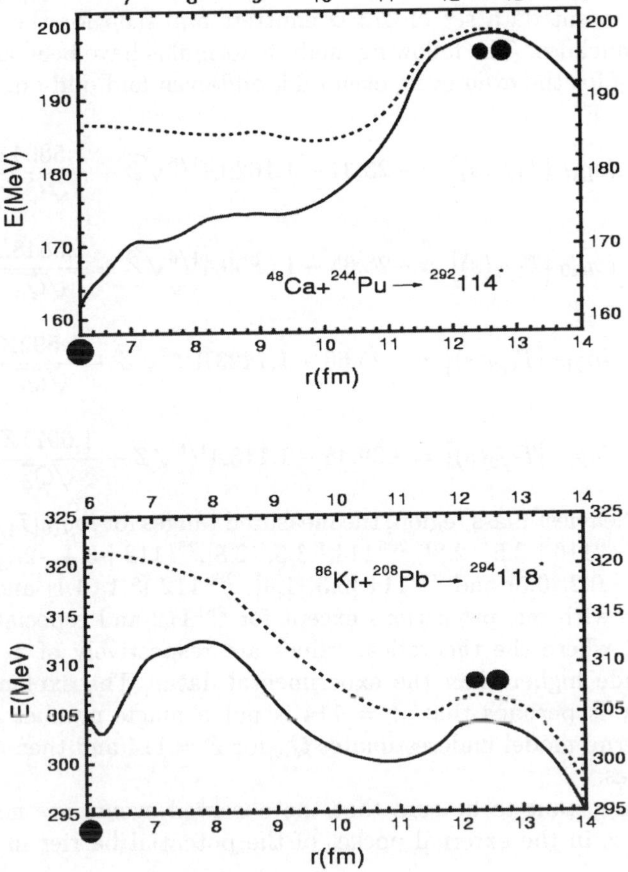

Figure 1. Macroscopic (dashed curve) and microscopic (solid one) fusion barriers versus the mass-centre distance.

2 α decay of the heaviest elements

The α decay of these very heavy nuclear systems have been studied within the same GLDM and adjusted to reproduce the experimental Q_α value or the value predicted by the Thomas-Fermi model. Preliminarily the α emission half-lives deduced from this model within the WKB barrier penetration penetrability had been compared with the experimental data. The rms deviation between the theoretical and experimental values of $log_{10}[T_{1/2}(s)]$ is only 0.63 for a recent data set of 373 α emitters and 0.35 for the subset of the even-even nuclides. The following analytic formulas have been also extracted, respectively for the even-even, even-odd, odd-even and odd-odd nuclei.

$$log_{10}\left[T_{1/2}(s)\right] = -25.31 - 1.1629A^{1/6}\sqrt{Z} + \frac{1.5864Z}{\sqrt{Q_\alpha}}, \tag{3}$$

$$log_{10}\left[T_{1/2}(s)\right] = -26.65 - 1.0859A^{1/6}\sqrt{Z} + \frac{1.5848Z}{\sqrt{Q_\alpha}}, \tag{4}$$

$$log_{10}\left[T_{1/2}(s)\right] = -25.68 - 1.1423A^{1/6}\sqrt{Z} + \frac{1.592Z}{\sqrt{Q_\alpha}}, \tag{5}$$

$$log_{10}\left[T_{1/2}(s)\right] = -29.48 - 1.113A^{1/6}\sqrt{Z} + \frac{1.6971Z}{\sqrt{Q_\alpha}}. \tag{6}$$

In the heaviest mass region, the measured ranges for $log_{10}[T_{1/2}(s)]$; ^{293}118 [-4.2, -3.5],^{289}116 [-3.5, -2.8], ^{285}114 [-3.5, -2.8],^{281}112 [-3.4, -2.7],^{277}110 [-2.8, -2.1],^{273}Hs [-0.2, 0.5] and ^{289}114 [0.3, 1.4], ^{285}112 [3.1, 4.1] and ^{281}110 [1.8, 2.9] ; agree with our predictions except for ^{281}112 and especially for ^{285}114 and ^{289}114 where the theoretical values are respectively of 5 and 2 orders of magnitude higher than the experimental data. The explanation of this discrepancy is perhaps that $Z = 114$ is not a magic number and that the Thomas-Fermi model underestimates Q_α for $Z = 114$ and then overestimates the half-lives.

It is important to be sure that the observed events are not incomplete fusion events in the external pocket of the potential barrier in the entrance channel.

References

1. G. Royer, R. Moustabchir and R.A. Gherghescu, Proc. Int. Conf. on Nuclear Reaction Mechanisms, Varenna, Italy, (ed. Universita Degli Studi Di Milano, 2000).
2. G. Royer, J. Phys. G **26**, 1149 (2000).
3. R. Moustabchir and G. Royer, Nucl. Phys. A **683**, 266 (2001).

THE EFFECT OF THE ENTRANCE CHANNEL IN COMPLETE FUSION REACTIONS LEADING TO THE PRODUCTION OF HEAVY NUCLEI

R. N. SAGAIDAK, V. I. CHEPIGIN, A. P. KABACHENKO, O. N. MALYSHEV, YU. TS. OGANESSIAN, A. G. POPEKO, A. V. YEREMIN
Flerov Laboratory of Nuclear Reactions, JINR, Dubna 141980, Russia

F. P. HEßBERGER, S. HOFMANN
Gesellschaft für Schwerionenforschung mbH, Darmstadt D-64291, Germany

V. NINOV
Nuclear Science Division, LBNL, Berkeley, California 94720, USA

CH. STODEL
Laboratoire de Physique Corpusculaire, Caen 14000,France

G. GIARDINA, A. LAMBERTO
INFN, Sezione di Catania, and Dipartamento di Fisica dell'Università di Messina 98166, Italy

A. K. NASIROV
Bogoliubov Laboratory of the Theoretical Physics, JINR, Dubna 141980, Russia; Heavy Ion Physics Department, INP, Tashkent, Uzbekistan

The excitation functions for the evaporation residues (ER) produced in the complete fusion reactions induced by ^{86}Kr leading to the ^{190}Hg*, ^{209}Fr*, ^{216}Ra*, ^{216}Th* and ^{222}Th* compound nuclei (CN) have been analyzed in the framework of the HIVAP-code with the potential barrier passing and standard statistical model approximations and using similar data obtained in asymmetric combinations leading to the same CN. The excitation functions for ER obtained in the vicinity of the nominal fusion barrier for nearly symmetric projectile-target combinations show significant suppression of the production cross section for ER in comparison with those which can be predicted by the model. An attempt has been made to find correlations between the extracted values of the fusion probability and the values of the additional barrier, corresponding to the minima of the driving potential.

It is clear now that characteristics of heavy-ion fusion reactions go beyond the conventional potential-barrier passing model of two colliding spherical nuclei. The experimental observations [1−4] show that two very massive nuclei have a strongly reduced complete-fusion cross section at incident energies around the expected value of the nominal fusion barrier [5]. Macroscopic models [6,7] predict that the "extra-extra push" (EEP) energy is needed for the "true" fusion of these massive nuclei with a formation of the equilibrated compact mononucleus. The diabatic approach [8] leads to a fast dynamical rise in the fusion barrier which is predicted to be strongly dependent on the nuclear structure. In the diabatic limit no re-arrangement of nucleons takes place. The additional value to the fusion barrier is determined by the number of nucleon pairs distributed equally on pair states spreaded over some energy range. A good "scaling" for both measured extra energies and the associated fluctuations was found with taking into account the number of valence nucleons (holes) above (below) the nearest closed shell of the projectile and target nuclei [4].

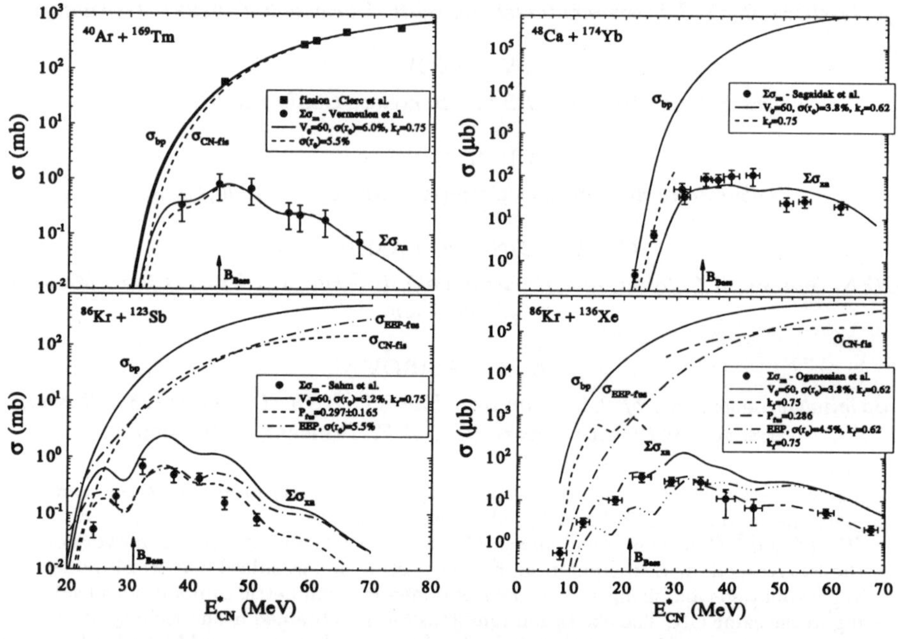

Figure 1. Excitation functions for the production of ER (filled circles — experimental data and lines — calculations) in the asymmetric and nearly symmetric combinations leading to the ^{209}Fr* (left panels) and ^{222}Th* (right panel) CN.

In continuation of these studies, we have measured the production cross sections for ER in the ^{86}Kr + ^{130}Xe and ^{86}Kr + ^{136}Xe reactions [9]. Production of ER is unambiguous signature of fusion. A strong suppression of fusion for the reaction on ^{130}Xe and a seeming absence of any suppression in the reaction on ^{136}Xe (fusion of two neutron-shell nuclei) [9,10], as it follows from our analysis with the potential-barrier passing model coupled with the standard statistical model [11], compelled us to look at the fusion probability in the reactions induced by ^{86}Kr once again. It is necessary from the point of view of the new data obtained recently in the ^{48}Ca + ^{174,176}Yb reactions [12]. In the reaction induced on ^{174}Yb leading to the same compound nucleus as in ^{86}Kr + ^{136}Xe, we observed relatively high production cross sections for ER corresponding to the 1n (α1n) and 2n evaporation channels at deep sub-barrier energies. The same ER are produced in the 3n and 4n evaporation channels observed in the ^{48}Ca + ^{176}Yb reaction. To describe both sets of data we had to use the same value of the scaling parameter at the liquid-drop fission barriers (k_f = 0.75). The same value of k_f was obtained in the analysis of excitation functions corresponding to the production of the deformed nuclei far from the N = 126 neutron shell, whereas the production of spherical ER in the vicinity of N = 126 (5n and 6n evaporation channels in the case of ^{48}Ca + ^{174}Yb), correspond to k_f = 0.55–0.65 [13]. This difference is the result of the collective enhancement in the level density [14], which is not taken into account in the HIVAP-code [11]. We applied these finding (obtained in the analysis of asymmetric combinations) to the ^{86}Kr induced reactions data. We fitted the parameter values of the nuclear potential (the strength, V_0, and the barrier fluctuation, $\sigma(r_0)$) as is shown in Fig. 1 for the ^{209}Fr* and ^{222}Th* CN.

Table 1. Fusion probabilities and additional potential barriers corresponding to re-sharing masses of fusioning nuclei in the ^{86}Kr induced reactions.

CN	V_0 (MeV/fm)	$\sigma(r_0)$ (%)		ΔB(GSM)	ΔB(LDM)
	$P_{fus}(E \lesssim B_{Bass})$		$P_{fus}(E>B_{Bass})$	(MeV)	(MeV)
	70	4.0±0.2			
^{190}Hg*	0.22±0.02			8.7	0.8
	60	3.2±0.2			
^{209}Fr*	0.30±0.17		0.30±0.17	11.2	2.5
	80	4.0±0.5			
^{216}Ra*	⩽ 0.2 (EEP)			14.9	1.2
^{216}Th*	60	2.0±1.0	0.007±0.003	15.0	3.4
	60	4.4±0.2			
^{222}Th*	Defined by EEP		0.29±0.01	17.0	1.2

The extracted values of the fusion probability (P_{fus}) at the energy below and above the Bass barrier [5] for the considered reactions are summarized in Table 1. It is important to note that almost all the reactions, as well as similar ones considered earlier [15], are described well with the constant value of P_{fus}

at energies below and above the nominal fusion barrier [5] without EEP. We attempted to find correlations between P_{fus} and the values of the additional barrier (ΔB) corresponding to the minima of the potential energy surface in the (Z,A)-coordinates [16] calculated for touching configurations of spherical nuclei (determined by their central radii [7]) and using ground-state (GSM) and liquid-drop (LDM) masses of nuclei. As it follows from the table, there are no evident correlations between the values of ΔB calculated in such simplified way and P_{fus}. Further considerations are needed for finding ingredients determining the fusion probability in nearly symmetric massive systems.

Acknowledgments

This work was performed partially under the financial support of the RFBR, contract 99–02–16447 and INTAS, contract 991–1344. The authors thank M. Morozova for her help in preparing the manuscript.

References

1. C.-C. Sahm *et al*, *Nucl. Phys.* A **441**, 316 (1985).
2. W.O. Shen *et al*, *Phys. Rev.* C **36**, 115 (1987).
3. W. Reisdorf *et al*, *Z. Phys.* A **342**, 411 (1992).
4. A.B. Quint *et al*, *Z. Phys.* A **346**, 119 (1993).
5. R. Bass, *Lect. Notes in Phys.* **117**, 281 (1980).
6. W.J. Swiatecki, *Nucl. Phys.* A **376**, 275 (1982).
7. J.P. Blocki *et al*, *Nucl. Phys.* A **459**, 145 (1986).
8. D. Berdichevsky *et al*, *Nucl. Phys.* A **499**, 609 (1989).
9. Yu.Ts. Oganessian *et al* in *JINR FLNR Scientific Report 1995-1996*, (JINR, E7-97-206, Dubna, 1997, p. 62).
10. R.N. Sagaidak *et al* in *Heavy Ion Physics. VI International School-Seminar, Dubna, Russia, 1997*, (WS, Singapore, 1998, p. 323).
11. W. Reisdorf, *Z. Phys.* A **300**, 227 (1981); W. Reisdorf, M. Schädel, *Z. Phys.* A **343**, 47 (1992).
12. R.N. Sagaidak *et al* in *JINR FLNR Scientific Report 1997-1998*, (JINR, E7-2000-232, Dubna, 2000, p. 60).
13. R.N. Sagaidak *et al* in *International Workshop "Fusion Dynamics at the Extremes", Dubna, Russia, 2000*, (WS, Singapore, 2001, p. 135).
14. A.R. Junghans *et al*, *Nucl. Phys.* A **629**, 635 (1998).
15. R.N. Sagaidak *et al* in *International Conference on Nuclear Physics "Shells–50", Dubna, Russia, 1999*, (WS, Singapore, 2000, p. 199).
16. H. Freiesleben and J.V. Kratz, *Phys. Rep.* **106**, 1 (1984).

FUSION CROSS SECTION FOR CHARGED PARTICLES BY USING THE OPTICAL THEOREM

A. SCALIA

Dipartimento di Fisica e Astronomia Università di Catania and INFN, Corso Italia 57, Catania I-95129, Italy
E-mail: scalia@catania.infn.it

One of the most remarkable results of collision theory is the Optical Theorem, an extension to charged particles has been often used in nuclear physics to extract the reaction cross section. In the present work the Optical Theorem is used to obtain the fusion cross section for charged particles at energies below the Coulomb barrier. In fact by using experimental evidences and the Optical Theorem it is possible to obtain the fusion cross section in terms of Rutherford cross section and experimental values of differential elastic cross section.

1 Elastic Scattering and Fusion

1.1 Optical Theorem

During the past several years elastic scattering angular distributions have become available for variety of targets and beam energies, so that at present it is possible to extract from experimental data the behaviour of elastic scattering angular distributions when the energy is varying. Elastic scattering differential cross section depicted as the ratio to Rutherford scattering exhibit the same general features. At forward angles the elastic scattering is Rutherford within experimental error, then the ratio to Rutherford elastic scattering angular distributions show oscillations, after the elastic scattering falls off rapidly increasing angle. However by inspection of experimental results [1-4] it is possible to see that the oscillations of the ratio to Rutherford elastic scattering angular distributions become more and more smooth when the energy decreases, so that at energy sufficiently low, $E < E_B$, E_B is the energy of the Coulomb barrier, it is reasonable to assume that

$$\frac{\sigma_{el}(\theta)}{\sigma_R(\theta)} = 1 \tag{1}$$

for $\theta < \theta_S$, where $\sigma_{el}(\theta)$ is the elastic scattering differential cross section, $\sigma_R(\theta)$ is the Rutherford cross section and θ_S is a suitable angle energy dependent and different for different reactions, which increases when the energy decreases [1-4]. By using the Optical Theorem for charged particles [5-11] it is possible to write the reaction cross section as

$$\sigma_r = \frac{4\pi}{\kappa} Imf'(0) + 2\pi \int_0^{\pi} (\sigma_R(\theta) - \sigma_{el}(\theta)) \sin(\theta) d\theta \tag{2}$$

where κ is the wave number and $Imf'(0)$ is the imaginary part of the residual (nuclear) amplitude $f'(\theta)$ calculated for $\theta = 0$, defined as

$$f(\theta) = f_R(\theta) + f'(\theta) \tag{3}$$

$f(\theta)$ is the elastic scattering amplitude, $f_R(\theta)$ is the Coulomb scattering amplitude and $f'(\theta)$ is the residual (nuclear) amplitude. Now if we consider the elastic scattering process for charged particles at energies sufficiently low, $E < E_B$, from eq.(1) it follows

$$Imf'(0) = 0 \tag{4}$$

and

$$\int_0^{\theta_s} (\sigma_R(\theta) - \sigma_{el}(\theta)) \sin(\theta) d\theta = 0 \tag{5}$$

so that eq.(3) becomes

$$\sigma_r = 2\pi \int_{\theta_s}^{\pi} (\sigma_R(\theta) - \sigma_{el}(\theta)) \sin(\theta) d\theta. \tag{6}$$

At low energy, $E < E_B$, we can assume $\sigma_r = \sigma_f$, where σ_f is the fusion cross section, so that finally we have

$$\sigma_f = 2\pi \int_{\theta_s}^{\pi} (\sigma_R(\theta) - \sigma_{el}(\theta)) \sin(\theta) d\theta. \tag{7}$$

References

1. J.E. Poling, E. Norbeck, R.R. Carlson,*Phys. Rev.* C **13**, 648 (1976).
2. H. Wojciechoski, D.E. Gustafson, L.R. Medskert, R.H. Davis,*Phys. Lett.* B **63**, 413 (1976).
3. F. Videbaek, R.B. Goldstein, L. Grodzins, S.G. Steadman,*Phys. Rev.* C **15**, 954 (1977).
4. K. Daneshvar, D.G. Kovar, S.J. Krieger, K.T.R. Davis,*Phys. Rev.* C **25**, 1342 (1982).

5. J.T. Holdeman, R.M. Thaler,*Phys. Rev. Lett.* **14**, 81 (1965).
6. J.T. Holdeman, R.M. Thaler,*Phys. Rev.* B **139**, 1186 (1965).
7. R. Giordano, F. Porto, S. Sambataro, A. Scalia,*Lett.Nuovo Cimento* **31**, 189 (1981).
8. S. Costa, R. Giordano, F. Porto, S. Sambataro, A. Scalia,*Nuovo Cimento* **72**, 40 (1982).
9. C. Marty,*Z. Phys.*A **309**, 261 (1983).
10. C. Marty,*Z. Phys.*A **322**, 499 (1985).
11. A.N. Ostrowski, W. Tiereth, H. Voit,*Phys. Rev.* C **44**, 2082 (1991).

DECAY OF COMPOUND NUCLEI *

C. SCHMITT AND J. BARTEL
IReS, Univ. L. Pasteur and IN2P3/CNRS, Strasbourg, France
E-mail: schmitt@ires.in2p3.fr

A. SUROWIEC AND K. POMORSKI
Katedra Fizyki Teoretycznej, UMCS, Lubin, Polska
E-mail: angasow@tytan.umcs.lublin.pl

Fission dynamics is investigated by solving the Langevin equation in a multi-dimensional deformation space. The influence of a neck degree of freedom, mass asymmetry and non-axiallity on fission times, and consequently, on prefission particle mutiplicities, is discussed. Differences of fission dynamics at low and high excitation energy, as well as deformation dependence of light particle emission width in Weisskopf and Thomas-Fermi models are pointed out.

1 Introduction

The model we are using [1] describes the dynamical evolution of an excited, rotating, deformed nucleus and its subsequent decay through fission, or ending up as an evaporation residue, taking into account particle evaporation channels. As the systems we considered until now were highly excited, we restricted ourselves to a semi-classical picture, and consequently to symmetric fission, and performed the study in an one-dimensional deformation space. In this context, the comparison between theory and experiment for prefission particle multiplicities is satisfactory over a wide range of nuclear masses [2]. Recently, a lot of experimental data became available dealing with systems at low excitation energy and exhibiting competition between symmetric and asymmetric fission. The system : $^{18}O +^{209} Bi \longrightarrow^{227} Pa$ was investigated at the JINR in Dubna at an excitation energy of $E^* = 26$ MeV. We are therefore aiming at extending our model to asymmetric fission.
To start with, let us temporarily omit shell effects which are important at low excitation energies, and simply include the asymmetry degree of freedom into the description. We won't include particle emission either, but rather investigate, as a first step, the influence of different aspects of our model on fission times, and consequently on prefission particle multiplicities.

*THE WORK IS PARTLY SUPPORTED BY THE POLISH COMMITY OF SCIENTIFIC RESEARCH UNDER CONTRACT NO. 2P03B011 12, THE POLONIUM FELLOWSHIP NO. 01704UG (2001) AND THE $IN2P3$ CONVENTION 3995.

2 Dynamical description

In order to treat the competition between symmetric and asymmetric fission, at least two coordinates are necessary for describing the evolution of the compound nucleus (CN). In this context, we use the Funny-Hills shape parametrization [3] which is characterized by three deformation parameters : c connected to elongation, α to left-right asymmetry and h to the neck formation.
In a multi-dimensional space, the dynamical evolution of the compound system is studied by the resolution of the Langevin equations :

$$
\begin{aligned}
\frac{dq_i}{dt} &= \sum_j [M^{-1}(\vec{q})]_{ij}\, p_j \\
\frac{dp_i}{dt} &= -\frac{1}{2}\sum_{j,k} \frac{d[M^{-1}(\vec{q})]_{jk}}{dq_i}\, p_j\, p_k - \frac{dV(\vec{q})}{dq_i} - \sum_{j,k} \gamma_{ij}(\vec{q})\, [M^{-1}(\vec{q})]_{jk}\, p_k + F_L(t)
\end{aligned}
\qquad (1)
$$

where M^{-1} is the inverse of the irrotational flow mass tensor in the Werner-Wheeler approximation, γ the wall and window friction tensor, V the collective potential determined through the finite temperature Liquid Drop Model (LDM) and $F_L(t)$ the fluctuating Langevin force related to diffusion (see ref. [1] for details).

3 Influence of several parameters on fission times

3.1 Neck degree of freedom

Choosing $\vec{q} = (c, \alpha)$ in eqs.(1) we present on figure 1, the LDM energy landscape taking $h = 0$: one can notice the deep symmetric fission valley ($\alpha = 0$). On this figure, we also show a typical fission trajectory : as no shell effects are presently included, the compound system ends up as symmetric fission fragments after a more or less long trajectory in the (c, α) plane.
The (c, α) energy landscape obtained after minimization with respect to the third deformation parameter h differs only slightly from the one with $h = 0$. Moreover, the mean fission time obtained after minimization with respect to h ($\bar{t} = 5.96 \cdot 10^{-17}s$) and the one calculated with $h = 0$ ($\bar{t} = 5.27 \cdot 10^{-17}s$) are quite similar. We will therefore from now on only consider (c, α) energy landscapes with $h = 0$.

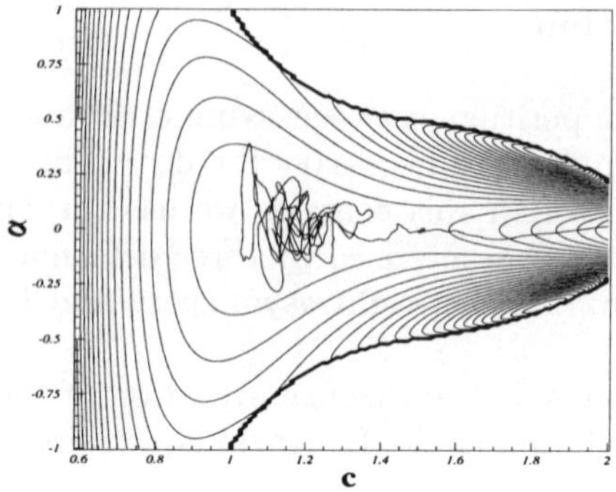

Figure 1. Liquid Drop energy landscape (c, α) with $h = 0$ for the compound system ^{227}Pa at $E^* = 26$ MeV and $L = 60\hbar$, and a typical fission trajectory

3.2 One-dimensional versus two-dimensional treatment

For the description of symmetric fission we solved an one-dimensional Langevin equation. In the two-dimensional description presented above, this corresponds to a trajectory along $\alpha = 0$ which is close to the bottom of the symmetric fission valley. As can be seen from the trajectory in figure 1, "excursions" into the asymmetric landscape are present, even if the dynamics finally ends up in the symmetric fission channel. The mean fission times, without particle emission, are quite comparable ($\bar{t} = 5.27 \cdot 10^{-17} s$ for the two-dimensional versus $\bar{t} = 6.36 \cdot 10^{-17} s$ for the one-dimensional equation), which gives us some confidence in the previously predicted particle multiplicities obtained with the one-dimensional description.

It could seem astonishing that the one-dimensional fission time is larger than the two-dimensional, i.e. that it takes more time for the CN to follow the bottom of the valley rather than to oscillate in the asymmetric landscape. But one could think of the Langevin equation as an approximation to the Focker-Planck equation which deals with probability distributions : the larger the dimensionality, the less constraints, or more flexibility, one has.

3.3 Presence of oblate deformation

Many systems, and in particular the system considered here, fission at large angular momenta L since fission barriers decrease with increasing L. In that case the semi-classical ground state can show an oblate deformation. This is the case for the CN ^{227}Pa. The resolution of the Langevin equation in the (c, α) space obtained after inclusion of this oblate minimum leads to a slightly higher fission time ($\bar{t} = 5.72 \cdot 10^{-17} s$) compared to the one calculated without this correction ($\bar{t} = 5.27 \cdot 10^{-17} s$), due to a slight increase of the fission barrier heigth. Nevertheless, the difference being small (8 %), we expect only little influence of this effect on prefission multiplicities.

3.4 Diffusion at low excitation energy

At high excitation energy, the diffusion tensor D can be related to the friction tensor γ and nuclear temperature T via the Einstein relation $D = \gamma T$. However, zero-point motion makes diffusion not to go to zero with temperature, so that this expression is no longer valid at low energy. Thus, we will use a modified Einstein relation $D = \gamma T^*$ where $T^* = (\hbar\omega/2)tgh(\hbar\omega/2T)$ and $\hbar\omega$ (≈ 1 MeV) is the energy of a typical collective excitation [4]. The fission time obtained is about 7 % smaller than the one calculated with $D = \gamma T$.

4 Particle emission width

In the case of highly excited nuclei, particle emission widths can be estimated using the statistical Weisskopf evaporation theory [5]. At low energy, however, this model seems to be questionable. Recently, we developed a more microscopic description of particle emission rates in the framework of the Thomas-Fermi theory : this new formulation is based on the picture of a Fermi gas of nucleons at temperature T, confined to a deformed rotating mean field [6]. On figure 2, we compare the emission widths obtained in the two approaches. One notices that, if for neutrons both models yield similar results, the Weisskopf theory underestimates substantially the emission widths for protons at large deformation.

5 Summary and conclusions

The average fission times evaluated in the one and two-dimensional description, without taking into account particle evaporation, are quite similar. The fission trajectories starting at an oblate minimum, present for rotating nuclei, yield fission times about 8 % larger than those beginning from a prolate minimum. The mean fission time obtained with the diffusion tensor evaluated with the use of the modified Einstein relation is about 7 % shorter than the one estimated with the standard Einstein relation. The proton emission widths obtained within the Thomas-Fermi model are significantly larger for

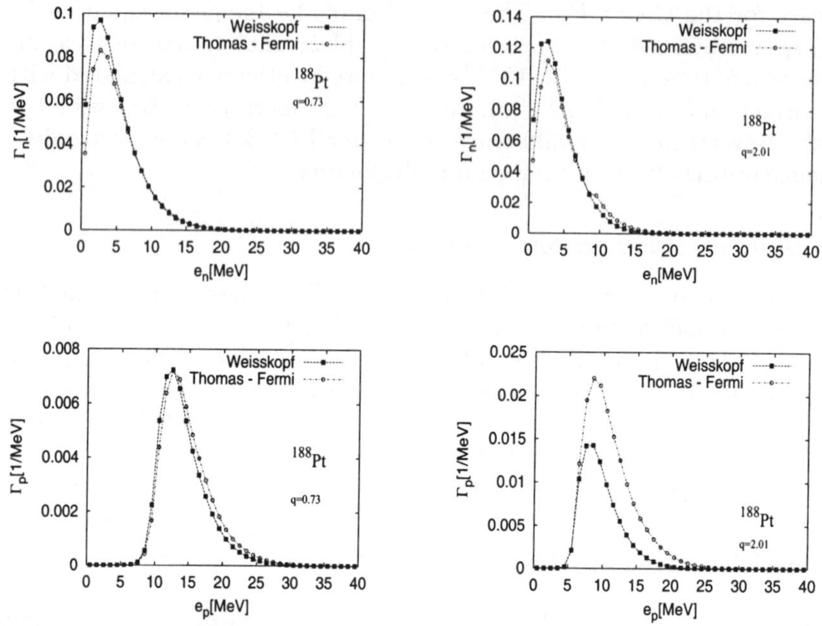

Figure 2. Emission width as function of particle energy : for neutrons (top) and protons (bottom) for almost spherical (left) and very elongated shapes (right)

very deformed shapes compared to those estimated with the Weisskopf theory. This could turn out to be important for our predictions of prefission charged particle multiplicities.

References

1. K. Pomorski et al., Nucl. Phys. **A605** (1996) 87.
2. K. Pomorski et al., Nucl. Phys. **A679** (2000) 23.
3. M. Brack et al., Rev. of Modern Phys. **vol.44** (1972) 320.
4. H. Hofmann, F.A. Ivanyuk, Phys. Rev. Lett. **vol.82** (1999) 4603 .
5. V. Weisskopf, Phys. Rev. **52** (1937) 295.
6. K. Dietrich et al., Z. Phys. **A351** (1995) 397.

STUDY OF TRANSFER INDUCED FISSION AND FUSION-FISSION REACTIONS FOR 28SI + 232TH SYSTEM AT 340 MEV

D. FABRIS, M. LUNARDON, G. NEBBIA, A. SAMANT, G. VIESTI, V. RIZZI.
Dipartimento di Fisica and INFN, Padova, Italy

E. FIORETTO, G. PRETE, M. CINAUSERO, D. V. SHETTY, S. PESENTE
INFN, Laboratori Nazionali di Legnaro, Legnaro, Italy

BRONDI, G. LA RANA, R. MORO, E. VARDACI, A. BOIANO, A. ORDINE
Dipartimento diScienze Fisiche and INFN, Napoli, Italy

N. GELLI, F. LUCARELLI
Dipartimento di Fisica and INFN, Firenze, Italy

A. SAXENA, B. K. NAYAK, D. C. BISWAS, R. K. CHOUDHURY, S. S. KAPOOR
Bhabha Atomic Research Centre, Mumbai, India

Transfer induced fission probabilities P_f have been measured in the collisions of 340 MeV ^{28}Si on ^{232}Th target as a function of the detected projectile-like fragment (PLF) charge. The measured P_f shows an increase with the net charge transfer ΔZ, in the range ΔZ =0-4. For larger ΔZ values (ΔZ =5-10), the fission probability exhibits first a saturation in the range of values P_f =0.4-0.6 and than a drop. Coincidences between PLF and light charge particles reveal that the charged particle emission is strongly damped when the fission fragments is required. An important component in the charged particle emission is localized in the direction of the recoiling target-like fragment (TLF); which is, in some cases, directly detected. Therefore results indicate a significant survival against fission of the populated TLF fragments.

1 Introduction

Transfer reactions on heavy targets offer the possibility of populating a wide range of target-like (TLF) fissioning nuclei at various excitation energies, which can prove useful to study in a systematic way the fission reaction mechanisms. In particular, the study of the fission probability P_f is not only interesting by itself, for the contained information on the fission dynamics [1], but also because it gives a direct measure of the possibility of survival (P_s=1- P_f) of the populated target like

fragments and thus the possibility of producing, by means of transfer reactions, specific nuclear species [2]. This is particularly appealing when TLF lies in mass regions of peculiar interest for the nuclear structure studies, as the super-heavy element region. Dynamical model studies for the production of super-heavy elements have quantitatively demonstrated that the fission hindrance factor plays a relevant role in defining the survival probability against fission of the populated composite systems [3]. Therefore, the detailed knowledge of the fission dynamics and the direct measures of the fission probability carry valuable information to establish the optimal conditions for the population of those rare nuclear species.
An experimental project aimed to study transfer-fission reactions has been started at the accelerator facilities of the Laboratori Nazionali di Legnaro, by using the newly completed 4π charged particle detector system. As a first step of this program, an exploratory study has been completed for the reactions of 340 MeV ^{28}Si on ^{232}Th. Results obtained in this work are presented here.

2 Experimental results

In the present work, we have carried out measurements of multi-nucleon transfer induced fission reactions in the ^{28}Si+ ^{232}Th system. The experiment has been performed at the accelerator complex of the Laboratori Nazionali di Legnaro consisting in the XTU Tandem and the Super-conductive Linac ALPI used as a post-accelerator. The 340 MeV ^{28}Si beam, having about 1 pnA intensity, was used to bombard a self-supporting ^{232}Th target of 1.5 mg/cm^2 thickness.
The 8πLP detector set-up [4] was employed. It consists of two detection sub-systems made of ΔE-E telescopes: the WALL, placed at 60 cm from the target, covering angles from $\Theta_{lab}= \pm 2.5°$ to $\Theta_{lab}=\pm 26°$, and the BALL having 15 cm internal radius, placed around the target. Each telescope uses a transmission 300 μm thick Si detector followed by a CsI(Tl) crystal, 15 mm (WALL) or 5 mm thick (BALL) . The telescopes stop light charged particles with energies up to 64 AMeV (WALL) and 34 AMeV (BALL), respectively. The silicon detectors of the BALL are mounted in flipped mode and allow the identification of light particles and fission fragments by Pulse Shape Discrimination method. Projectile like fragments (PLF) were detected using 3 ΔE-E telescopes, made by 40 μm + 300 μm silicon detectors, placed around the grazing angle (Θ_{lab}=39.5°, 45.5° ,45.5°) with a solid angle of 1.3 msr each. PLF in single mode and in coincidence with Fission Fragments (FF) and light Charged Particles (LCP) have been analyzed. From the PLF energy spectra the experimental mean energy-loss as a function of transferred Z (ΔZ) was obtained and excitation energy and spin of target-like fragments (TLF) was evaluated in the frame of two-body inelastic collision. It resulted in a constant increase of TLF excitation energy and spin as a function of transferred Z. The fission probability P_f was deduced for each ΔZ from the ratio between the yields of transfer induced fission events (defined as the coincidence between the PLF and at least one fission fragments detected in the Ball detectors) and the single yields of

PLF.

The coincidence rate was corrected first by tacking into account the fraction of the total solid angle open for the fission decay covered by the BALL detectors used to detect fission fragments. Secondly, kinematical effects were accounted for, associated with the changes in direction of the recoiling TLF when different ΔZ are considered. It is seen in fig.1 that the fission probability P_f first increases with increasing the net transferred charge ΔZ up to $\Delta Z=4$ and then shows a kind of saturation around the values $P_f=0.5$ with a final decrease for the higher Z-transfers. The rise of the fission probability at increasing ΔZ is qualitatively understood as due from one side to the strong increase of the estimated average TLF excitation energy and angular momentum in increasing the net transferred charge ΔZ and, on the other side, to the decrease in the fission barrier in populating higher Z_{TLF}.

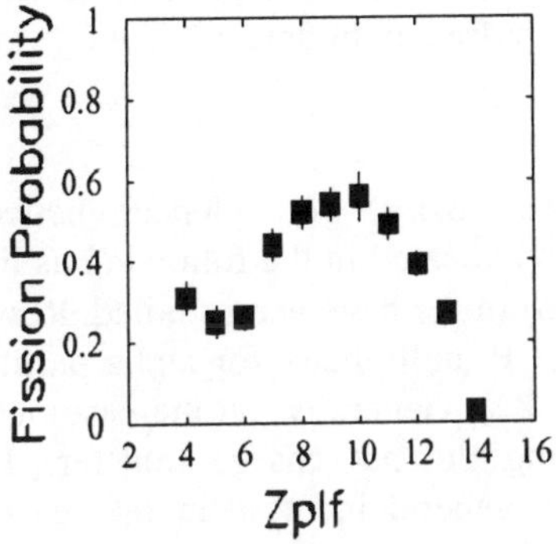

Fig.1 Fission probability vs. Z_{PLF}

Results from the statistical model calculations (PACE2) point out that TLF fragments should fission with very high probability already in the case of $\Delta Z=1$. In the calculations, the fission probability rises and saturates immediately to the value $P_f=1$ for $\Delta Z>2$. The experimental decrease of P_f at large ΔZ, which corresponds to large PLF excitation energy, can be generally explained only considering the opening of new decay channels at high excitation energy and spin that compete strongly with fission. This can be the case when fission times becames longer than typical particle emission times. In this case the dissipation of excitation energy and spin associated by the particle emission might leave the residual nucleus is a state for which the fission probability is smaller, giving rise to the apparent decrease of the measured fission probability if fission is strongly dependent on excitation energy and spin. This seems not to be the case for the nuclei populated in this reaction in which the low fission barrier make fission highly probable. A second explanation for this effect is that the measured energy loss is hardly correlated to the excitation energy and spin of the TLF fragment. This is certainly the case of more complex reaction mechanisms as those due to projectile break-up.

On the other end the fact that a sizeable fraction of the primary TLF fragments survives fission, as suggested by our fission probability measurement, can be experimentally proved by searching directly for heavy, low energy fragments detected in coincidence with PLF. Despite the reduced coverage of the Ball detectors ($\Theta_{lab}>51^0$), TLF could be directly detected in the case of $Z_{PLF}=12$ ($\Delta Z=2$). The Statistical Model predicts a very large fission probability for these events, on the contrary, we observe directly signals of heavy mass ions in kinematical coincidence with PLF. The events, characterized by very low energy (~5 MeV), are supposed to be TLF and are well separated from the position of fission fragments. The measured kinetic energy is compatible with that estimated from kinematics, corrected for the energy loss in the target and the pulse height defect.

3 Light Charged Particle results

To explore more in detail the reaction mechanisms, coincidences between projectile-like fragments and protons and alphas (indicated in the following as light charges particles LCP) detected in the 8πlp spectrometer have been studied. Results are reported in Fig.2 in term of the measured LCP multiplicity for alpha particles (protons have the same trend) as a function of Z_{plf}. Generally, an increase of the particle multiplicities is evidenced in increasing the net charge transfer. This increase is somewhat expected, considering the general increase in the reaction inelasticity as a function of ΔZ. The impressive fact in Fig.2 is that the quite large particle multiplicity for proton and alpha particles measured for $Z_{PLF}<8$ is completely damped when a further coincidence with a fission fragment is required. This indicates clearly that the events for which we do not detect a fission fragment in coincidence with PLF are really a different class of events respect to the detected PLF-fission coincidences, characterized by a higher charged particle multiplicity.

To better qualify the PLF-LCP coincidences, we have also reconstructed their angular correlation functions. The Φ correlation is dominated by a pronounced peak centered in the reaction plane (Φ=180°) on the opposite side of the PLF trigger demonstrating that a sizeable part of the LCP yield is coming from the sequential decay of the complementary TLF.

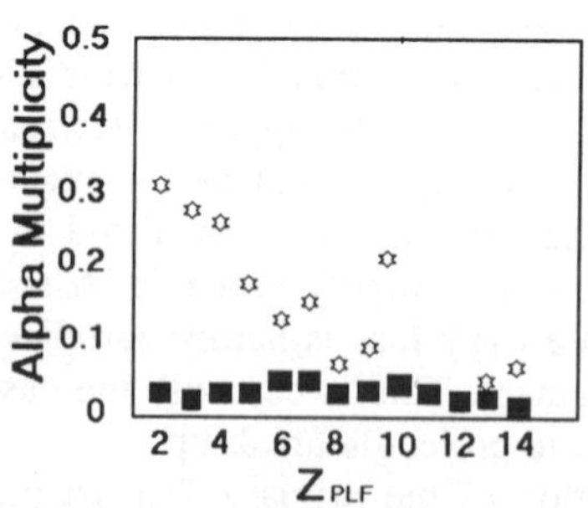

Fig. 2 Alpha particle multiplicity in coincidence with PLF(stars) and Fission Fragments (squares).

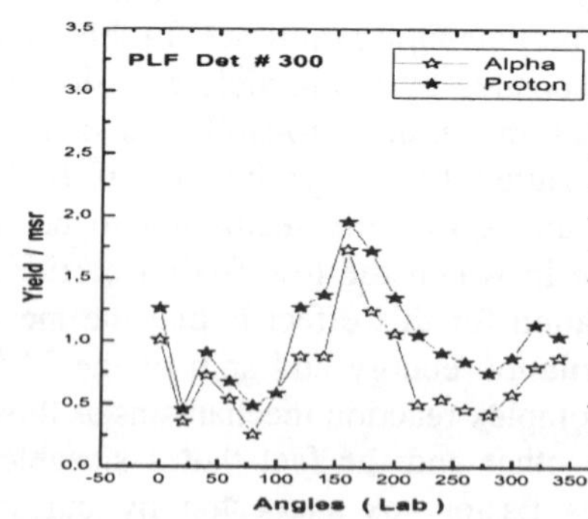

Fig.3 PLF-LCP out-of –plane angular correlation function (Φ), integrated on Θ

A second, weaker, component is located also in plane, from the same side of the trigger, at Φ=0° and Φ=360°. This component is certainly associated with the PLF sequential decay. It appears from the correlation functions that the bulk of the LCP emission is consistent with the assumption of a sequential decay of the fragments produced in a two-body collision.

4 Conclusions

The multi-nucleon transfer reaction induced by 340 MeV ^{28}Si on a ^{232}Th target have been studied by using a large area detection system for fission fragments and light charged particles triggered by projectile-like fragments detected at an angle of about Θlab=40°, close to the grazing angle. It was found that the measured ratio between PLF in coincidence with at least one fission fragment and the PLF singles maintains, on average, around the value 0.4-0.6, suggesting that the survival probability of TLF fragments against fission is not as small as predicted by the standard Statistical Models. The analysis of the light charged particles emitted in coincidence with PLF demonstrates that PLF in anti-coincidence with fission fragments are associated to a sizeable light charged particle emission that is greatly reduced when the coincidence with fission fragments is required. The evidenced fission hindrance (or survival probability) can be explained in details only by knowing the regions of the excitation energy-angular momentum plane where the TLF are populated (with or without the presence of projectile break-up) and the dynamical effects associated to the nuclear viscosity. The latter effects are known to play a crucial role in tuning the fission-evaporation competition in this mass region [3]. Quantitative modeling of the TLF decay appears, however, to be very difficult at the present level of knowledge in the field. On the other side, the effect reported here, when confirmed by systematic measurements, is quite interesting from the point of view of the production of nuclei with Z>92 for nuclear structure studies.

References

1. J.B. Natowitz et al., Phys. Lett. B247 (1990) 242.
2. J.F.C. Cooks et al., J. Phys. G 26 (2000) 23 and references therein.
3. Y. Aritomo et al., Phys. Rev. C59 (1999) 796.
4. G. Prete et al. ,Nucl Inst. and Meth. A422 (1999) 263.

STATISTICAL DECAY OF HEAVY NUCLEUS-NUCLEUS SYSTEMS

K. SIWEK-WILCZYŃSKA

Institute of Experimental Physics, Warsaw University, Hoża 69, 00-516 Warsaw, Poland

E-mail: siwek@npdaxp.fuw.edu.pl

A hybrid model combining deterministic dynamics with evaporation of light particles and/or dissipative fission is discussed. The model is applied in analysis of neutron multiplicities measured in fusion-fission and fast fission reactions in collisions of heavy nuclear systems. Results of simulations are compared with the correlation between the prescission and postscission multiplicities deduced with the backtracing method by the DéMoN Collaboration for a heavy composite system Z=110.

1 Introduction

Fusion-fission and fast fission reactions are the large-scale mass rearrangement processes best suited for measurements of the reaction time scale by using the "neutron clock" during the prescission stage of the reaction. Measurements of the prescission neutron multiplicities revealed already more than a decade ago that the fusion-fission processes seem to proceed slower than expected on the grounds of the standard transition-state theory of fission proposed by Bohr and Wheeler [1]. An alternative view of nuclear fission, formulated in terms of the Kramers [2] transport theory, gives an explanation of this phenomenon. Another framework, of a similar physical background as the latter approach, stems from employing Langevin dynamics to the description of fission reactions.

In the Kramers model of fission, the coupling between fluctuations and dissipation is assumed, leading to the dependence of the time scale of fission on nuclear dissipation. In such a way, measurements of the prescission neutron multiplicities, the "neutron clock", can be used for determination of viscosity of nuclear matter.

Similarly, evaporation of light particles in strongly damped collisions and "fast fission" reactions (when fusion does not occur) may also play the role of a clock measuring the time scale of nuclear interaction. These non-fusion reactions can be described in terms of deterministic dynamical models [3] involving dissipative (frictional) forces.

Both theoretical tools, deterministic dissipative dynamics and the transport theory approach, have to be employed in analysis of prescission neutron multiplicities in fusion-fission reactions because the latter formalism applies only until the saddle point is passed. Beyond the saddle point, when fission is decided, the deterministic dynamics become the most convenient framework for calculating post-saddle evaporation.

In this presentation I will discuss the general scheme of a hybrid model for analysis of neutron multiplicities measured in both fusion-fission and fast fission reactions in collisions of heavy nuclear systems.

2 Structure of the Model Calculations

For heavy nuclear systems, overcoming the interaction barrier and formation of the composite system does not guarantee that the system will eventually fuse. Therefore, in our simulations we consider two scenarios: the fusion-fission scenario (in which the compound-nucleus fission is assumed), and the fast fission scenario, in which the time evolution of the combined system is entirely described by deterministic dynamics. For distinction between these two types of reactions, we use predictions of the dynamical code HICOL [3], which is based on the concept of one-body dissipation [4] and was proved to reasonably describe essential features of nucleus-nucleus collisions. Formation of the compound nucleus is predicted for all partial waves below a certain limiting value l_{fus}. For all higher partial waves, fast-fission-like processes are predicted. The boundary between these two types of processes strongly depends on the details of the nucleus-nucleus interaction. Consistency between predicted cross sections of the fusion-fission and fast-fission components with observed mass distributions of the fission fragments is required.

For both, compound-nucleus fission and fast-fission scenarios, we calculate pre-scission and post-scission multiplicities, ν_{pre} and ν_{post}, respectively. In the case of compound-nucleus fission (the left branch in Fig. 1), ν_{pre} consists of two differently calculated components: pre-saddle and saddle-to-scission. The pre-saddle neutron multiplicity in a fusion-fission process is calculated [5] with a time-dependent Monte Carlo program in which evaporation of light particles competes with dissipative fission described with a modified [6] Kramers formalism. The number of neutrons emitted during the pre-saddle stage depends on the value of the dissipation coefficient in the Kramers hindrance factor, which in the stationary limit reduces the fission width Γ^K relative to the nonviscous Bohr-Wheeler width Γ^{BW}:

$$\Gamma^K = \Gamma^{BW}(\sqrt{1+\gamma^2} - \gamma), \tag{1}$$

where γ is the dimensionless dissipation coefficient. Following the Grangé-Weidenmüller model [6], a time dependence of Γ^K is introduced into Eq. (1)

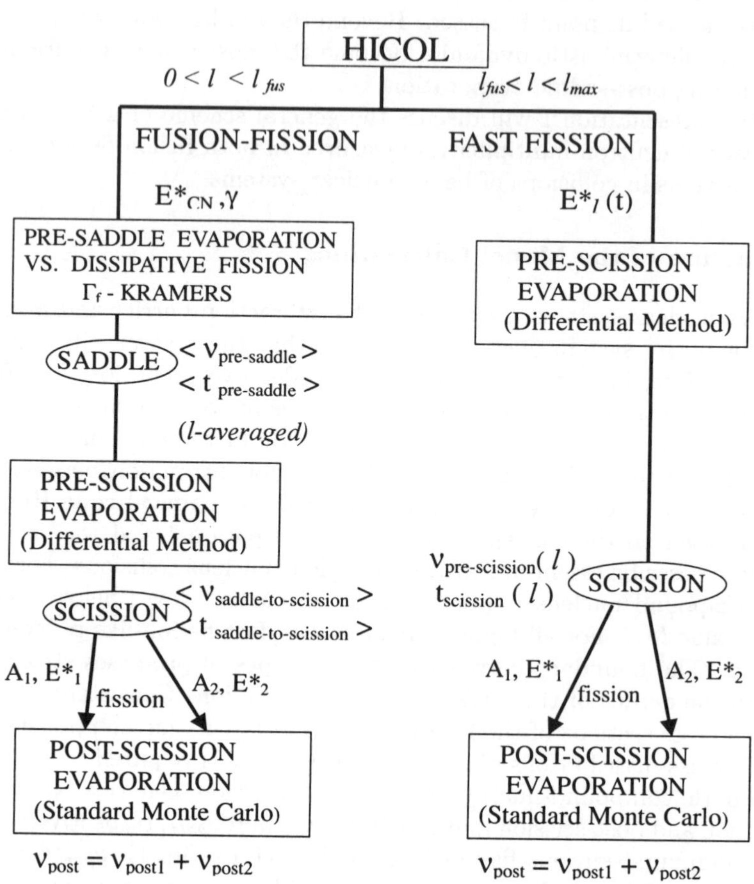

Figure 1. Calculation scheme of pre- and postscission neutron multiplicities in fusion-fission and fast fission reactions.

accounting for the transient time needed to build up the quasistationary probability flow over the fission barrier.

When the saddle point is reached (see the left branch in Fig. 1), fission is decided, and the system slides towards the scission point. Evaporation of neutrons (and other light particles) during the saddle-to-scission stage is simulated with a time-dependent "differential" Monte Carlo code, in which light-particle emission or non-emission probability is being drawn in small consecutive time intervals. (Since fission is decided, during the post-saddle cascade $\Gamma_f = 0$.) The "differential" approach is essential in description of statistical cascades from rapidly evolving systems, in which excitation energy is generated in a time scale comparable or faster than the decay rate. Time dependence of the excitation energy generated in the fissioning system along the fission trajectory is calculated using the code HICOL.

The same method, as that for calculating the saddle-to-scission component of ν_{pre} in fusion-fission reactions outlined above, is used to calculate the pre-scission neutron multiplicity in the fast-fission scenario, see the right branch in Fig. 1. The rate of excitation energy generated along the whole fast fission trajectory is calculated for a given angular momentum using the code HICOL. This information is then supplied to the evaporation cascade calculations. The upper limit of the angular momentum range for fast fission processes, l_{max}, is determined by the maximum value of the mass asymmetry of the fragments, used in a given experiment to trigger the prescission neutrons, and otherwise predicted in HICOL calculations.

Both scenarios, fusion-fission and fast fission, end at scission of the composite system. At this point, each individual Monte Carlo cascade is finished and information on the final value of the thermal- and deformation excitation energies, mass and charge of the fragments is stored and can be used for calculating deexcitation cascades of the postscission fragments.

Results of simulations described above are compared with the correlation between the prescission and postscission multiplicities deduced with the backtracing method for the ^{58}Ni + ^{208}Pb and ^{40}Ca + ^{232}Th reactions by the DéMoN group [7]. Results of the neutron-multiplicity calculations show that fission of heavy composite systems is slowed down by intense dissipative forces comparable with one-body dissipation. The dominating part of the observed fission-like processes seems to correspond to fast fission reactions, consistently described with the HICOL model implying one-body dissipation.

References

1. N. Bohr and J.A. Wheeler, *Phys. Rev.* **56**, 426 (1939).
2. H.A. Krammers, *Physica* **7**, 284 (1940).
3. H. Feldmeier, *Rep. Prog. Phys.* **50**, 915 (1987).
4. J. Błocki *et al., Ann. Phys. (NY)* **113**, 330 (1978).
5. K. Siwek-Wilczyńska *et al., Acta Phys. Pol.* **B29**, 451 (1998).
6. P. Grangé and H.A. Weidenmüler, *Phys. Lett.* **96B**, 26 (1980).
7. L. Donadille *et al., Nucl. Phys.* **A656**, 259 (1999).

STRUCTURE AND PROPERTIES OF SUPERHEAVY NUCLEI

I. MUNTIAN[1,2] AND A. SOBICZEWSKI[1]
[1] Soltan Institute for Nuclear Studies, Hoża 69, PL-00-681 Warsaw, Poland
[2] Institute for Nuclear Research, Kiev, Ukraine
E-mail: adam.sobiczewski@fuw.edu.pl

A few problems connected with the structure and properties of superheavy nuclei are considered. These are the problems of experimental confirmation of deformed shapes of nuclei situated around the nucleus ^{270}Hs and the localization, in nuclear chart, of spherical superheavy nuclei with the largest (negative) shell correction to energy. Question of existence of superheavy nuclei in nature is also shortly discussed.

1 Introduction

The objective of this article is a short discussion of a few problems connected with the structure and properties of superheavy nuclei. One of them is the problem of experimental confirmation of deformed shapes of nuclei situated around the nucleus ^{270}Hs, predicted by theory. Another one is the question, how accurate is theoretical localization of spherical superheavy nuclei (with neutron number around $N = 184$) in proton number Z. Still another question is the existence of superheavy nuclei in nature.

2 Regions of superheavy nuclei

Expected regions of increased stability of superheavy nuclei are the regions of the lowest (negative and the largest in absolute value) values of shell correction to energies of these nuclei, E_{sh}. Figure 1 shows contour map of the energy E_{sh} plotted for a large region of proton, Z=82-120, and neutron, N=126-190, numbers.

The energy was calculated [1] in a macroscopic-microscopic approach. The result suggests the existence of two regions of superheavy nuclei: one around the nucleus ^{270}Hs ($Z = 108$, $N = 162$) and the other around the nucleus $^{298}114$ ($Z = 114$, $N = 184$). Theoretical analysis of shapes of nuclei in the two regions indicates that those of the first region should be deformed, while those of the second one should be spherical. This is illustrated in Fig. 2 (Ref. [2]).

At present, a number of nuclei of the first region have already been observed, and nuclei being presently synthesized approach to the second region (cf. e.g. Refs. [3,4]). It has not been exeperimentally shown yet, however, that

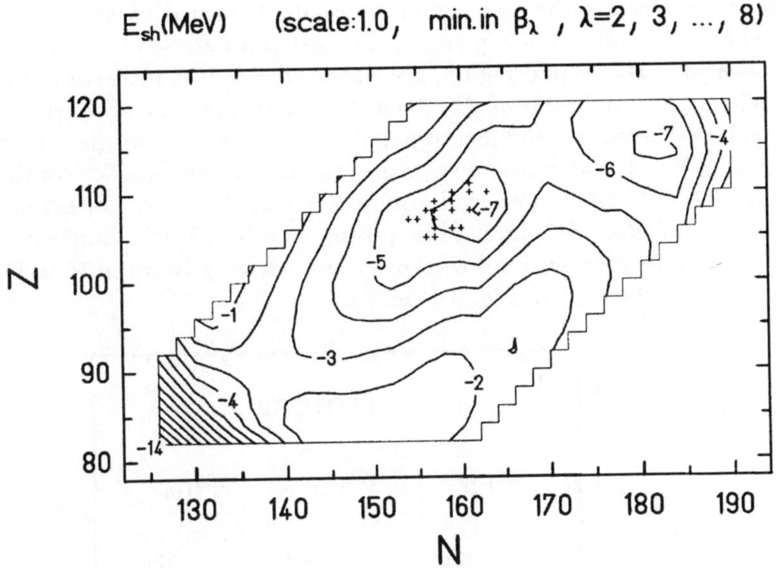

Figure 1. Contour map of the ground-state shell correction energy E_{sh} [1].

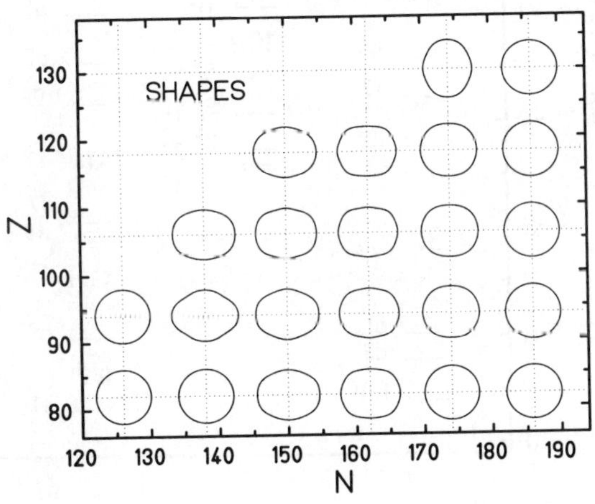

Figure 2. Calculated shapes of nuclei plotted for a wide region of Z=82-130 and N=126-190 [2].

the observed nuclei of the first region are really deformed. A short discussion of the chance of doing this is given in the next chapter.

Concerning the second region, not observed yet, it is interesting to learn, how well is the minimum of E_{sh} localized in the proton number Z. The problem is discussed in a number of papers (e.g. Refs. [6,7]). In the calculations presented in Fig. 1, the minimum is rather shallow in the "west", "south-west" and "north-west" directions, while quite steep in the "east" direction. The slow change of E_{sh} with increasing proton number Z of a nucleus, above $Z = 114$, is connected with a low proton level density between $Z = 114$ and $Z = 126$, as can be seen in Fig. 3 (Ref. [5]).

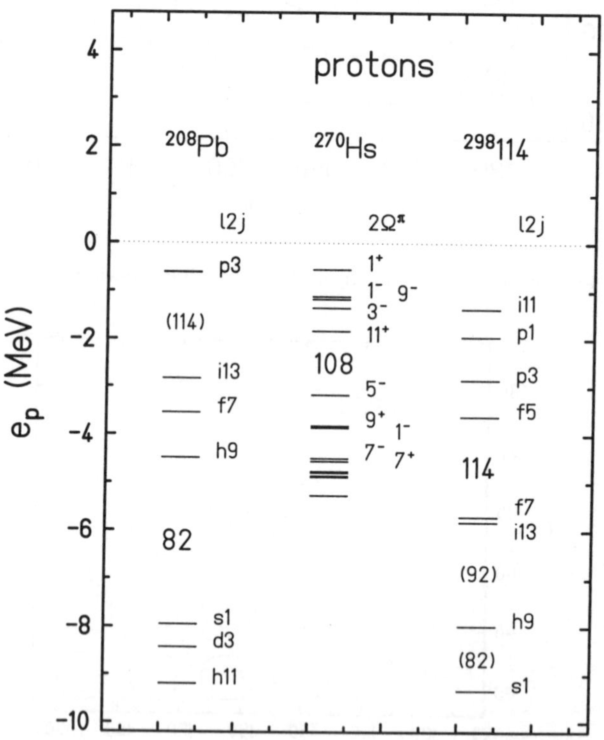

Figure 3. Proton single-particle energy levels calculated for the doubly magic nuclei: ^{208}Pb, ^{270}Hs and ^{298}114. Spectroscopic symbol for the orbital angular momentum l and total spin (multiplied by two) $2j$ are given at each level of the spherical nuclei ^{208}Pb and ^{298}114. Projection of spin on the symmetry axis of a nucleus (multiplied by two) 2Ω and parity π are shown at each level of the deformed nucleus ^{270}Hs [5].

3 Problem of shapes of supeheavy nuclei around ^{270}Hs

To show experimentally that nuclei around ^{270}Hs are really deformed, it is enough to observe a rotational band in their spectra. The heaviest nuclei for which such bands have been observed are ^{254}No [8,9] and ^{252}No [10]. There is, however, a small chance to observe such a band for superheavy nuclei, because cross sections to synthesize them are very low and the effectiveness of γ spectroscopy is relatively small. A more promising way is to see the first excited state 2+ of an even-even nucleus in α-decay or electron spectra. If the energy of such state is found to be of about 40-50 keV, as is expected, the state cannot be of other nature than rotational. An indication of this kind for deformation of the nucleus ^{256}Rf [11] was the observation of two close lines in the α-decay spectrum of ^{260}Sg [12]. A wider and more systematic study of this kind is, however, needed. The role of theoreticians in such a study is to predict energies of the first rotational state 2+, and the probabilities of α-decay to this state for even-even superheavy nuclei in the region around ^{270}Hs. Such predictions have been done in Refs. [13,14,15]. The energy E_{2+} calculated for nuclei with Z=102-112 is shown in Fig. 4, taken from Ref. [15]. One can see that

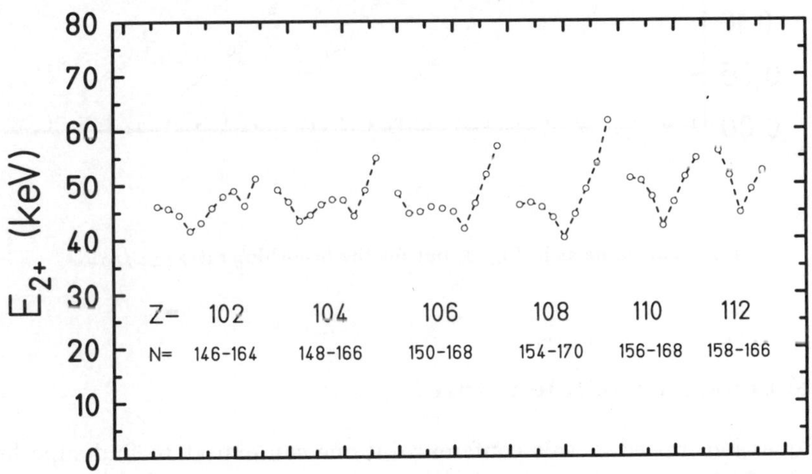

Figure 4. Dependence of the energy E_{2+} on neutron number N, calculated for elements with proton number Z=102-112. For each element, values of considered N are specified below the value of Z.

the energy is very low, about 40-50 keV, for many of these nuclei. Figure 5, taken from Ref. [15], shows the branching ratio p_{2+}/p_{0+} calculated for the same nuclei as in Fig. 4. Here, p_{I+} is the transition probability from the ground state 0+ of an even-even nucleus to the I+ state (I=2 and 0) of its daughter. One can see that the ratio strongly decreases with increasing neutron number N. It also decreases with increasing proton number Z. The obtained values indicate that the observation of the 2+ state and measurement of its energy is possible already presently.

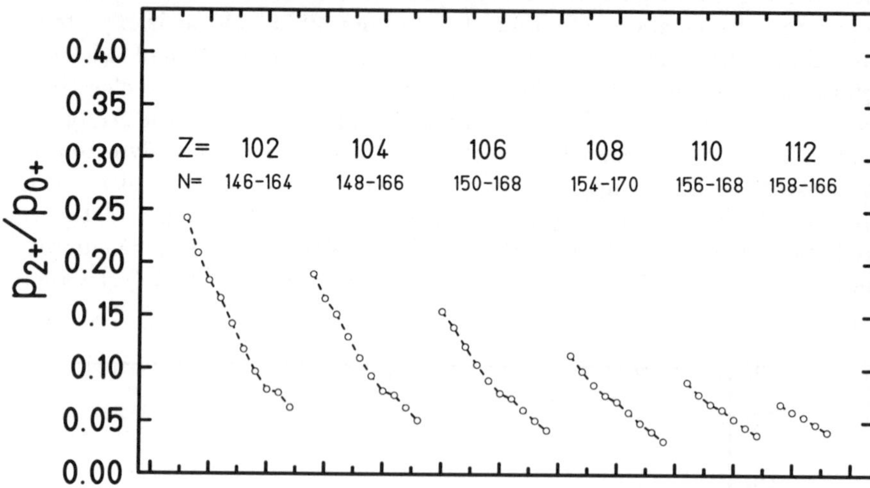

Figure 5. Same as in Fig. 4, but for the branching ratio p_{2+}/p_{0+}.

4 Superheavy nuclei in nature?

The question, raised at this conference, if one can expect to find superheavy nuclei in nature, is quite natural. What one can answer to it on the basis of theoretical calculations done up to present? Results obtained for (deformed) superheavy nuclei around the nucleus ^{270}Hs suggest that there is practically no chance to find in nature nuclei from this region. The largest calculated half-lives of even-even nuclei in this region are of the order of few hours, as can be seen in Fig. 6, taken from Ref. [16].

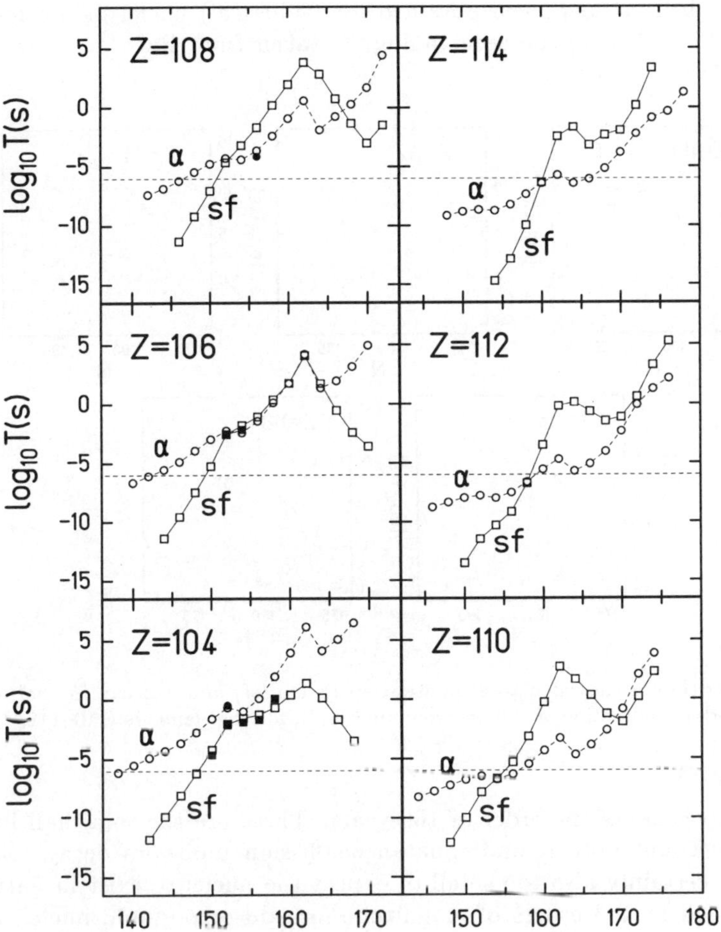

Figure 6. Dependence of the logarithm of the calculated spontaneous-fission (sf) half-lives, given in seconds, on the neutron number N, for elements 104–114. The α-decay half-lives (α), taken from Ref. [1], are also shown, for comparison. Experimental values are given as full symbols. The horizontal dashed line indicates about the lowest half-life (1 μs) of a nucleus, which can be detected in a present-day set-up, after its synthesis [16].

Larger half-lives have been obtained for (spherical) nuclei in the region around $^{298}114$. This can be seen in Fig. 7, taken from Ref. [17]. Here, the

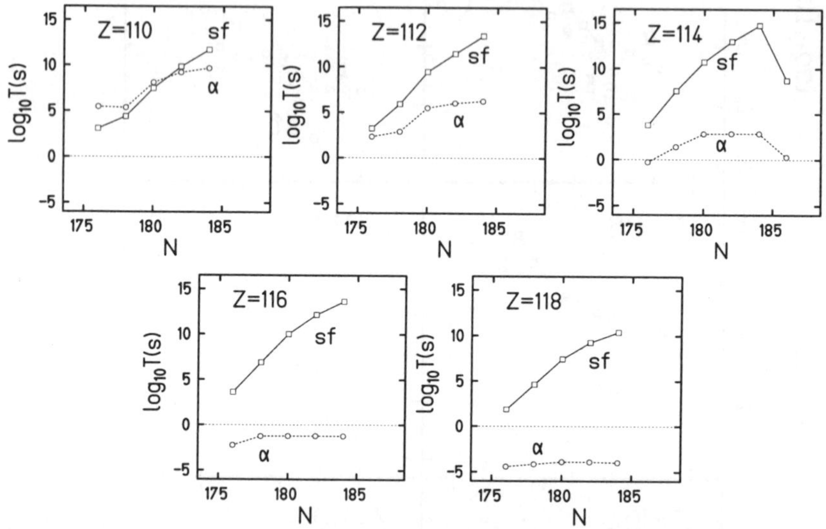

Figure 7. Logarithm of the calculated spontaneous-fission (sf) and α-decay (α) half-lives (given in seconds), as function of the neutron number N, for the elements: 110–118 [17].

largest half-lives are of the order of 100 years. These are the total half-lives, taking into account both α and spontaneous-fission modes of decay. Such half-lives are certainly also too small to expect the nuclei to exist in nature, even if enlarged by 2-3 orders of magnitude for odd or odd-odd nuclei, due to odd-particle effects. However, as the nuclei are in some distance from the region known experimentally, and as there is a sensitivity of the calculated half-lives to various, not well established, factors, the problem of existence of superheavy nuclei in nature may be still considered as being open.

Acknowledgements

The authors would like to acknowledge a support by the Polish State Committee for Scientific Research (KBN), Grant No. 2 P03B 117 15.

References

1. R. Smolańczuk and A. Sobiczewski, in *Proc. of the XV Nuclear Physics Conference: "Low Energy Nuclear Dynamics"*, St. Petersburg (Russia) 1995, edited by Yu.Ts. Oganessian, W. von Oertzen and R. Kalpakchieva (World Scientific, Singapore, 1995) p. 313.
2. I. Muntian and A. Sobiczewski, *Acta Phys. Pol.* B **32**, 629 (2001).
3. S. Hofmann and G. Münzenberg, *Rev. Mod. Phys.* **72**, 733 (2000).
4. Yu.Ts. Oganessian, *Phys. At. Nuclei* **63**, 1315 (2000).
5. A. Sobiczewski, I. Muntian and Z. Patyk, *Acta Phys. Slov.* **49**, 83 (1999).
6. S. Ćwiok, J. Dobaczewski, P.-H. Heenen, P. Magierski and W. Nazarewicz, *Nucl. Phys.* A **611**, 211 (1996).
7. K. Rutz, M. Bender, T. Bürvenich, T. Schilling, P.-G. Reinhard, J.A. Maruhn and W. Greiner, *Phys. Rev.* C **56**, 238 (1997).
8. P. Reiter, T.L. Khoo, C.J. Lister *et al.*, *Phys. Rev. Lett.* **82**, 509 (1999).
9. M. Leino, H. Kankaanpää, R.-D. Herzberg *et al.*, *Eur. Phys. J.* A **6**, 63 (1999).
10. R.-D. Herzberg and M. Leino, private communication.
11. G. Münzenberg and P. Armbruster, in *Exotic Nuclear Spectroscopy*, edited by W.C. McHarris (Plenum Press, New York, 1990) p.181.
12. G. Münzenberg, S. Hofmann, H. Folger *et al.*, *Z. Phys.* A **322**, 227 (1985).
13. I. Muntian, Z. Patyk and A. Sobiczewski, *Acta Phys. Pol.* B **30**, 689 (1999); *Phys. Rev.* C **60**, 041 302 (1999).
14. I. Muntian, Z. Patyk and A. Sobiczewski, *Phys. Lett.* B **500**, 241 (2001).
15. A. Sobiczewski, I. Muntian and Z. Patyk, *Phys. Rev.* C **63**, 034 306 (2001).
16. R. Smolańczuk, J. Skalski and A. Sobiczewski, *Phys. Rev.* C **52**, 1871 (1995).
17. R.A. Gherghescu, Z. Patyk, J. Skalski and A. Sobiczewski, in *Proc. of the Intern. Workshop: Research with Fission Fragments*, Benediktbeuern (Germany) 1996, eds. T. von Egidy, D. Habs, F.J. Hartmann, K.E.G. Löbner and H. Nifenecker (World Scientific, Singapore 1997) p. 116.

COULOMB EXCITATION OF NI AND ZN ISOTOPES AROUND THE N=40 SUBSHELL CLOSURE

O. Sorlin[1], S. Leenhardt[1], C. Donzaud[1], F. Azaiez[1], F. Amorini[1,*], A. Astier[2], D. Baiborodin [3], M. Belleguic[1], K. Bienczak[4], C. Bourgeois[1], C. Borcea[5], H. Brenning[1], D. M. Cullen[13], I. Deloncle[10], Z. Dlouhy[3], Zs. Dombrádi[6], E. Dragulescu[5], J. Duprat[1], M. Gorska[7], H. Grawe[7], S. Grévy[8], D. Guillemaud-Mueller[1], G. Hagemann[9], B. Herskind[9], J. Kiener[10], R. Lemmon[11,#], M. Lewitowicz[4], D. Pantalica[6], S.M. Lukyanov[12], Yu.-E. Penionzhkevich[12], M.G. Porquet[10], P. Mayet[7], F. de Oliveira Santos[4], L. Petizon[1], F. Pougheon[1], N. Redon[2], M.G. Saint-Laurent[4], J.A. Scarpaci[1], G. Sletten[9], M. Stanoiu[4], O. Tarasov[12], Ch. Theisen[14]

[1] *Institut de Physique Nucléaire, IN2P3-CNRS, F-91406 Orsay Cedex, France*
[2] *Institut de Physique Nucléaire Lyon, F-69622 Villeurbanne Cedex, France*
[3] *Nuclear Physics Institute, AS CR, CZ 25068, Rez, Czech Republic*
[4] *GANIL, B. P. 5027, F-14076 Caen Cedex, France*
[5] *IFIN-HH, P. O. Box MG-6, 76900 Bucarest-Magurele, Rumania*
[6] *Institute of Nuclear Research, H-4001 Debrecen, Pf. 51, Hungary*
[7] *GSI, Postfach 110552, D-64200 Darmstadt, Germany*
[8] *LPC, ISMRA, F-14050 Caen Cedex, France,*
[9] *NBI, University of Copenhagen, Copenhagen, Denmark*
[10] *CSNSM, F-91405 Orsay Cedex, France*
[11] *Department of Physics, University of Surrey, Guilford GU2 7XH, U.K.*
[12] *FLNR, JINR, 141980 Dubna, Moscow region, Russia*
[13] *Oliver Lodge Laboratory, University of Liverpool, Liverpool L69 7ZE, U.K.*
[14] *CEA-Saclay, DAPNIA-SPhN, F-91191 Gif sur Yvette Cedex, France*
[*] *present address: INFN -Lab. del Sud, via S.Sofia 44, 95123 Catania, Italy*
[#] *present address: CRLC, Daresbury Lab., Daresbury, Warrington, WA4 4AD, U.K.*

The reduced transition probabilities $B(E2: 0_1^+ \rightarrow 2^+)$ of $^{66,68}_{28}$Ni and $^{70,72}_{30}$Zn have been measured at GANIL by Coulomb excitation at intermediate energy in a ^{208}Pb target. A remarkable change in the evolution of the B(E2) is seen while crossing the N=40 closed shell in Ni and Zn isotopic chains.

1 Experimental results

1.1 Coulomb excitation of $^{66,68}Ni$ and ^{70}Zn

^{66,68}Ni nuclei have been produced at the GANIL facility in the fragmentation of a 65.9·A MeV ^{70}Zn^{29+} beam onto a 140μm-thick ^{58}Ni target. They have been selected by the LISE3 spectrometer in two different magnetic rigidity settings. A wedge-shaped Be-foil of 217μm thickness was inserted between the two dipoles of LISE3 to perform an additional energy-loss selection. Beam rates of 18000s^{-1} and 15000s^{-1} have been obtained for ^{66}Ni and ^{68}Ni isotopes respectively. Coulomb excitation of these two nuclei at $v/c \sim 0.3$ was induced by a 99% pure ^{208}Pb target of 220 mg/cm^2 thickness located at the final focal plane of LISE3. The target was surrounded by four segmented clover Ge detectors placed at 90 degrees at a distance of 5.5 cm of the target, yielding a photopeak efficiency of ϵ_γ=4.0% at 1.3 MeV. Two annular silicon detectors were mounted 50 cm behind the lead target in order to identify the deflected nuclei by their energy losses and residual energies. Nuclei coming out of the target with angles smaller than 1.5 degree were detected in a plastic scintillator two meters downstream. This detector also served to monitor the secondary beam intensity and determine the total number of implanted nuclei N_n. Photons emitted in flight in coincidence with the scattered nuclei were detected in segmented Ge clover-detectors. Each Ge crystal representing a 'leaf" of the clover is longitudinally segmented in four quadrants, reducing the angular acceptance of the individual detector elements. The add-back mode has been used for the detection of the high-energy γ-lines, increasing the efficiency by 40% for a γ at 2.033 MeV, leading to ϵ_γ=3.4%. The Doppler-corrected spectra for the Coulomb excitation of ^{66}Ni and ^{68}Ni are shown in Fig.1. From the total number of γ rays with respect to the number N_n of nuclei impinging onto the target, B(E2) values have been determined [1] (Fig. 2). $B(E2)\uparrow$ of ^{66}Ni and ^{68}Ni are 625(100) e^2fm^4 and 265(60) e^2fm^4 respectively.

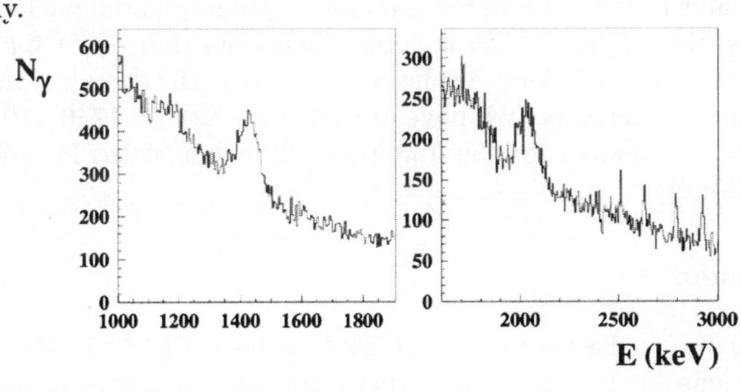

Figure 1. Doppler-corrected γ-energy spectra obtained in the Ge-clover detectors from the Coulomb excitation of ^{66}Ni(left) and ^{68}Ni(right).

By use of the primary beam ^{70}Zn at lower intensity, we have remeasured the B(E2) of ^{70}Zn. We found a value of $B(E2)\uparrow= 1640(280)e^2fm^4$ which is in accordance with that of [2], $B(E2) \uparrow= 1600(140)e^2fm^4$, obtained at a beam-energy close to the Coulomb barrier.

1.2 Coulomb excitation of ^{72}Zn

In another experiment, we have determined the B(E2) of ^{72}Zn. It has been produced at a mean β(=v/c) value of 0.28 via reactions of a 50 A.MeV ^{86}Kr beam in a ^{58}Ni target and selected by the LISE3 spectrometer. Other even-even isotopes, as ^{76}Ge, were transmitted in the same setting. The rates of ^{76}Ge and ^{72}Zn produced were about 100 and 120 s^{-1} respectively. Their Coulomb excitation has been induced by the field of Pb nuclei in a 220mg.cm^{-2} target placed in the center of a large gamma-array detector. They were identified event-by-event by two large-area 25cm^2 Si-detectors mounted at a distance of 50 cm downstream the Pb target. The Si-detectors covered angular deflections up to 3 degrees in the referential frame of the projectiles. This value is well below the grazing angle of the system (6.5 degrees), insuring that process leading to the excitation of the first 2^+ of the nucleus is of Coulomb origin only. The target was surrounded by a γ-array composed with 70 BaF_2 detectors of the Chateau de Cristal" which were mounted in a 4π geometry at a distance of 32 cm form the target. Each BaF_2 crystal is an hexagonal cylinder of 9cm diameter and a length of 14cm. The photopeak efficiency of this array at 600 keV was 6.5%. ^{76}Ge is a well-known deformed nucleus, exhibiting a low 2^+ energy of 563 keV and a large reduced transition probability, $B(E2 : 0_1^+ \rightarrow 2^+) = 2680(80)e^2fm^4$ [3]. It has been used as a reference nucleus to determine the B(E2) value of ^{72}Zn. From the numbers N_i of implanted nuclei, and N_γ of detected γ's, the B(E2) of ^{72}Zn is determined from that of ^{76}Ge after having corrected from the ϵ_γ efficiency of the BaF_2 array at the energy of the $2^+ \rightarrow 0^+$ transition in each nucleus. We have obtained $B(E2) \uparrow= 1740(210)e^2fm^4$ for ^{72}Zn, which is measured for the first time. This translates in a deformation of $\beta_2 = 0.233(14)$.

2 Discussion

The "bell-shape" behaviour of the $B(E2 \uparrow)$ values (Fig. 2) in the Ni isotopes provides a clear illustration of the structural evolution between the two magic nuclei ^{56}Ni and ^{68}Ni. As neutrons are added to ^{56}Ni, possibilities of creating 2^+ excitations via particle(s)-hole(s) excitations (p-h) are gradually increased until the number of p-h reaches a maximum value at the mid fp-shell, e.g. at ^{62}Ni. While filling the fp orbits up to N=40, the amount of quadrupole excitations is drastically reduced to a value which is about three times smaller than that of ^{56}Ni [4,5]. The low B(E2) of ^{68}Ni comes from two reasons: The first is the presence of an energy gap across N=40 which hampers the excitation of neutrons. The second is that a change in parity is occuring across N=40 between the fp and gd orbitals since these two groups of levels originate from different Harmonic-Oscillator shells. Therefore, excitations which preserve the parity-symmetry across the gap, as quadrupole excitations B(E2:$0^+ \rightarrow 2^+$), are strongly hindered. The size of the N=40 gap is however very fragile since it is of the order of the pairing energy-difference between $p_{1/2}$ and $g_{9/2}$.

In the Zn isotopic chain, the B(E2) curve deviates from that of Ni isotopes at N=38, reaching a maximum value at ^{72}Zn. The two protons added to a Ni core may deform the nuclei, the spherical shell gap being not strong enough to stabilize the nuclei. As a consequence, positive parity single particle levels penetrate in the region of negative parity states due to the splitting of the orbitals as a function of deformation. Even with a moderate deformation, $\beta_2 \sim 0.25$, the density of levels is very high, providing lots of possibilities to create 2^+ states via particle(s)-hole(s) excitations. The $B(E2 : 0_1^+ \rightarrow 2^+)$ is therefore large at N=40. The Zn isotopic chain is a transitional region between spherical Ni and deformed Ge for which there is a steady increase of collectivity up to N=42, without any kink at N=38.

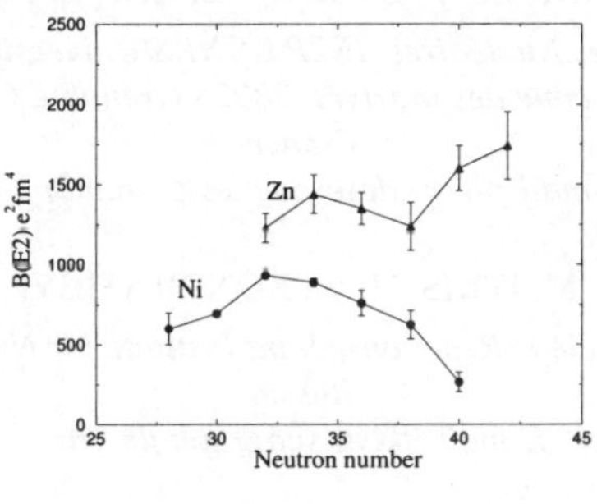

Figure 2. Experimental $B(E2)\uparrow$ values in the Ni (filled circles) and Zn (triangles) isotopic chains.

References

1. O. Sorlin *et al*, to be published and S. Leenhardt thesis, IPN Orsay, IPNO T 00-03.
2. P. H. Stelson and F. K. McGowan, *Nucl. Phys.* A **32**, 652 (1962).
3. S. Raman *et al* in *At. Data and Nucl. Data Tables* 36, 1 (1987).
4. G. Kraus *et al*, *Phys. Lett.* B **73**, 1773 (1994).
5. Y. Yanagisawa *et al*, in *proc. of the ENAM98 conference*, eds B. M. Sherrill et al., p. 610.

BACKTRACED NEUTRON MULTIPLICITIES AND CAPTURE DYNAMICS IN THE SUPERHEAVY REGION

L. STUTTGÉ, O. DORVAUX, C. SCHMITT
Institut de Recherches Subatomiques, IN2P3-CNRS/Université Louis Pasteur, 67037 Strasbourg Cedex 2, France
E-mail: stuttge@in2p3.fr

F. HANAPPE, B. BENOIT, E. DE GÓES BRENNAND, T. MATERNA
Université Libre de Bruxelless, CP229, 1050 Bruxelles, Belgium
E-mail: fhanappe@ulb.ac.be

K. SIWEK-WYLCZYNSKA, .J. WYLCZYNSKI
Nuclear Physics Department, Institute of Physics, Warsaw University, Warsaw, Poland
E-mail: siwek@npdaxp.fuw.edu.pl

L. DONADILLE, P. DÉSESQUELLES, E. LIATARD
Institut des Sciences Nucléaires, IN2P3-CNRS/Université Joseph Fourier, 53, Avenue des Martyrs, 38026 Grenoble Cedex, France
E-mail:pierre.desesquelles@isn.in2p3.fr

M. G. ITKIS, YU. M. ITKIS, N. A. KONDRATIEV, E. M. KOZULIN
Flerov Laboratory of Nuclear Reactions, Joint Institute for Nuclear Research, Dubna, Russia
E-mail:itkis@sungraph.jinr.ru

The dynamics of the fusion-fission process is being investigated over a wide range of mass, energy and angular momentum by the DEMON-CORSET collaboration. Results on the superheavy system Z=110 produced by different entrance channels at different excitation energies are presented. A new analysing procedure, the backtracing, gives access to the neutron multiplicity distributions as well as to the correlation between the pre- and post-scission emission, allowing for the first time to separate fusion-fission from fast fission.

1 Introduction

The dynamical aspects of the fusion-fission process have been ignored for many years, the main characteristics of the process being interpreted in the frame of the liquid drop model established already in 1939 by Bohr and Wheeler [1]. Only recently the concepts of dissipation, friction and viscosity have been introduced although they had already been established by Kramers in 1940 [2] after their importance was revealed by the discovery in the 70's of the deep inelastic collisions and the fusion hindrance [3]. Since then a huge amount of experimental as well as theoretical work has been performed to try to explain all the aspects of the fission process and the vanishing of the fusion cross-section in heavy ion reactions. The major limit in the comparison of experimental data with theoretical predictions lies in the difficulty to separate fusion-fission from other dissipative processes, mainly from fast-fission which presents almost the same characteristics in mass, angular and energy distributions. As a consequence the extraction of a meaningful dissipation coefficient becomes meaningless and thus there is still a lot of controversy about the nature (one body or two body), the intensity of the dissipation, as well as on its dependence with temperature. In the heavy region, a large set of data has been obtained by Hinde et al [4]. Friction coefficients deduced with the dynamical model including the entrance channel of Siwek-Wylczynska et al [5] and plotted in figure 1 show that if for the lighter systems one body dissipation reproduces well the data, for the superheavy region the very high values could reflect a two body dissipation process

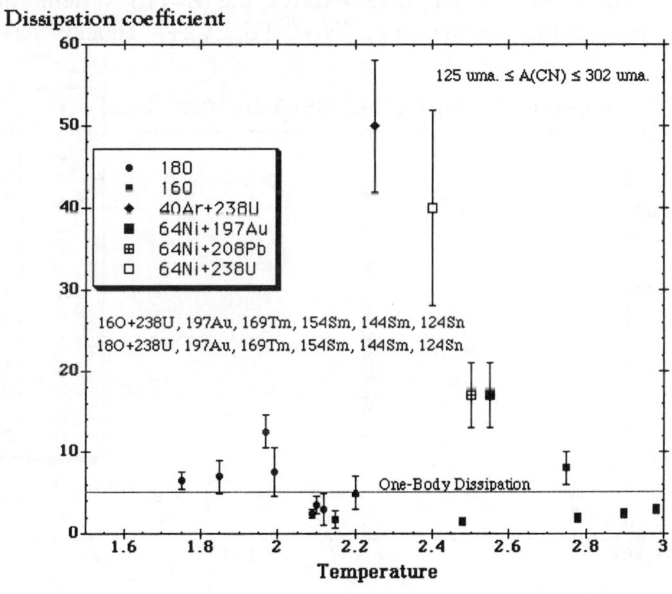

Figure 1. Dissipation coefficients for different heavy systems deduced by the Siwek-Wilczynska et al model [5] from the Hinde et al data [4].

As mentioned above, one has to remind that in the Hinde at al data there was no way to separate fusion-fission from fast fission, which is supposed to dominate the cross-section in the superheavy region.

2 Pre- and post-scission neutron multiplicities in the Z=110 system

A vast campaign of measurements has been undertaken by the DEMON-CORSET collaboration [6], which aims to get a detailed overview on all aspects of the fusion-fission dynamics. Many systems ranging over the whole fissility domain have been studied at different energies, angular momenta and entrance channels.

In these experiments the two fission fragments were detected by the CORSET setup developed in Dubna, which consists in two arms composed of micro channel plates characterized by very good time and angular resolutions (Δt~150 ps, ΔΘ~0.1°) leading to a mass resolution of about three mass units. The angular and energy distributions of the neutrons were measured in coincidence by the DEMON neutron multidetector [7].

Figure 2 shows the systems already measured in the superheavy region. Excitation functions have been measured for the Z=110 system through four different entrance channels, $^{58}Ni+^{208}Pb$, $^{64}Ni+^{208}Pb$, $^{40}Ca+^{232}Th$ and $^{40}Ar+^{238}U$ [8,9].

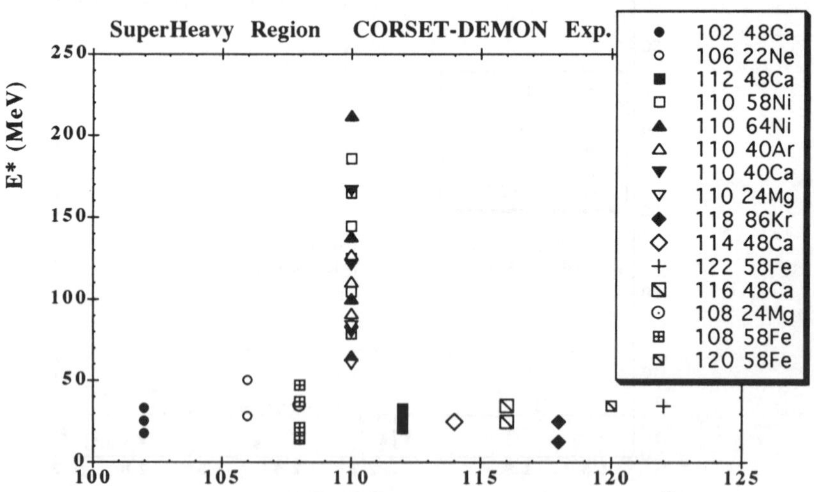

Figure 2. Measured systems in the superheavy region by the DEMON-CORSET collaboration in the frame of the study of the fusion-fission process .

A conventional analysis with a χ^2 minimization using a three Maxwellian source deconvolution has been performed to separate the pre-scission neutrons from the post-scission ones. Figure 3 shows the multiplicities obtained for the different entrance channels as a function of the excitation energy. One can observe the known behaviour of the pre-scission multiplicity which increases with increasing excitation energy whereas the post-scission multiplicity increases very slowly showing that the fission is a cold process. The χ^2 minimization giving only access to the mean values of the multiplicities, it doesn't allow a separation between fast-fission and fusion-fission which are mixed up in the deduced multiplicities.

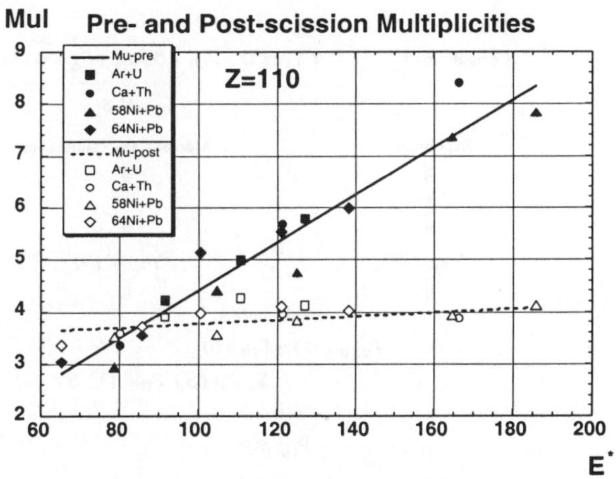

Figure 3. Pre- and post-scission multiplicities of the neutrons emitted by the Z=110 system deduced with a χ^2 minimization using three Maxwellian sources, as a function of the excitation energy for different entrance channels.

3 Backtraced neutron multiplicities

To overcome this limitation a new protocol, the backtracing, introduced in nuclear physics by P. Désesquelles [10] has been developed for our purpose. The backtracing method consists in finding the distributions of the model input variables which provide an optimal description of the experimental observables. This is done by a recursive procedure. The flow chart of the protocol is shown in figure 4.

A physical process is completely described by the probability distribution $P_\Phi(O/S)$ of an observable O generated by an initial condition called source variable S. A theory is completely described by the corresponding theoretical probability $P_{th}(O/S)$. The theory is compatible with the process if, whatever the source S and whatever the observable O, the process and the theory give the same distribution. Thus, on one hand, the experimental data filtered by the experimental setup and the analysis selections produce the experimental observable distributions $P_{exp}(O)$. On the other hand, a Monte-Carlo simulation is performed for any source value S. The theoretical outputs are filtered by the same filter as the experimental data. The obtained probability distributions $P_{th}(O)$ are compared to the experimental ones in order to extract the source distributions $P_{th}(S)$. This is done by an iterative and recursive matrix inversion, the backtracing. This method presents the advantage to give not just mean values of the theoretical input parameters, but their distribution as well as the correlation between them.

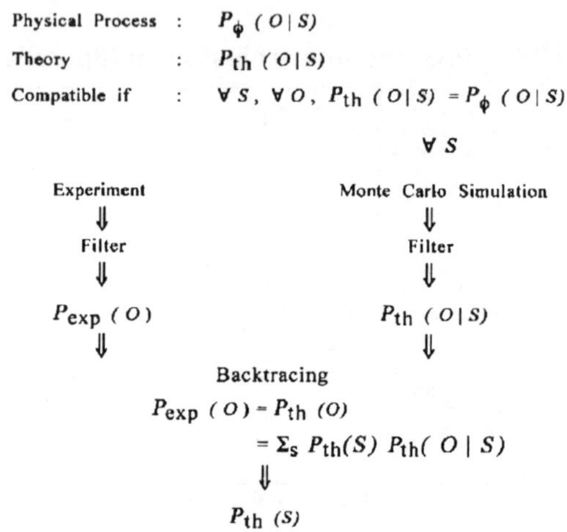

Figure 4. Protocol flow chart of the backtracing procedure

In theory one can introduce as many input parameters as desired but the backtracing procedure becomes very quickly highly computer and time consuming and so one has to restrict to the minimum number of observables containing the maximum of information and discrimination of the process. The backtracing procedure also requires global variables. As DEMON is far from being a 4π detector, this is not the case for the measured observables. For these reasons the procedure was performed in two steps as following. The first step was dedicated to the correlation between the total excitation energy and the total neutron multiplicity leading to the thermal excitation energy of the system. Three global observables, O_1, O_2 and O_3, related to the neutron energy were used defined as following:

- $O_1=O_2=O_3=0$ for 0-fold events
- $O_1=E_{n1}$, $O_2=O_3=0$ for 1-fold events
- $O_1=0$, $O_2= E_{n1}+E_{n2}$, $O_3=0$ for 2-fold events
- $O_1=0$, $O_2=0$, $O_3=E_{n1}+E_{n2}+E_{n3}$ for 3-fold events,

where E_{ni} is the energy of neutron i.

In the second step the obtained thermal excitation energy and the total neutron multiplicity were used to get the correlation between the pre- and the post-scission multiplicities by taking this time as observables the angles θ, φ and the energy of the neutron considering only one-fold events.

To validate this new method it has been applied to a simple case, ^{126}Ba, for which only fusion-fission occurs and thus the neutron emission is not contaminated by fast-fission. This system had been measured through two entrance channels ^{19}F+^{107}Ag and ^{28}Si+^{98}Mo at several incident energies [11]. Figure 5 shows the pre-scission multiplicities as a function of the excitation energy obtained by the χ^2 minimization. The results are compared to the predictions of the dynamical model of Pomorski et al [12] which considers the fusion-fission process only but without any adjustable parameter. The straight line on the figure which is the best fit of the theoretical points shows the good agreement with the experimental data. Figure 6 shows the pre-scission multiplicity distribution obtained with the backtracing procedure at the highest excitation energy, E^*=132 MeV, for the ^{28}Si+^{98}Mo interaction. The comparison with the predicted distribution by the Pomorski et al model shows a good agreement. Also the mean value obtained by the backtracing, $<\mu_{pre}>_{Back}=2.54$, is the same as that obtained by the χ^2 minimization, $<\mu_{pre}>_{\chi 2}=2.52$ and in good agreement with the predicted one, $<\mu_{pre}>_{Pom}=2.29$. These results validate the backtracing as well as the Pomorski et al model which uses a one-body dissipation.

The backtracing procedure has been applied to the Z=110 system for the ^{58}Ni+^{208}Pb and ^{40}Ca+^{232}Th interactions at the highest energies. For this system there is an important competition between fusion-fission and fast fission. Figure 7 shows the correlation obtained between the pre- and post-scission multiplicities. One can identify clearly two components in both cases which can be intuitively attributed to the fast fission for the lowest one at $\mu_{pre}\sim4$ and to the fusion-fission for the highest at $\mu_{pre}\sim7$.

Calculations for these two systems have been performed with the dynamical model of K. Siwek-Wilczynska et al. This model uses the HICOL code developed by Feldmeier [13] which calculates the trajectories leading to the different processes: deep inelastic, fast fission and fusion-fission and predicts the excitation energy created in the composite system. It is followed by a statistical code which calculates the deexcitation and gives the pre- and post-scission multiplicities. For the fast fission a dissipation of γ=5 corresponding to a one body dissipation has been used. For the fusion-fission two values γ=5 and γ=11, which is twice the one body dissipation, have been used. The calculated values are plotted on the experimental correlations of figure 7 as white and black rectangles. The angular momenta are deduced by HICOL from the experimental masses taken into account. The agreement is perfect for the fast fission and the two values of γ cover the experimental values for the fusion-fission.

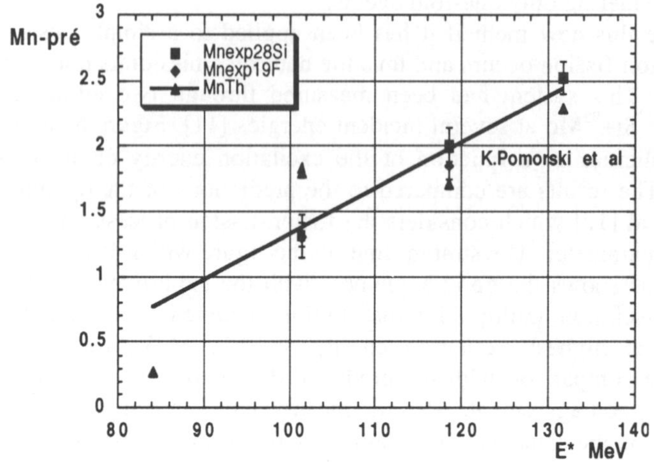

Figure 5. Pre-scission multiplicities obtained by a χ^2 minimization of ^{126}Ba formed through the $^{19}F+^{107}Ag$ and $^{28}Si+^{98}Mo$ interactions as a function of the excitation energy.

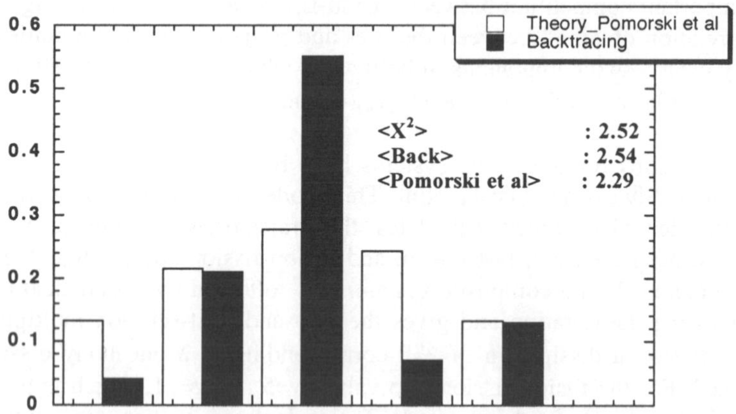

Figure 6. Pre-scission multiplicity distribution obtained with the backtracing for ^{126}Ba through the $^{28}Si+^{98}Mo$ interaction at E^*=132 MeV, compared to the distribution predicted by Pomorski et al.

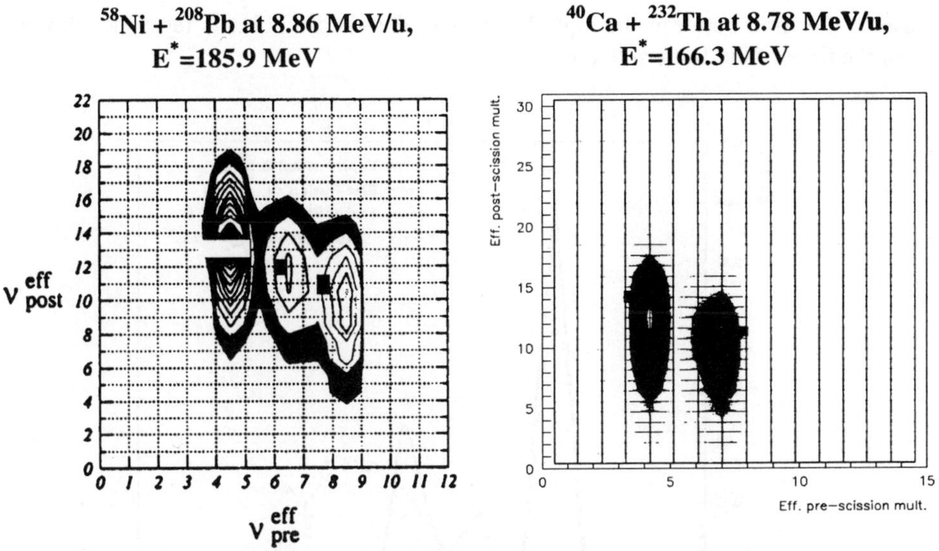

Figure 7. Correlation between pre- and post-scission multiplicities obtained by the backtracing for the ^{58}Ni+^{208}Pb and ^{40}Ca+^{232}Th leading to the Z=110 system.

The backtraced pre-scission multiplicity distribution has also been compared to the predictions of the Pomorski et al model and is shown in figure 8 for the ^{58}Ni+^{208}Pb case.

On this figure the two components obtained with the backtracing have been fitted by two independent Gaussians (gray and dark gray). The predicted distribution (dashed line and points) is in good agreement with the fusion-fission Gaussian of the backtracing although the predicted mean value as well as the width are slightly higher than the backtracing ones. This can be partly due to the fact that in its present stage the Pomorski model considers only strictly symmetric fission whereas the experimental masses range over $A_{CN}/2\pm30$, where A_{CN} is the total mass of the system. The model neither considers the entrance channel, which in this case is very important. It takes the excitation energy deduced from the incident energy, thus dissipation effects in the first step of the interaction are not included. This can also explain the discrepancy with the experimental value. The model is presently being developed to include these considerations.

The model on the other hand gives access to the different time scales of the process and it shows that one has to be very careful when using the neutron clock. The calculations show that if for heavy systems like Z=110, the fission time mainly due to the time between the saddle and the scission points is equal to the time of the last emitted neutron, for light systems like ^{126}Ba, the fission time mainly due in this case to the time between the equilibrium and the saddle point is ten times longer than the time of the last emitted neutron.

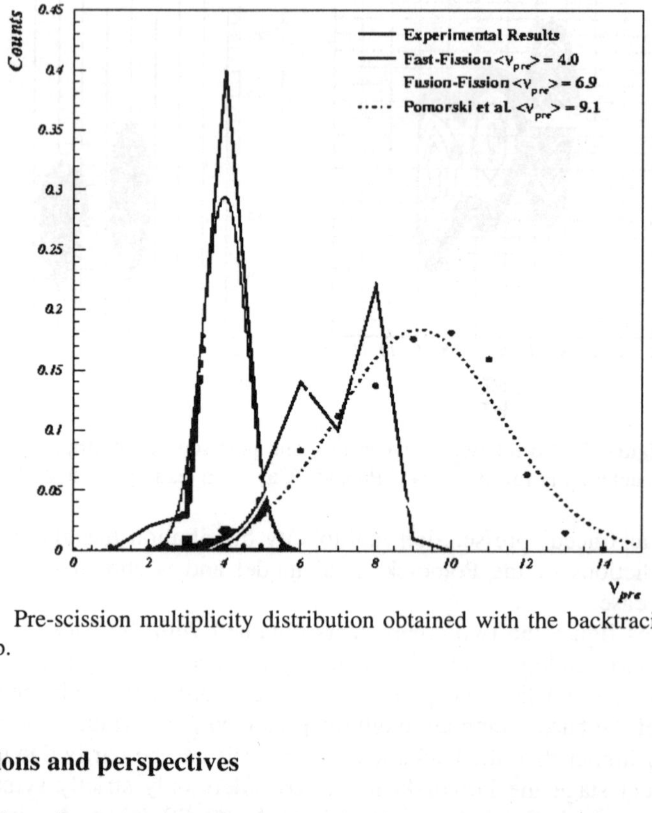

Figure 8. Pre-scission multiplicity distribution obtained with the backtracing for $^{58}Ni+^{208}Pb$.

4 Conclusions and perspectives

The new backtracing protocol allows for the first time to discriminate between fusion-fission and fast fission. The analysis with this procedure shows that there is no need to introduce a two body dissipation in the fusion-fission process. Once the different processes are separated a one body dissipation reproduces the experimental data over the whole mass region. The backtracing has been used for two systems at a high excitation energy, E^*>150 MeV. The weak point of the backtracing at its present stage of development is that it requires a huge statistics, experimental as well as simulated. The procedure is presently under development to allow also the analysis of systems where the statistics is presently too low. This is the case at lower energy where in addition the multiplicities are much lower and the two components of fast fission and fusion-fission probably much less separated. On the theoretical side, many developments are in progress aiming to include namely the entrance channel with dissipation, the mass asymmetry, shell effects and the temperature dependence of the dissipation.

References

1. Bohr N. and Wheeler J. A., *Phys. Rev.* **53** (1939) pp. 426.
2. Kramers H. A., *Physica* **7** (1940) pp. 284.
3. Tamain B, *PhD Thesis*, Université de Paris Sud, 1976
   Hanappe F. et al, *Phys. Rev. Lett.* **32** (1974) pp. 738
   Le Brun C. et al, *Nucl. Phys.* **A321** (1979) pp. 207.
4. Hinde D. et al, *Phys. Rev.* **C45** (1992) pp. 1229
   Hinde D. et al, *Nucl. Phys.* **A592** (1995) pp. 271.
5. Siwek-Wilczynska K. et al, *Phys. Rev.* **C51** (1995) 2054 and *this conference.*
6. Hanappe F. et al, INTAS Project 97-11929.
7. Tilquin I. et al, *NIM* **A365** (1995) pp. 446.
8. Donadille L., *Nucl. Phys.* **A656** (1999) pp. 256.
9. Benoit B., *PhD Thesis*, **ULB**, 2000.
10. Désesquelles P. et al, *Nucl. Phys.* **A604** (1996) pp. 183.
11. Góes de Brennand E., *PhD Thesis*, **ULB**, 2000.
12. Pomorski K. et al, *Nucl. Phys.* **A605** (1996) pp. 87
    Pomorski K. et al, *Nucl. Phys.* **A679** (1) (2000) pp. 25.
13. Feldmeier H., *Rep. Prog. Phys.* **50** (1987) pp. 915.

MICROSCOPIC CLUSTERS IN THE GROUND STATES OF LITHIUM AND BERYLLIUM ISOTOPES

M. TOMASELLI, T. KÜHL, P. EGELHOF, W. NÖRTERSHÄUSER, C. KOZHUHAROV, A. DAX, H. WANG, S.R. NEUMAIER, D. MARX, AND H.-J. KLUGE

Gesellschaft für Schwerionenforschung (GSI), D-64291 Darmstadt, Germany
e-mail: m.tomaselli@gsi.de

I. TANIHATA

RIKEN 2-1 Hirosawa, Wako-Shi, Saitama 351-0198, Japan

S. FRITZSCHE

Institute of Physics, Kassel University, D-34132 Kassel, Germany

M. MUTTERER

Institute of Nuclear Physics, Darmstadt University, D-64289 Darmstadt, Germany

The matter and charge distributions of lithium and beryllium isotopes were calculated in terms of interacting microscopic clusters: the valence- (one, two, three particles) and the intrinsic vacuum-clusters (collective-excitations of reference vacuum). Root-mean-square matter and charge radii have been extracted and are in good agreement with experimental results.

1 Introduction

Elastic proton scattering experiments in inverse kinematics recently performed at GSI [1] have reopened important, partially unresolved questions concerning the physics of the halo nucleus ^{11}Li. In order to obtain a deeper insight into the halo structure of light exotic nuclei and to understand the nature of the difference between the matter and the charge distributions, microscopic calculations for the ground states of the lithium and beryllium isotopes have been performed within the Dynamic-Correlation Model (DCM). The DCM describes the ground states of nuclei in terms of interacting clusters: valence particles and intrinsic vacuum states. The amplitudes of the mixed-mode wave-functions are derived in the framework of non-perturbative solutions of the Equation of Motion (EOM). The matter and charge distributions obtained by this microscopic approach can be used to predict experimentally accessible quantities. In the case of lithium this was utilized for elastic proton scattering cross sections (p-Li) at an energy of 0.7 GeV/u as measured recently at GSI. Similar results were already presented for ^{6}He and ^{6}Li [2]. Furthermore (rms) matter and charge radii of lithium and beryllium isotopes are calculated.

While there are experimental values for the matter radii of all lithium isotopes under investigation [3], the charge radii are only known for the two stable isotopes ^{6,7}Li [4]. Experimental and theoretical values are in good agreement. However, for a better understanding on the influence of the neutron halo on the core nucleus, it is desirable to determine the charge radii particularly for ^{11}Li. For this purpose an experiment is being prepared at GSI and ISOLDE, CERN to determine this value by means of an optical isotope shift measurement [5].

2 Theoretical Approach, Results and Discussion

In dynamic-correlation calculations the single-particle model is redefined by describing nuclei with mixed-mode states which take care of the excitations of the valence particles and their interaction with the reference core. A detailed description of the DCM has been given symbolically in Fig. 1 left. The most remarkable difference between the dinamical models and the perturbation calculations is the exact treatment of higher-order diagrams which in the DCM are considered via the collective mean field potential, while in perturbative approximations result difficult to be evaluated without introducing additional approximations.

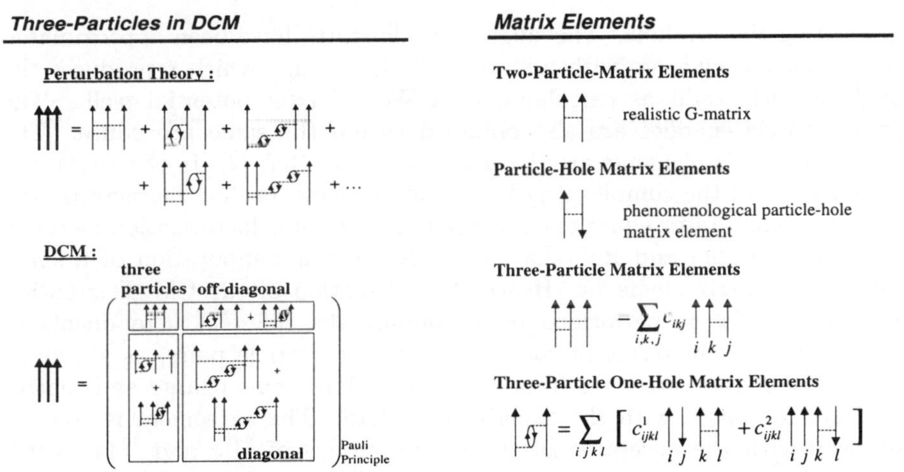

Figure 1. Left: Symbolic representation of the DCM; Right: cluster factorization of the matrix elements.

The model spaces for the ground states of the ^{6,7,9,11}Li and ^{7,9}Be isotopes are constructed by allowing valence particles to be scattered to higher configuration states ($2\hbar\omega$) and to interact with the intrinsic core states formed

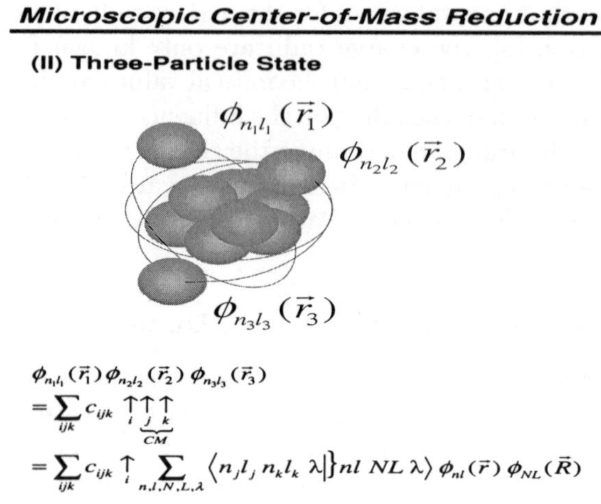

Figure 2. Centre of mass reduction: three particle state.

by exciting the s,p-hole. The single particle states have been approximated by harmonic oscillators with a state-dependent range which reproduces the single particle radii as calculated in a Wood-Saxon potential well. The single-particle energies are also obtained within the same procedure. The two-body matrix elements are the same as used in Ref. [2]. In the DCM, the computation of the complex $\{n\}$ body matrix elements that characterize the different commutator equations is simplified by cluster factorization methods (see Fig. 1 right) and Ref. [2] which involve the computation of a small number of matrix elements. Hence the calculation of the CM distribution of $\{A\}$ particles is performed by expanding the different components of the CMWFs [2] in terms of consistent sums of particle pairs, as shown in Fig. 2 for mass A=11. In Tab. 1 the calculated rms charge and matter radii are compared with the experimental data. The agreement is good in all cases with the exception of the matter radius of ^{9}Be and ^{11}Li, which are considerably larger than the experimental value of Ref. [3]. It should be mentioned that experimental values are model-dependent and that new proton-scattering data from GSI [6] indicate a larger matter radius than the one given previously. The DCM matter distributions of the ^{11}Li is presented in Fig. 3 left and compared with the best experimental fit. Furthermore in Fig. 3 right the calculated proton scattering cross section for the p-^{11}Li elastic scatteriing is compared with the preliminar results of Ref. [6].

References

1. S.R. Neumaier et al., Nucl. Phys. A (2000), to be published; G.D. Alkhazov et al., Phys. Rev. Lett. 78, 2313 (1997).
2. M. Tomaselli et al., Phys. Rev. C62, 67305 (2000).
3. I. Tanihata et al., Phys. Lett. B 206, 592 (1988).
4. C.W. de Jaeger et al., At. Data Nucl. Data Tables 14, 479 (1974).
5. A. Dax et al., CERN/INTC 2000-006 INTC/P118.
6. A.V. Dobrowolsky et al., GSI Report (1999) and Verhdlg. DPG 35, 5 (2000); M. Mutterer, and the IKAR collaboration, this conference.

	R^{calc}_{matter}	R^{exp}_{matter}	R^{calc}_{charge}	R^{exp}_{charge}
^7Be	2.38 fm	2.31(2)[3] fm	2.38 fm	?
^9Be	2.46 fm	2.38(1)[3] fm	2.62 fm	2.47(1)[3] fm
^6Li	2.41 fm	2.44(7)[6] fm	2.55 fm	2.55(4)[4] fm
		2.35(3)[3] fm		
^7Li	2.36 fm	2.35(3)[3] fm	2.41 fm	2.39(2)[5]
^9Li	2.42 fm	2.42(7)[6] fm	2.42 fm	?
		2.32(2)[3] fm		
^{11}Li	3.64 fm	3.65(19)[6] fm	2.67 fm	?
		3.10(17)[3] fm		

Table 1. Calculated and experimental rms matter and charge radii of Be and Li isotopes.

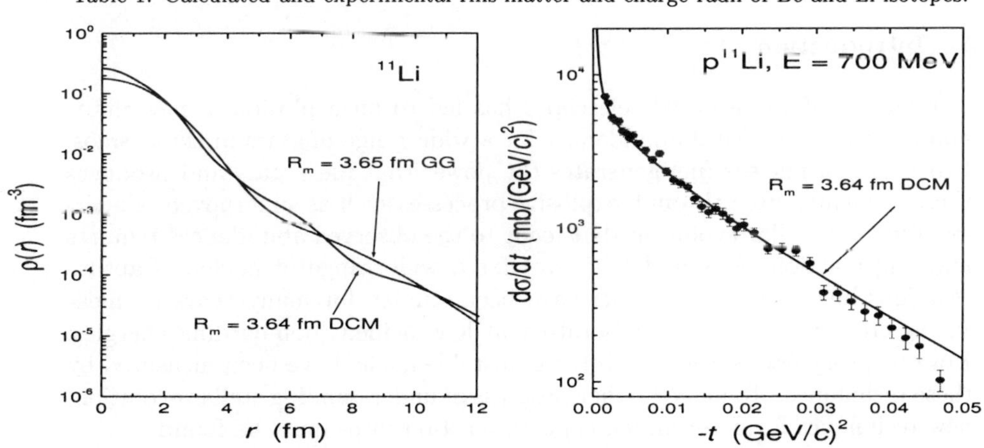

Figure 3. Left: Calculated (full line) and experimental (dash line) mass densities of ^{11}Li; Right: calculated and experimental p-^{11}Li cross section.

ASTROPHYSICAL S FACTORS FROM ASYMPTOTIC NORMALIZATION COEFFICIENTS

R.E. TRIBBLE[1], A.AZHARI[1], P. BEM[2], V. BURJAN[2], F. CARSTOIU[3], J. CEJPEK[2], H.L. CLARK[1], C.A. GAGLIARDI[1], V. KROHA[2], Y.-W. LUI[1], A.M. MUKHAMEDZHANOV[1], J. NOVAK[2], S. PISKOR[2], A. SATTAROV[1], E. SIMECKOVA[2], X. TANG[1], L. TRACHE[1], J. VINCOUR[2]

[1] *Cyclotron Institute, Texas A&M University, College Station, Texas 77843 USA*
[2] *Institute for Nuclear Physics, Prague-Řež, Czech Republic*
[3] *Institute for Atomic Physics, Bucharest, Romania*

Peripheral transfer reactions can be used to determine asymptotic normalization coefficients (ANC). These coefficients, which provide the normalization of the tail of the overlap function, determine S factors for direct capture reactions at astrophysical energies. A variety of proton transfer reactions have been used to measure ANC's. As tests of the technique, ANC's for $^{10}B \leftrightarrow {}^{9}Be + p$ and $^{17}F \leftrightarrow {}^{16}O + p$ have been measured and their values were used to predict the S factors for $^{9}Be(p,\gamma)^{10}B$ and $^{16}O(p,\gamma)^{17}F$. In both cases the predictions are in excellent agreement with measured S factors. Following the same technique, the $^{10}B(^{7}Be,^{8}B)^{9}Be$ and $^{14}N(^{7}Be,^{8}B)^{13}C$ reactions have been used to measure the ANC appropriate for determining $^{7}Be(p,\gamma)^{8}B$. The results from the two transfer reactions provide an indirect determination of $S_{17}(0)$. Recent measurements have been completed on the $^{14}N(^{11}C,^{12}N)^{13}C$ reaction which will allow us to define the astrophysical S factor for the $^{11}C(p,\gamma)^{12}N$ reaction.

1 Introduction

The advent of space based telescopes has led to an explosion of new information about elemental abundances for a wide range of stars in the cosmos. Processing of nuclear fuel generates the power that fuels stars and produces these elements, often through explosive processes such as supernovae. Understanding the stellar evolution that leads to the observed abundances requires knowing reaction rates and half lives for a wide range of nuclei. Capture reactions involving stable nuclei have been studied for many years by measuring cross sections in the laboratory at low incident bombarding energies. However, very few reactions involving unstable nuclei have been measured by these techniques. To make further progress in understanding stellar evolution, new techniques for determining capture reaction rates must be found.

Direct capture reactions of astrophysical interest usually involve systems where the binding energy of the captured proton is low. At stellar energies, the capture proceeds through the tail of the nuclear overlap function whose shape is completely determined by the Coulomb interaction. Consequently, the amplitude of the overlap function defines the capture reaction rate. The ^{7}Be(p,γ)^{8}B reaction is an excellent example of such a direct capture process.

The asymptotic normalization coefficient (ANC) C for $A + p \leftrightarrow B$ specifies the amplitude of the tail of the overlap function for the system. In previous communications[1,2], we have pointed out that astrophysical S factors for peripheral direct radiative capture reactions can be determined through measurements of ANC's using traditional nuclear reactions such as peripheral nucleon transfer. Direct capture S factors derived with this technique are most reliable at the lowest incident energies in the capture reaction, precisely where capture cross sections are smallest and most difficult to measure directly. It is extremely important to test the reliability of the ANC technique in order to know the precision with which it can be applied. Below we give a brief description of this technique and discuss two tests that we have performed. We then describe our ^{10}B(^{7}Be,^{8}B)^{9}Be, ^{14}N(^{7}Be,^{8}B)^{13}C, and ^{14}N(^{11}C,^{12}N)^{13}C radioactive beam studies and their use to determine the astrophysical S factors for ^{7}Be(p,γ)^{8}B and ^{11}C(p,γ)^{12}N.

2 ANC's from Proton Transfer Reactions

Traditionally spectroscopic factors have been obtained from proton transfer reactions by comparing experimental cross sections to DWBA predictions. For peripheral transfer, the ANC is better determined and is the more natural quantity to extract. Consider the proton transfer reaction $a + A \to c + B$, where $a = c + p$, $B = A + p$. As was previously shown[3] we can write the DWBA cross section in the form

$$\frac{d\sigma}{d\Omega} = \sum_{j_B j_a} (C^B_{Apl_Bj_B})^2 (C^a_{cpl_aj_a})^2 \frac{\sigma^{DW}_{l_Bj_Bl_aj_a}}{b^2_{Apl_Bj_B}\, b^2_{cpl_aj_a}}, \tag{1}$$

where $\sigma^{DW}_{l_Bj_Bl_aj_a}$ is the reduced DWBA cross section and j_i, l_i are the total and orbital angular momenta of the transferred proton in nucleus i. The factors $b_{cpl_aj_a}$ and $b_{Apl_Bj_B}$ are the ANC's of the normalized single particle bound state proton wave functions in nuclei a and B. If the reaction under consideration is peripheral, the ratio in Eq. (1) is independent of the single particle ANC's $b_{Apl_Bj_B}$ and $b_{cpl_aj_a}$. Thus for surface reactions the DWBA cross section is best parametrized in terms of the product of the square of the ANC's of the initial and final nuclei $(C^B)^2(C^a)^2$ rather than spectroscopic factors.

We have used this formulation to extract ANC's for ^{10}B $\leftrightarrow$ ^{9}Be + p, ^{14}N $\leftrightarrow$ ^{13}C + p, and ^{17}F $\leftrightarrow$ ^{16}O + p from the reactions ^{9}Be(^{10}B,^{9}Be)^{10}B, ^{13}C(^{14}N,^{13}C)^{14}N, ^{13}C(^{3}He,d)^{14}N, and ^{16}O(^{3}He,d)^{17}F. The first two reaction studies were carried out with beams from the K500 superconducting cyclotron at Texas A&M University. The ^{13}C(^{3}He,d)^{14}N reaction was investigated at the U-120M isochronous cyclotron of the Nuclear Physics Institute of the Czech Academy of Sciences. The ^{16}O(^{3}He,d)^{17}F reaction was studied at both laboratories. Details of the experiments can be found in [3,4,5,6].

In order to extract ANC's, DWBA calculations were carried out with the finite range code PTOLEMY[7], using the full transition operator. A check on the extracted ANC's versus Woods-Saxon well radial parameters indicated that the calculated DWBA cross sections are insensitive to assumptions about the wave functions in the nuclear interior. For the (^{3}He,d) reactions, normalizing the DWBA calculations to the data and dividing by the ANC's for the single particle orbitals yields the product of the ANC's for $^{3}\mathrm{He} \to d + p$ and either $^{14}\mathrm{N} \to {}^{13}\mathrm{C} + p$ or $^{17}\mathrm{F} \to {}^{16}\mathrm{O} + p$. Dividing this product by the known ANC for $^{3}\mathrm{He} \to d + p$ provides C^2 for the other vertex of the reaction[8]. For proton transfer between ground states in ^{9}Be(^{10}B,^{9}Be)^{10}B and ^{13}C(^{14}N,^{13}C)^{14}N, the two vertices are identical and each cross section depends on C^4.

The two independent measurements of the $^{14}\mathrm{N} \leftrightarrow {}^{13}\mathrm{C} + p$ ANC's provide a consistency check on our ability to extract ANC's from proton transfer reactions. We find the ANC for the dominant $p_{1/2}$ component of the ^{14}N ground state to be $C^2 = 18.6 \pm 1.2$ fm^{-1} from the ^{13}C(^{14}N,^{13}C)^{14}N reaction[4] and 17.8 ± 1.3 fm^{-1} from the ^{13}C(^{3}He,d)^{14}N reaction[5]. We adopted the weighted average of these two results when analyzing the ^{14}N(^{7}Be,^{8}B)^{13}C and ^{14}N(^{11}C,^{12}N)^{13}C reactions described below. This would not have been practical if we had utilized spectroscopic factors determined from the DWBA fits to ^{13}C(^{14}N,^{13}C)^{14}N and ^{13}C(^{3}He,d)^{14}N instead of ANC's, because the inferred spectroscopic factors are very sensitive to the choice of single-particle bound state orbitals in the DWBA calculations.

3 Using ANC's to Predict Astrophysical S Factors: Test Cases

The ANC's found from the proton transfer reactions can be used to determine direct capture rates at astrophysical energies. The relation of the ANC's to the direct capture rate at low energies is straightforward to obtain. The cross section for the direct capture reaction $A + p \to B + \gamma$ can be written as

$$\sigma = \lambda \left| < I^B_{Ap}(\mathbf{r}) \mid \hat{O}(\mathbf{r}) \mid \psi_i^{(+)}(\mathbf{r}) > \right|^2, \tag{2}$$

where λ contains kinematical factors, I^B_{Ap} is the overlap function for $B \to A+p$, $\hat{O}$ is the electromagnetic transition operator, and $\psi_i^{(+)}$ is the scattering wave in the incident channel. If the dominant contribution to the matrix element comes from outside the nuclear radius, the overlap function may be replaced by

$$I^B_{Ap}(r) \approx C\frac{W_{-\eta,l+1/2}(2\kappa r)}{r}, \tag{3}$$

where C defines the amplitude of the tail of the radial overlap function I^B_{Ap}, W is the Whittaker function, η is the Coulomb parameter for the bound state $B = A + p$, and κ is the bound state wave number. The required C's are just the ANC's found above from transfer reactions. Thus, the direct capture cross sections are directly proportional to the squares of these ANC's.

Using the results for the ANC's discussed above, the S factors describing the capture to both the ground and first excited states for $^{16}O(p,\gamma)^{17}F$ were calculated, with no additional normalization constants. The results are shown in Fig. 1 compared to the two previous measurements of $^{16}O(p,\gamma)^{17}F$[9,10]. Both E1 and E2 contributions have been included in the calculations, but the E1 components dominate the results. The theoretical uncertainty in the S factors, not including the uncertainty due to the ANC's, is less than 2% for energies below 1 MeV. Above 1 MeV the nuclear interaction begins to be important in the evaluation of the scattering wave function. The agreement between the measured S factors and those calculated from our $^{17}F \rightarrow {}^{16}O + p$ ANC's is quite good, especially for energies below 1 MeV where the approximation of ignoring contributions from the nuclear interior should be very reliable and the optical potential uncertainties are negligible. Overall, the results verify that the technique is valid for determining S factors to accuracies of at least 9%.

The S factor for $^9Be(p,\gamma)^{10}B$ has contributions from both resonance and direct capture at stellar energies. Thus the connection between the ANC and the capture cross section is more complicated and is discussed elsewhere[12]. Our measurement of the ANC's for $^{10}B \rightarrow {}^9Be + p$ fixes the nonresonant capture component. With this result, an R-matrix fit was made to the existing data[13] using the known locations of the resonance states and their widths. The fit[12], which is shown in Fig. 2, does an excellent job of reproducing the data. Prior to our determination of the direct capture contribution, similar attempts to fit the data required substantial changes in the known resonance positions and widths.

4 ANC's for stellar processes

We have measured the (^{7}Be,^{8}B) reaction on a 1.7 mg/cm^2 ^{10}B target[14] and a 1.5 mg/cm^2 Melamine target[15] in order to extract the ANC for $^8B \rightarrow {}^7Be +$

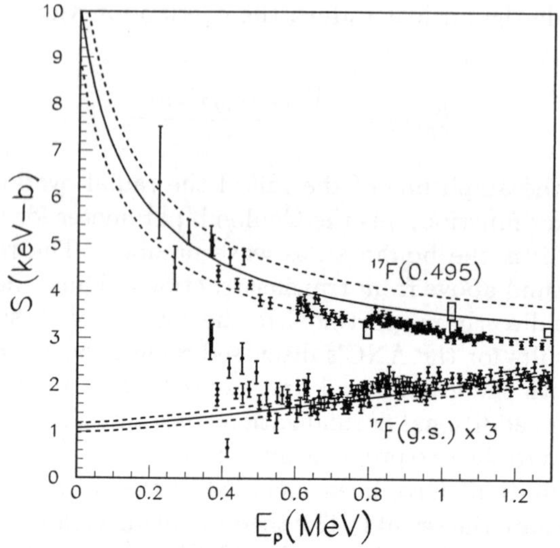

Figure 1. A comparison of the experimental S factors to those determined from the ANC's found in $^{16}O(^{3}He,d)^{17}F$. The solid data points are from[9], and the open boxes are from[10]. The solid lines indicate our calculated S factors, and the dashed lines indicate the $\pm 1\sigma$ error bands. Note that the experimental ground state S factor may be contaminated by background[11] at energies below 500 keV.

p. The radioactive ^{7}Be beam was produced at 12 MeV/A by filtering reaction products from the $^{1}H(^{7}Li,^{7}Be)n$ reaction in the recoil spectrometer MARS, starting with a primary ^{7}Li beam at 18.6 MeV/A from the TAMU K500 cyclotron. The beam was incident on an H_2 cryogenic gas target, cooled by LN_2, which was kept at 1 atm (absolute) pressure. Reaction products were measured by 5 cm × 5 cm Si detector telescopes consisting of a 100 μm ΔE strip detector, with 16 position sensitive strips, followed by a 500 or 1000 μm E counter. Details of the experiments can be found elsewhere[14,15].

In order to extract ANC's from reactions involving radioactive-ion beams, it is necessary to have optical model parameters. We have carried out a series of elastic scattering measurements with stable beam and target combinations that are close to those for our radioactive beam measurements. We used the folding model prescription to calculate the potential parameters and then renormalized the real and imaginary parts of the potentials to fit the data. Several different interactions were tried but the most consistent results were found to be from the JLM interaction[16]. The renormalization coefficients were found to be 0.366 ± 0.014 and 1.000 ± 0.087 for the real and imaginary

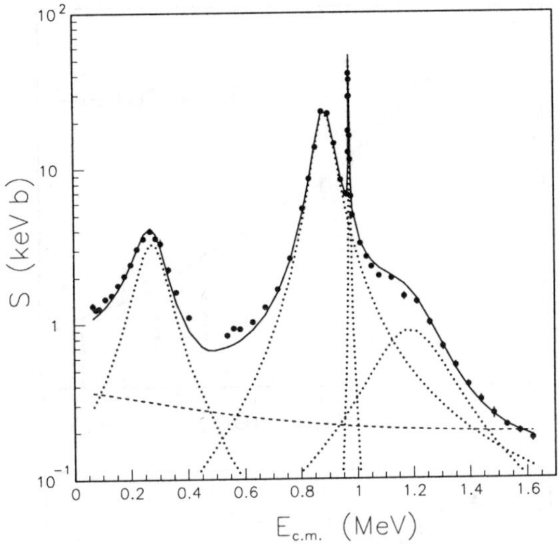

Figure 2. The result of an R-matrix fit to the direct plus resonance contributions to the S factor for ^{9}Be(p,γ)^{10}B. Standard resonance widths and positions were used in the fit. The direct capture contribution to the S factor is the dashed line.

potentials. The uncertainties are from the dispersion in the renormalization coefficients that we found for the different reactions[17]. With these potential parameters, it is now possible to calculate elastic scattering and transfer reactions for loosely bound p-shell systems.

Angular distributions for both elastic scattering on the ^{10}B and ^{14}N targets and the (^{7}Be,^{8}B) transfer reactions populating the ground states of ^{9}Be and ^{13}C were extracted from the data. A Monte Carlo simulation of the beam and detector system was used to generate the solid angle factor for each angular bin and the smoothing needed for the DWBA calculations to account for the finite angular resolution of the beam. We obtained excellent agreement between the optical model calculations, using the parameters discussed above, and the elastic scattering yields from both targets without additional normalization coefficients. The angular distributions for the transfer reactions are compared to DWBA calculations in Fig. 3.

The ANC for $^8\text{B} \to {}^7\text{Be} + p$ was extracted based on the fit to the angular distributions and the ANC's for $^{10}\text{B} \to {}^9\text{Be} + p$ and $^{14}\text{N} \to {}^{13}\text{C} + p$ following the procedure outlined above in our test case. Two ^{8}B orbitals, $1p_{1/2}$ and $1p_{3/2}$, contribute to the transfer reaction but the $1p_{3/2}$ dominates in both

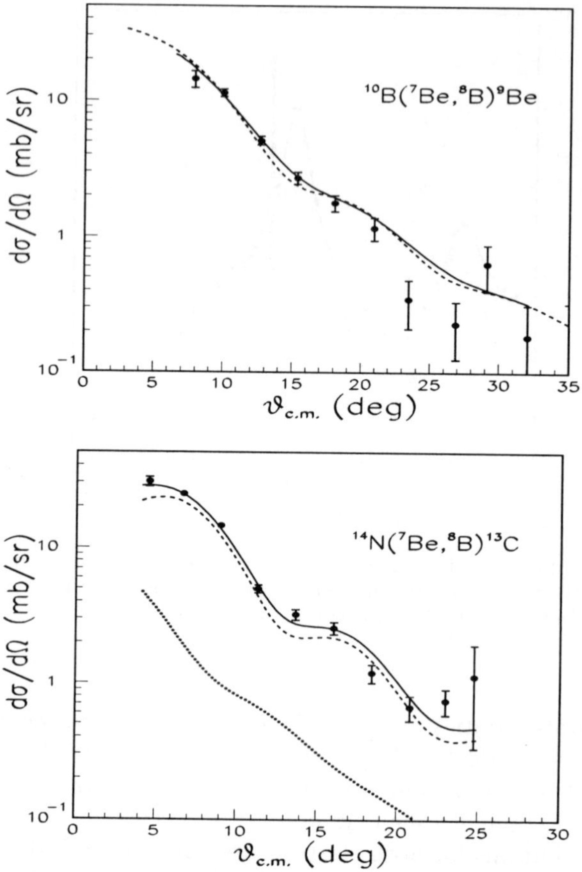

Figure 3. Angular distributions for ^8B populating the ground state of ^9Be from the ^{10}B target and ^{13}C from the Melamine target. In the top figure, the dashed curve shows the result of a DWBA calculation for the dominant component that contributes to the cross section. Two components are shown in the bottom figure. In both cases, the solid curve is smoothed over the angular acceptance of each bin.

cases. In calculating the angular distributions, we used the ratio for the two orbitals as given by a microscopic description of the ^8B ground state[2]. The optical model parameters were obtained from renormalized microscopic folding potentials using the JLM effective NN interaction described above. We have checked the sensitivity of the calculations by varying the normalization parameters. As in previous studies, the results are insensitive to bound state

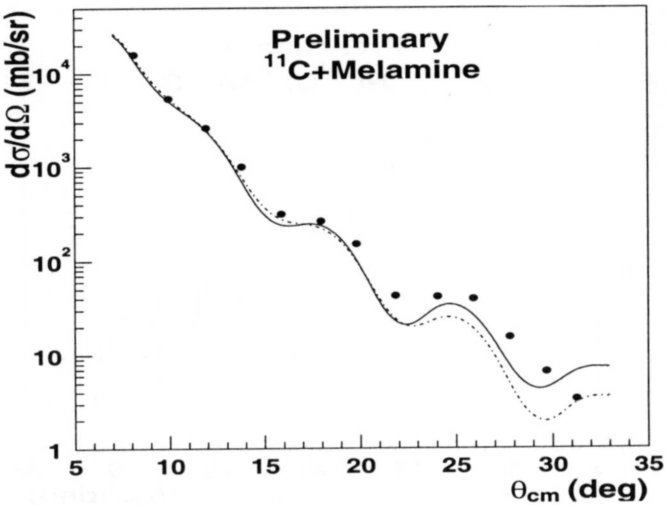

Figure 4. Preliminary angular distribution for ^{11}C elastic scattering from the Melamine target. The curves are two optical model predictions, as described in the text.

single particle well parameters in the DWBA calculations.

The astrophysical S factor for ^{7}Be(p,γ)^{8}B has been determined from the ANC for each target. The quoted uncertainties include 8.1% for optical model parameters. Experimental uncertainties and the ANC of the other vertex contribute an additional 10.9% for the ^{10}B target and 7.6% for the ^{14}N target. The values that we find are $S_{17}(0) = 18.4 \pm 2.5$ eV b for the ^{10}B target and 16.6 ± 1.9 eV b for the ^{14}N target, which are in good agreement with the recommended value[18] of 19^{+4}_{-2} eV b. Since the optical model parameters for the two different targets are derived by the same technique, this introduces correlations in the uncertainties between the two results. Accounting for this, we find $S_{17}(0) = 17.3 \pm 1.8$ eV b from the combined measurements.

We have measured the ^{14}N(^{11}C,^{12}N)^{13}C reaction to determine the ANC for $^{11}\mathrm{C} + p \leftrightarrow {}^{12}\mathrm{N}$ and, in turn, the direct capture rate for ^{11}C(p,γ)^{12}N at astrophysical energies. The experiment used a ^{11}C radioactive beam, with an intensity of $\sim 4.2 \times 10^5$ particles/sec at an energy of 10 MeV/u, incident on a 1.5 mg/cm^2 Melamine target. The radioactive beam was produced in the reaction ^{1}H(^{11}B,^{11}C)n and filtered by MARS, using procedures similar to those in the (^{7}Be,^{8}B) measurements.

Both ^{11}C elastic scattering off the Melamine target and (^{11}C,^{12}N) proton transfer reactions were observed simultaneously. Figure 4 shows the prelimi-

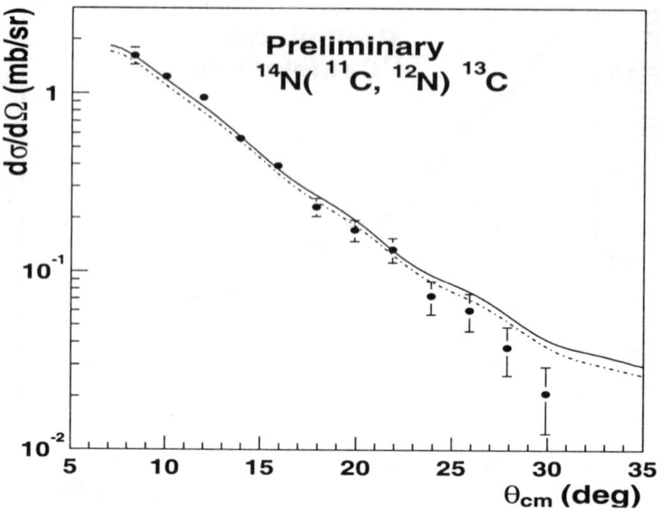

Figure 5. Preliminary angular distribution for the $^{14}N(^{11}C,^{12}N)^{13}C$ reaction. The curves represent DWBA predictions of the angular distribution obtained using the same optical potentials as in Fig. 4. The solid curve has been normalized to the data. The dot-dashed curve uses the same normalization constant as the solid curve

nary results for the measured elastic scattering distribution, together with two theoretical predictions, each of which has already been smoothed to account for finite angular acceptance. As above, the absolute cross sections predicted by the optical potentials are compared to the data without additional normalization coefficients. The dot-dashed curve uses the double-folding optical potential for loosely-bound $1p$ shell nuclei described above and provides a poor description of the data beyond 20°. However, one should note that ^{11}C is not loosely bound, so the renormalization of the real potential should not be as great. The solid curve uses the same normalization factors for the double-folding optical potential that provided the best fit to $^{13}C + ^{14}N$ elastic scattering[17], and provides a better description of the data.

Figure 5 shows our preliminary angular distribution for the $^{14}N(^{11}C,^{12}N)^{13}C$ reaction. The solid curve is a DWBA prediction of the proton transfer reaction that uses the same optical potential for ^{11}C + ^{14}N as in the solid curve in Fig. 4, and uses the optical potential for loosely bound nuclei for $^{12}N + ^{13}C$. The normalization of this DWBA prediction to the data determines the ANC for $^{12}N \leftrightarrow {}^{11}C + p$. The dot-dashed curve is a DWBA prediction that uses the optical potential for loosely bound nuclei in both the entrance and exit channels and the same normalization as the solid curve. It demonstrates that the sensitivity of the elastic scattering distribution to the choice of double-folding potential renormalization factors illustrated in Fig. 4 leads to an uncertainty in the inferred ANC for ^{12}N $\leftrightarrow$ ^{11}C + p of only 4%. Our preliminary results indicate that the direct capture contribution to the ^{11}C$(p,\gamma)^{12}$N S factor is larger than previous estimates[19,20].

Acknowledgments

Supported in part by the U.S. Department of Energy (DE-FG03-93ER40773), the National Science Foundation (INT 9909787), the Czech Academy of Sciences and by the Robert A. Welch Foundation.

References

1. H.M. Xu *et al.*, Phys. Rev. Lett. **73**, 2027 (1994).
2. A.M. Mukhamedzhanov and N.K. Timofeyuk, JETP Lett. **51**, 282 (1990).
3. A.M. Mukhamedzhanov *et al.*, Phys. Rev. C **56**, 1302 (1997).
4. L. Trache *et al.*, Phys. Rev. C **58**, 2715 (1998).
5. P. Bem *et al.*, Phys. Rev. C **62**, 024320 (2000).
6. C.A. Gagliardi *et al.*, Phys. Rev. C **59**, 1149 (1999).
7. M. Rhoades-Brown, M. McFarlane and S. Pieper, Phys. Rev. C **21**, 2417 (1980); Phys. Rev. C **21**, 2436 (1980).
8. A.M. Mukhamedzhanov, R.E. Tribble and N.K. Timofeyuk, Phys. Rev. C **51**, 3472 (1995).
9. R. Morlock *et al.*, Phys. Rev. Lett. **79**, 3837 (1997).
10. H.C. Chow, G.M. Griffith and T.H. Hall, Can. J. Phys. **53**, 1672 (1975).
11. R. Morlock, private communication.
12. A. Sattarov *et al.*, Phys. Rev. C **60**, 035801 (1999).
13. D. Zhanow *et al.*, Nucl. Phys. **A589**, 95 (1995).
14. A. Azhari *et al.*, Phys. Rev. Lett. **82**, 3960 (1999).
15. A. Azhari *et al.*, Phys. Rev. C **60**, 035801 (1999).
16. J.P. Jeukenne, A. Lejeune and C. Mahaux, Phys. Rev. C **16**, 80 (1977).
17. L. Trache *et al.*, Phys. Rev. C **61**, 024612-1 (2000).
18. E.G. Adelberger *et al.*, Rev. Mod. Phys. **70**, (1998) 1265.
19. M. Wiescher *et al.*, Astrophys. J. 343 (1989) 352.
20. A. Lefebvre *et al.*, Nucl. Phys. A 592 (1995) 69.

NEUTRON INDUCED REACTIONS ON RADIOACTIVE BERYLLIUM AND ARGON ISOTOPES

C. WAGEMANS AND G. GOEMINNE

University of Gent, Dept. of Subatomic and Radiation Physics, B-9000 Gent, Belgium
E-mail: wagemans@irmm.jrc.be

J. WAGEMANS

EC, JRC, Institute for Reference Materials and Measurements, B-2440 Geel, Belgium

U. KÖSTER

CERN-ISOLDE, CH-1211 Geneva, Switzerland

M. LOISELET AND M. GAELENS

University of Louvain, Cyclotron Research Center, B-1348 Louvain-la-Neuve, Belgium

P. GELTENBORT

Institute Laue-Langevin, BP 156, F-38042 Grenoble, France

The preparation of radioactive ^{10}Be, ^{37}Ar and ^{39}Ar samples is discussed. Investigation of the ^{10}Be(n_{th},γ)^{11}Be, ^{37}Ar(n_{th},α)^{34}S, ^{37}Ar(n_{th},p)^{37}Cl and ^{39}Ar(n_{th},α)^{36}S reactions is reported, and resonances in the ^{37}Ar(n,α)^{34}S reaction have been observed and analysed.

1 Introduction

Radioactive beams are used for a large variety of nuclear reaction studies, but a direct interaction with a neutron beam has not been realised so far. In the present paper we discuss a two-step method, in which rare radioactive nuclides are produced, ionised, accelerated and implanted in a foil, which then is bombarded with neutrons.

This method was applied recently at the cyclotron of the UCL (Louvain-la-Neuve, Belgium), where ^{37}Ar (half-life 35.04 days) was produced, and at CERN-ISOLDE (Geneva, Switzerland), where ^{10}Be (half-life 1.6 million years) and ^{39}Ar (half-life 269 years) samples were made. These samples were then irradiated with thermal neutrons at the high flux reactor of the ILL (Grenoble, France); some of the ^{37}Ar samples were also used for measurements with resonance neutrons at the GELINA neutron facility in Geel (Belgium).

2 Sample preparation and characterisation

The ^{37}Ar samples were produced via the $^{37}Cl(p,n)^{37}Ar$ reaction using a 30 MeV proton beam of the CYCLONE 30 cyclotron of the UCL. Up to 7x10^{15} ^{37}Ar ions could be implanted in thin aluminium foils. The number of atoms was determined by measuring the 2.62 keV KX-rays emitted during the EC-decay of ^{37}Ar. A detailed description of the procedure can be found in [1].

^{10}Be and ^{39}Ar on the other hand were produced via spallation reactions on graphite and titaniumoxide respectively, induced by 1 GeV protons of CERN's PS booster. Then the reaction products were ionised and mass-analysed at ISOLDE and implanted in a Ge and Al strip, respectively. In this way, a ^{10}Be sample containing 10^{15} atoms and a ^{39}Ar sample containing 3x10^{14} atoms were obtained. The number of atoms was determined by counting the 0.6 MeV β-particles emitted in the decay of both ^{10}Be and ^{39}Ar.

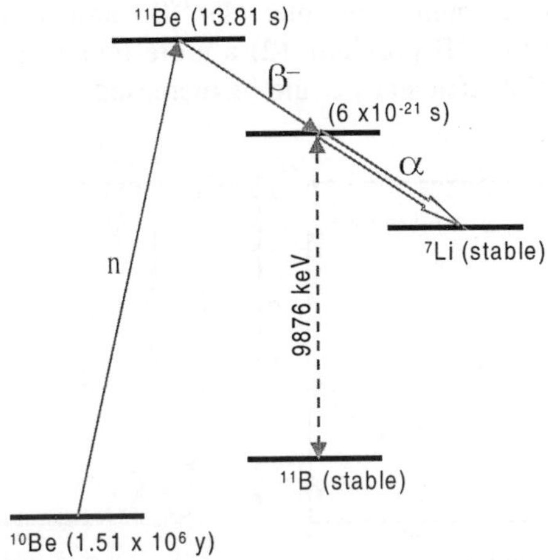

Figure 1. Schematic representation of the $^{10}Be(n_{th},\gamma)^{11}Be(\beta^-)^{11}B^*(\alpha)^7Li$ process.

3 Measurements and results

All measurements with thermal neutrons were performed at the ILL's curved neutron guide H22D, where a vacuum chamber with suited surface-barrier detectors was installed. A very clean thermal neutron beam with 3.5x10^8 neutrons/cm^2.s was hitting the sample. Measurements with resonance neutrons were only performed for ^{37}Ar. These experiments were carried out at a 8m long flight-path of the GELINA neutron facility. The charged particles were detected using a Frisch-gridded ionisation chamber with a methane gas-flow.

3.1 ^{10}Be

The ^{10}Be sample was used to determine the cross section of the $^{10}Be(n_{th},\gamma)^{11}Be$ reaction leading to the halo nucleus ^{11}Be. Due to the small number of atoms, this was done in an indirect way, i.e. via the detection of the 0.77 MeV α-particles emitted in the $^{10}Be(n_{th},\gamma)^{11}Be(\beta^-)^{11}B^*(\alpha)^7Li$ processes (Figure 1). We know indeed that the α-decay from the short-living 9.876 MeV level in ^{11}B to the ^{7}Li ground state has a branching ratio of 2.7 % [2], so the $^{10}Be(n_{th},\gamma)^{11}Be$ cross section can immediately be deduced from this "quasi (n,α)" experiment. The measurement was hampered by the presence of a small amount of ^{10}B in the Ge strip, the $^{10}B(n_{th},\alpha)^7Li$ reaction having a very large cross section (3840 b). So a disturbing ^{7}Li peak appeared at 0.83 MeV, its tail hampering a clean detection of the 0.77 MeV α's. As a result, we could only determine an upper-limit of 5 b for the $^{10}Be(n_{th},\gamma)^{11}Be$ cross section. This measurement however demonstrated the feasibility of the method, significant improvements being possible: (1) ^{10}Be atoms could be implanted in ultra-pure Al, solving the ^{10}B problem; (2) a more intense neutron beam could be used and (3) the detection efficiency could be increased.

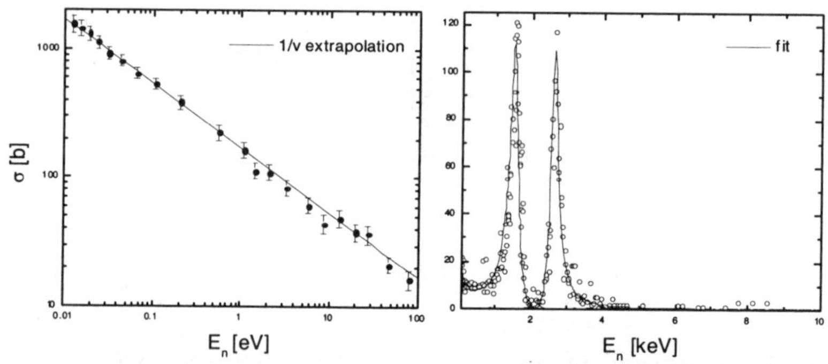

Figure 2. The $^{37}Ar(n,\alpha)^{34}S$ cross section from 0.01 eV to 10 keV.

3.2 ^{37}Ar

The $^{37}Ar(n,p)^{37}Cl$ and $^{37}Ar(n,\alpha)^{34}S$ reactions were investigated with thermal and with resonance neutrons, ^{37}Ar being one of the few nuclides with positive reaction energies for both (n,p) and (n,α) reactions. Moreover, the capture of a neutron leads to the semi-double-magic nucleus ^{38}Ar, for which no resonance data exist.

The measurements with thermal neutrons resulted in the following cross sections: $^{37}Ar(n_{th},\alpha_0)^{34}S$ (1070 $\pm$ 80)b; $^{37}Ar(n_{th},\alpha_1)^{34}S$ (0.31 $\pm$ 0.10)b; $^{37}Ar(n_{th},p)^{37}Cl$ (37 $\pm$ 4)b [3]. These values are two times smaller than those of Asghar et al. [4].

Experiments performed at GELINA covered the neutron energy region from 0.01 eV up to 100 keV. The $^{37}Ar(n,\alpha)^{34}S$ cross section was shown to have a 1/v-shape (figure 2 left), and in the keV region several resonances were observed (figure 2 right) and analysed.

3.3 ^{39}Ar

For the $^{39}Ar(n_{th},\alpha)^{36}S$ reaction cross section, a theoretical estimate of 3.5 b can be deduced from the data of Woosley *et al.* [6]. In the present work, a much lower upper-limit of 260 mb is obtained.

4 Discussion

The present study of the $^{37}Ar(n,\alpha)^{34}S$, $^{37}Ar(n,p)^{37}Cl$ and $^{39}Ar(n,\alpha)^{36}S$ reactions yields valuable information for nuclear spectroscopy (e.g. width and spin of nuclear levels), but these reactions play also an important role in the nucleosynthesis network in the sulfur region. Due to the absence of experimental data, theoretical values were used so far in the network calculations. We transformed our $\sigma(E_n)$ values in reaction rates [5]. Strong discrepancies with the theoretical values [6,7,8] were observed. Replacing these theoretical values by our experimental data significantly modifies the branchings at ^{37}Ar and ^{39}Ar. In the case of carbon burning *e.g.*, 87.3 % of the reaction flow at ^{37}Ar passes now via the (n,α) channel compared to 45.5 % with the theoretical values. This results in a modification of the nucleosynthesis flow, influencing the s-process abundances of *e.g.* ^{36}S and ^{40}Ar.

References

1. C. Wagemans, M. Loiselet, R. Bieber, B. Denecke, D. Reher and P. Geltenbort, *Nucl. Instr. Meth. A* **397** (1997) pp. 22-25.
2. R. Firestone, Table of Isotopes (J. Wiley, New York, 1996).
3. R. Bieber *et al.*, *Nucl. Phys. A* **647** (1999) pp. 3-11.
4. M. Asghar, A. Emsallem, E. Hagberg, B. Jonson and P. Tidemand-Petersson, *Z. Phys. A* **288** (1978) pp. 45-48.
5. G. Goeminne, C. Wagemans, J. Wagemans, O. Serot, M. Loiselet and M. Gaelens, *Nucl. Phys. A* **678** (2000) pp. 11-23.
6. S. Woosley and T. Weaver, *Ann. Rev. Astr. Astrophys.* **24** (1986) pp. 205-253.
7. T. Rauscher, Proc. Int. Symp. Nuclei in the Cosmos, Volos (Greece), Ed. Frontieres, Paris (1999) pp. 484-490.
8. S. Goriely, Proc. Int. Symp. on Capture Gamma-Ray Spectroscopy and Related Topics, Santa Fe (USA), AIP Conf. Proc. 529 (2000) pp. 287-294.

PRE-FUSION CAPTURE CROSS SECTIONS FOR HEAVY NUCLEUS-NUCLEUS SYSTEMS

J. WILCZYŃSKI
Institute for Nuclear Studies, 05-400 Otwock-Swierk, Poland
E-mail: januszw@iriss.ipj.gov.pl

K. SIWEK-WILCZYŃSKA
Institute of Experimental Physics, Warsaw University, 00-681 Warsaw, Poland
E-mail: siwek@npdaxp.fuw.edu.pl

Excitation functions for fusion and/or capture reactions have been analysed in a phenomenological model assuming the Gaussian distribution of fusion barriers. Preliminary systematics of the barrier-distribution parameters show the role of nuclear structure effects, which we propose to account for by relating values of the dispersion parameter with fusion energy thresholds calculated assuming the fusion adiabatic potential.

1 Introduction

Modern experiments aimed at production of new superheavy elements require precise predictions of the optimum value of bombarding energy, at which the production cross section attains its maximum. Very accurate knowledge of the fusion- or rather capture excitation function is one of the essential factors needed for such predictions. The term "capture" can be identified with the overcoming of the interaction barrier in a nucleus-nucleus collision. For light and medium systems, the overcoming of the interaction barrier leads to fusion in nearly every event, but for very heavy systems, overcoming of barrier guaranties only formation of a composite system that most probably will reseparate into two fragments in a process called "fast fission". Only very rarely will the system fuse forming a compound nucleus.

The most promising way to precisely predict excitation functions in capture reactions for very heavy systems is to study fusion reactions for somewhat lighter systems and then extrapolate the knowledge about conditions of the overcoming of the interaction barrier to the range of very heavy non-fusing systems.

In the first part of this presentation, we will show results of analysis of about 50 precisely measured fusion excitation functions, for which a new criterion of determining fusion energy thresholds was applied, and so-determined "experimental fusion thresholds" have been compared with predictions of the barrier heights calculated with a proposed adiabatic fusion potential [1].

Later, some preliminary results of an analysis of the fusion- and also capture excitation functions, assuming the Gaussian shape of the fusion barrier distribution, will be presented. Results of this analysis, combined with the information on the adiabatic fusion thresholds, give the possibility of predicting the mean position and width of the barrier distribution for any fusion- or capture reaction including reactions of very heavy systems. In such a way, fusion- and capture excitation functions can be accurately predicted for reactions not yet studied experimentally.

2 Experimental fusion energy thresholds

It is well known that fusion excitation functions cannot be explained assuming a single value of the fusion barrier. To obtain a reasonable agreement with experimental data, it is necessary to assume coexistence of different barriers, a situation that is naturally accounted for in description of fusion reactions in terms of coupled channel calculations. As it was demonstrated by Rowley, Satchler and Stelson, [2] an effective fusion-barrier distribution can be deduced from a precisely measured fusion excitation function by taking the double derivative of the product of the cross section multiplied by energy, $d^2(\sigma E)/dE^2$. The low-energy edge of the barrier distribution, "the lowest barrier", tells us where the onset of above-the-barrier fusion processes is, and therefore this energy is a good reference point that we can call an effective fusion-energy threshold.

Since determination of the fusion-barrier distributions and location of their low-energy edges require very precise measurements of the fusion cross sections at near-threshold energies, we have proposed in Ref.[1] an alternative, easy to apply definition of the fusion-energy threshold, which turns out to give threshold values almost exactly coinciding with those resulting from the original definition of "the lowest barrier". We propose to use the following operational definition of the fusion energy threshold: this threshold is the energy $E = E_{thr}$ at which the measured fusion cross-section is equal to the s-wave absorption cross section:

$$\sigma_{fus} = \pi \lambda^2 = \frac{\pi \hbar^2}{2\mu E_{thr}}, \tag{1}$$

where λ is the wavelength of the fusing system, and μ-its reduced mass. Note that in semi-classical language, $\pi\lambda^2$ is a "minimum portion" of above-the-barrier fusion cross section, and therefore seems to be a quantity appropriate for consistent definition of the easy-to-determine fusion energy threshold.

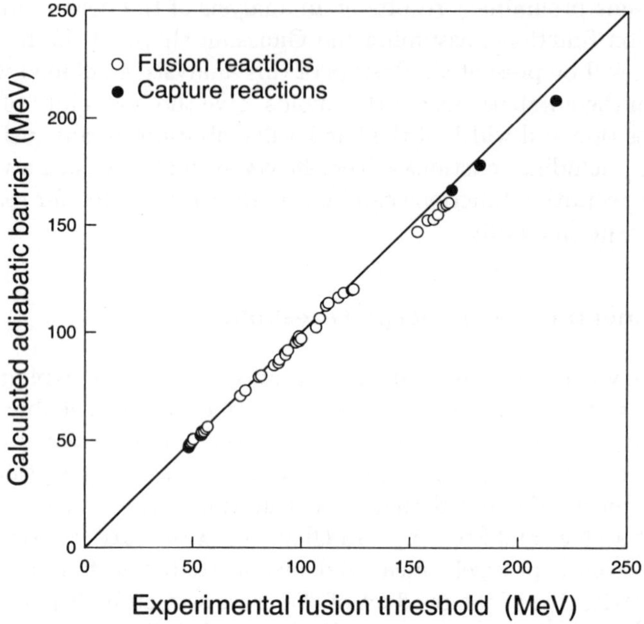

Figure 1. Fusion barriers calculated with the adiabatic fusion potential, Eqs.(2)-(5), compared with experimental fusion thresholds deduced from measured [3-16,20] fusion excitation functions.

In Ref.[1] we have demonstrated that "experimental" energy thresholds determined according to the criterion given by Eq.(1) perfectly agree with positions of the low-energy edge in the fusion-barrier distributions, for an ample set of about 50 excitation functions [3-16], for which the barrier distributions had been determined.

3 Adiabatic fusion barriers

In Ref.[1] we gave a prescription for a simple nucleus-nucleus fusion potential that has been constructed avoiding the question of specific shape parametrization of the fusing nucleus-nucleus system. We have used only the known characteristics of the system at the onset of nuclear interaction (the contact force) and in the final state of the equilibrated compound nucleus. Smooth interpolation between these two reference points is done essentially without free parameters, assuming that the effective one-dimensional nuclear potential has the Saxon-Woods shape,

$$V_n(r) = \frac{-V_0}{1 + exp(\frac{r-R_0}{a})} , \tag{2}$$

where

$$V_0 = (M_1 + M_2 - M_{cn})c^2 + C_{cn} - C_1 - C_2 + S_{cn} , \tag{3}$$

$$R_0 = r_0(A_1^{1/3} + A_2^{1/3}) , \tag{4}$$

$$a = \frac{V_0(R_1 + R_2)}{16\pi\gamma R_1 R_2} . \tag{5}$$

Here M_1, M_2 and M_{cn} are the ground-state masses of the two fusing nuclei and the final compound nucleus, respectively, and similarly, C_1, C_2 and C_{cn} represent intrinsic Coulomb energies of these nuclei, $C_i = 0.7054 Z_i^2/A_i^{1/3}$ MeV, with the constant taken from the standard liquid-drop-model fit to nuclear masses [17]. The diffuseness parameter a is determined by the strength of the nucleus-nucleus attractive force in the contact configuration $R_0 = R_1 + R_2$, calculated [18] with the liquid-drop model as a derivative of the surface energy,

$$\left(\frac{dV_n}{dr}\right)_{r=R_0} = \frac{V_0}{4a} = 4\pi\gamma\frac{R_1 R_2}{R_1 + R_2} , \tag{6}$$

where γ is the surface-tension coefficient [17] of the combined system. Calculating the liquid-drop-model contact force, we scaled the radii R_1 and R_2 as $r_0 A_i^{1/3}$, allowing for some fine tuning of the value of the r_0 parameter in order to obtain the best agreement of the predicted barrier heights with experimental values of the fusion thresholds. The radius parameter r_0, optimized in such a way, was found to be $r_0 = 1.15$ fm, a value that can be associated with the nuclear "equivalent sharp radius" [19].

For determination of the fusion barrier, it is sufficient to consider the Coulomb potential only in the outer region, $r > R_0$, where the point-charge approximation,

$$V_c(r) = \frac{Z_1 Z_2 e^2}{r} \tag{7}$$

gives sufficient accuracy. For completeness, in the inner region $r < R_0$, a simple one-dimensional parametrization joining the known value of the Coulomb energy in the contact configuration with that at the equilibrium shape, has been proposed [1].

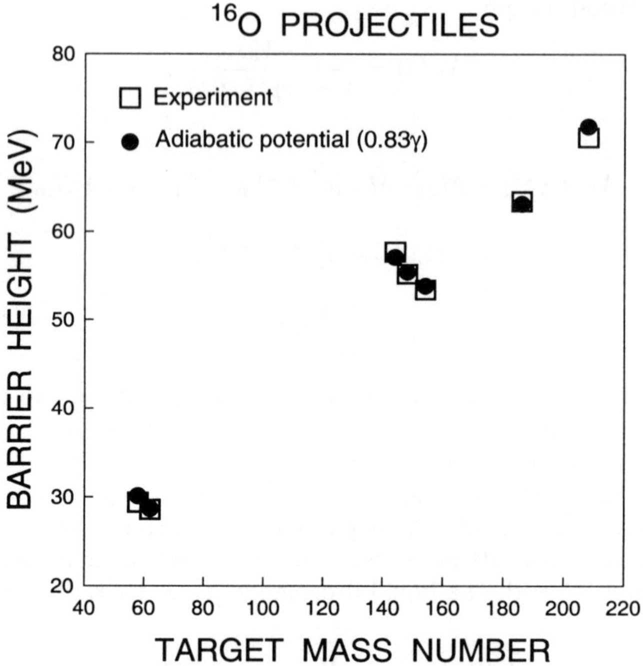

Figure 2. Nuclear structure (isotopic) effects in experimental fusion energy thresholds and calculated adiabatic fusion barriers in ^{16}O-induced fusion reactions on different target nuclei.

The proposed nucleus-nucleus potential, Eqs. (2)-(5), is expected to describe the adiabatic path of the colliding system towards fusion, leading over the lowest possible barrier. Therefore, we compare the experimental fusion thresholds determined by using the s-wave absorption criterion, Eq. (1), with our adiabatic barriers B_{adiab} defined by the condition $d(V_n + V_c)/dr = 0$. In Fig. 1, the calculated fusion barriers B_{adiab} are compared with the experimental fusion energy thresholds E_{thr}, determined [1] for about 50 reactions studied experimentally [3–16,20]. Very good agreement between these two quantities is observed, especially for medium-weight systems with E_{thr} below 100 MeV. For heavy systems, a good correlation between both quantities still remains, although a clear, systematic effect of the increasing difference $E_{thr} - B_{adiab}$ with increasing barrier height is observed. In Ref. [1] we have demonstrated that this effect can be identified with the "extra push" energy [21], mostly associated with dissipative losses of kinetic energy outside the barrier.

As seen from Fig. 1, we have included in our analysis recent data on capture reactions for very heavy systems [20]. They perfectly conform with the trend established for fusion reactions.

It should be emphasized that height of the adiabatic barriers B_{adiab} depends to some extent on structural properties of the fusing nuclei (shell effects, deformations, isospin, etc.) because parameters of the adiabatic potential, V_0 and a, depend on the ground-state masses of these nuclei. The influence of these structural effects is illustrated in Fig. 2, where both E_{thr} and B_{adiab} values for fusion of ^{16}O with different target nuclei are plotted as a function of the target mass number. All irregularities in the dependence of E_{thr} on A are well reproduced by the calculated adiabatic barriers.

4 Fusion- and capture excitation functions

As discussed above, by using the adiabatic potential we are able to predict energy thresholds and the corresponding cross sections for fusion and capture reactions. This information is unquestionably too limited for solving practical questions, usually requiring the knowledge of fusion or capture cross sections in a certain range of energies. Therefore we have attempted [22] to parametrize fusion- and capture excitation functions arbitrarily assuming the Gaussian shape of the barrier distribution:

$$p(B) = \frac{1}{\sigma_B\sqrt{2\pi}} \exp\left[-\frac{(B-B_0)^2}{2\sigma_B^2}\right], \tag{8}$$

with the hope of establishing the systematics of the parameters B_0 and σ_B for reasonable predictions of the cross sections for not yet studied systems.

Assuming Gaussian distribution of the barrier heights, one can quite precisely reproduce the measured fusion excitation functions, as shown in Fig. 3 on the example of four selected reactions. By minimizing χ^2, one can then determine the mean barrier B_0 and the dispersion σ_B for each reaction individually. With this procedure, using published data on fusion and/or capture excitation functions[3–6,8–16,20,23–25], we have determined the parameters B_0 and σ_B for about 50 reactions. Compilation of these results can be used for predictions of fusion- and capture cross sections for reactions not yet studied experimentally, especially for very heavy systems used to produce superheavy elements.

Fig. 4 shows the preliminary results of our analysis: the deduced values of the mean barrier B_0 plotted as a function of the parameter $x = Z_1Z_2/(A_1^{1/3} + A_2^{1/3})$. This dependence is very regular and can be approximated by a second

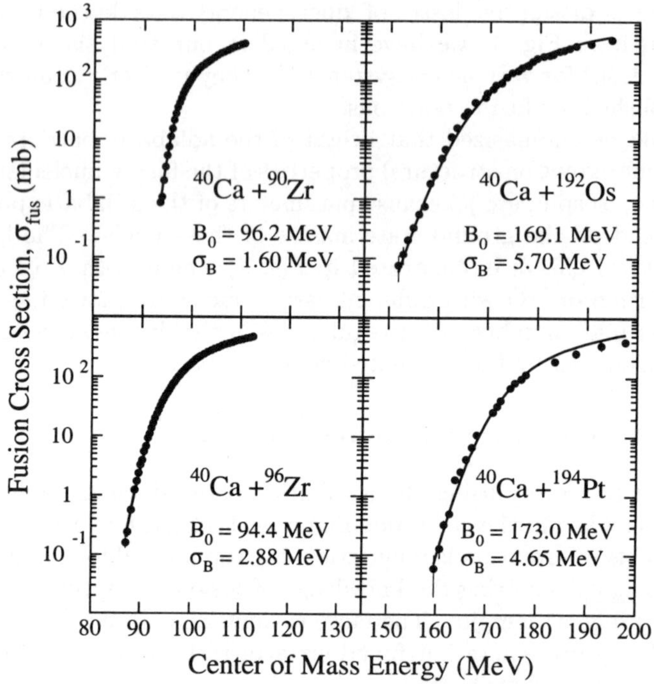

Figure 3. Examples of fusion excitation functions calculated assuming Gaussian distribution of the fusion barrier, fitted to experimental data taken from Refs. [11,12].

order polynomial function,

$$B_0 = 0.00136x^2 + 0.78x + 4.2 \text{ MeV}. \tag{9}$$

We emphasize the fact that the parametrization established for fusion reactions (full circles in Fig. 4) holds for capture reactions for very heavy systems (squares). This gives good argument for using Eq. (9) for predictions of the capture cross sections in reactions leading to production of superheavy nuclei.

Contrary to B_0, the dispersion parameter σ_B does not behave so regularly. This is obvious, taking into account the nature of fusion-barrier distributions, namely the coupling to the rotational and vibrational states in the fusing nuclei. Clearly, σ_B must depend not only on the global parameters, such as Z and A of the fusing nuclei, but also on structural characteristics of these nuclei. We observed, however, that knowing the fusion energy thresholds B_{adiab}, which reflect the structural effects, we can account to some extent

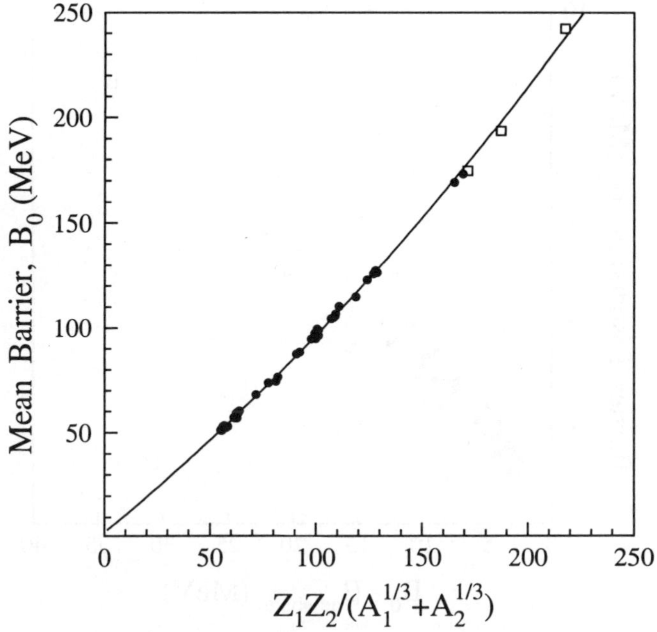

Figure 4. Systematics of the mean barrier parameter B_0 deduced from analysis of a set of published fusion- and capture excitation functions [3-6,8-16,20,23-25], presented as a function of the Coulomb barrier parameter $Z_1 Z_2/(A_1^{1/3} + A_2^{1/3})$. Results for capture reactions [20] are indicated by different symbols (squares). Solid line represents the parametrization given by Eq. (9).

for these effects in individual values of σ_B. Namely, we found that σ_B is correlated quite closely with the difference $B_0 - B_{adiab}$. This correlation is shown in Fig. 5. It presents a possibility to predict σ_B for any system using the following relation:

$$\sigma_B = 0.22(B_0 - B_{adiab}) + 0.7 \text{ MeV}, \tag{10}$$

where B_0 is given by Eq. (9) and B_{adiab} is calculated as described in Section 3. Again, similarly as in case of B_0 systematics, we emphasize the fact that the correlation established for fusion reactions holds perfectly also for capture reactions. This gives us the possibility to quite accurately predict the parameters of the barrier distribution for any system and use this information for calculating fusion- or capture excitation functions.

To summarize, we propose to use the $\pi\lambda^2$ criterion for determination of

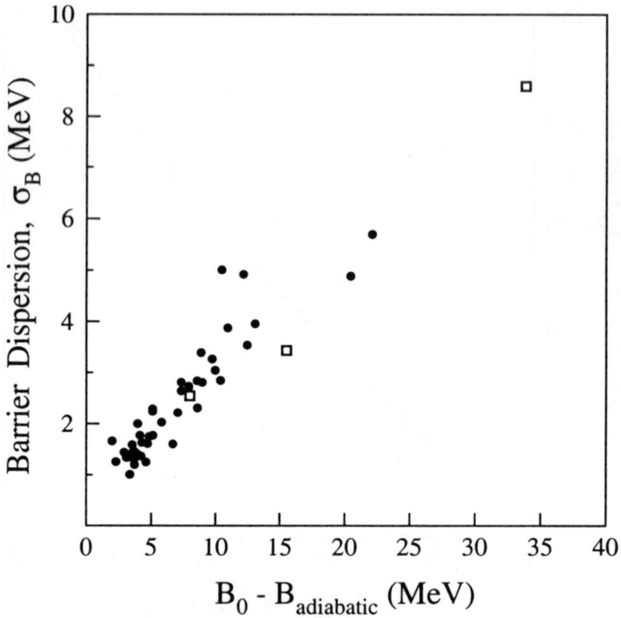

Figure 5. Systematics of the barrier dispersion parameter σ_B deduced from analysis of a set of published fusion- and capture excitation functions [3–6,8–16,20,23–25], presented as a function of $B_0 - B_{adiabatic}$ (see text). Results for capture reactions [20] are indicated by different symbols (squares).

the "experimental fusion energy thresholds" E_{thr}. So-defined fusion energy thresholds perfectly agree with positions of the low-energy edge in experimental fusion barrier distributions and thus can be identified with the lowest (adiabatic) fusion barriers. On the other hand, the experimental fusion thresholds are closely correlated with the adiabatic barriers B_{adiab} that can be calculated with the proposed nucleus-nucleus fusion potential. In order to be able to predict fusion and/or capture cross sections in a wider range of energies, we analysed published data on fusion- and capture excitation functions assuming the Gaussian distribution of fusion barriers. From such an analysis, we obtained preliminary [22] systematics of the parameters of barrier distributions enabling one to accurately predict fusion and/or capture excitation functions for all systems, including those that might be used for synthesis of superheavy nuclei.

Acknowledgments

We would like to thank W.J. Świątecki for many inspiring discussions and E. Siemaszko for her contribution to the yet unpublished work on analysis of fusion excitation functions. We are grateful to A.M. Stefanini for providing us with some tabulated data on fusion excitation functions and A.M. Itkis for sending us yet unpublished data on capture cross sections for very heavy systems. This work was supported by the Poland-U.S.A. Maria Skłodowska-Curie Joint Fund II, under Project No. PAA/DOE-98-34, and by the Polish State Committee of Scientific Research, KBN, Grant No. 2P03B05419.

References

1. K. Siwek-Wilczyńska and J. Wilczyński, *Phys. Rev.* C, in press.
2. N. Rowley, G.R. Satchler and P.H. Stelson, *Phys. Lett.* B **254**, 25 (1991).
3. M. Beckerman *et al.*, *Phys. Rev. Lett.* **45**, 1472 (1980).
4. M. Beckerman *et al.*, *Phys. Rev.* C **25**, 837 (1982).
5. H.A. Aljuwair *et al.*, *Phys. Rev.* C **30**, 1223 (1984).
6. W. Reisdorf *et al.*, *Nucl. Phys.* A **438**, 212 (1985).
7. W. Reisdorf *et al.*, *Nucl. Phys.* A **444**, 154 (1985).
8. A.M. Stefanini *et al.*, *Nucl. Phys.* A **456**, 509 (1986).
9. A.M. Stefanini *et al.*, *Phys. Rev. Lett.* **74**, 864 (1995).
10. A.M. Stefanini *et al.*, *Phys. Rev.* C **52**, R1727 (1995).
11. J.D. Bierman *et al.*, *Phys. Rev.* C **54**, 3068 (1996).
12. H. Timmers *et al.*, *Nucl. Phys.* A **633**, 421 (1998).
13. A.A. Sonzogni *et al.*, *Phys. Rev.* C **57**, 722 (1998).
14. F. Scarlassara *et al.*, *Nucl. Phys.* A **672**, 99 (2000).
15. A.M. Stefanini *et al.*, *Phys. Rev.* C **62**, 014601 (2000).
16. C.R. Morton *et al.*, *Phys. Rev.* C **62**, 024607 (2000).
17. W.D. Myers and W.J. Swiatecki, *Ark. Fys.* **36**, 343 (1967).
18. J. Wilczyński, *Nucl. Phys.* A **216**, 386 (1973).
19. W.D. Myers, *Nucl. Phys.* A **204**, 465 (1973).
20. M.G. Itkis *et al.*, Proc. International Workshop *"Fusion Dynamics at the Extremes"*, Dubna, Russia, 2000 (World Scientific, Singapore, in press).
21. W.J. Swiatecki, *Nucl. Phys.* A **376**, 275 (1982).
22. K. Siwek-Wilczyńska, W.J. Świątecki, and J. Wilczyński, to be published.
23. C.R. Morton *et al.*, *Phys. Rev.* C **52**, 243 (1995).
24. J.R. Leigh *et al.*, *Phys. Rev.* C **52**, 3151 (1995).
25. N. Keeley *et al.*, *Nucl. Phys.* A **628**, 1 (1998).

MAGNEX: A LARGE-ACCEPTANCE SPECTROMETER FOR RADIOACTIVE NUCLEAR BEAM EXPERIMENTS

A. CUNSOLO[1,2], F. CAPPUZZELLO[1], A. FOTI[2,3], A. LAZZARO[1,2], A.L. MELITA[1,2], C. NOCIFORO[1,2], V. SHCHEPUNOV[1,4], J.S. WINFIELD[1]

[1] *INFN - Laboratori Nazionali del Sud, Via S. Sofia 44, 95123 Catania, Italy*

[2] *Dipartimento di Fisica e Astronomia, Universita di Catania, Corso Italia 57, 95129 Catania, Italy*

[3] *INFN - Sezione di Catania, Corso Italia 57, 95129 Catania, Italy*

[4] *Flerov Laboratory of Nuclear Reactions, Joint Institute of Nuclear Research, 141980 Dubna, Russia*

We describe a large-acceptance (50 msr) high-resolution magnetic spectrometer "MAGNEX", which has been designed primarily for the radioactive beam facility EXCYT at INFN-LNS Catania. The configuration is a vertically-focusing quadrupole followed by a 55° bend angle dipole. The aberrations are corrected partly by shaping the entrance and exit effective field boundaries and partly by software compensation. Surface coils within the dipole pole-gap are used to compensate for the kinematic effect.

1 Introduction

A new magnetic spectrometer "MAGNEX" is under construction at INFN-LNS Catania.[a] It is designed primarily for the radioactive beam facility EXCYT[1], which is nearing completion. The relatively low beam energies (0.5 - 8 MeV/A) and excellent beam quality from the 15 MV Tandem accelerator will provide new opportunities for research in near Coulomb-barrier transfer and elastic and inelastic scattering reactions with exotic beams. However, the relatively low secondary beam intensities compared to stable beams (at least three orders of magnitude lower) means that a large acceptance spectrometer is required, not just in terms of overall counting rate, but also so that angular distributions can be measured with relatively few spectrometer angle settings. At the same time, one wishes to approach the high-resolving power achieved by smaller spectrometers, since this is both a partial compensation for low-statistics spectra (improving the signal to noise ratio) as well as an advantage in the separation of possible close-lying states. These two criteria – large solid angle of acceptance together with high resolving power – mean that the ion-optical aberrations must be very carefully treated.

[a] The magnets are being made by DANFYSIK A/S, Jyllinge, Denmark.

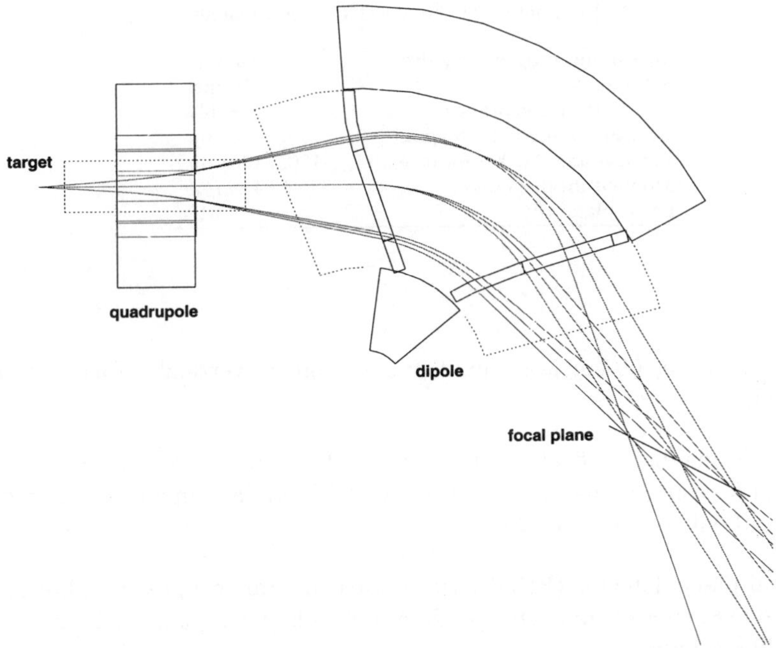

Figure 1. Plan view of the MAGNEX spectrometer. The ion trajectories correspond to three horizontal angles $\theta = 0, \pm 70$ mr, for three relative momenta ($\delta = 0, \pm 10\%$). The dotted lines enclose the regions of magnetic fields used in the ray-tracing programs.

2 General Description and Ion Optics

In order to minimize the number of large-volume magnetic elements, the spectrometer has a simple quadrupole-dipole (QD) configuration. The first-order layout of bend angle, Q-D separation, pole face angles, focal plane dimensions, maximum field, etc., has been determined by expressing such quantities as functions of transport matrix elements and performing optics calculations (see ref. [2]). A plan view of the magnetic elements and ion trajectories is shown in Fig. 1.

The principal components of MAGNEX are:

- An entrance position-sensitive detector (PSD) which provides the start signal for the TOF. The vertical angle of the particles is also measured

Table 1. Parameters of the MAGNEX spectrometer.

Maximum magnetic rigidity	1.80 T-m
Solid angle	50 msr
Momentum acceptance	±10%
Horizontal angular acceptance	(-90,+110) mr
Vertical angular acceptance	(-130,+130) mr
Momentum dispersion	4.04 cm/%
Focal plane tilt	61°

here.

- A 40-cm bore, 60-cm long quadrupole magnet, vertically focusing the particles.
- A dipole magnet, deflecting the particles by a mean angle of 55°. Horizontal focusing is mainly achieved by 18° boundary inclinations at the exit and entrance of the dipole.
- A focal plane detector (FPD) used to measure particle position, horizontal angle θ, energy, ionic charge, Z and A. The focal plane is 81 cm long and 30 cm high.

Details of the detectors may be found in ref. [3]. The main parameters of the spectrometer are given in Table 1. The first-order momentum resolution is 5400, although the actual resolution is limited by 2nd and higher order aberrations, which are partly corrected by shaping the entrance and exit effective field boundaries (EFBs). These shapes have been automatically optimised with a 2nd order spline in ZGOUBI.[4]

Because of the large θ acceptance, a correction for the kinematic effect (change of reaction product energy with scattering angle) is very important. Kinematical correction down to a k value of -0.5 ($k = +1/p\mathrm{d}p/\mathrm{d}\theta$) is provided by a system of surface correcting coils within the dipole together with a displacement of the focal plane detector. Such large values of k are a feature of inverse kinematics reactions, which have certain advantages for radioactive beam facilities.[5,6] To keep the shift of the FPD with respect to the dipole exit to a minimum (< 10 cm at maximum rigidity), we have optimised the pole profiles for $k = -0.25$ and introduced quadrupolar and sextupolar surface correction coils[7] inside the dipole gap. The remaining important aberrations will be corrected by software, with the use of the COSY INFINITY iterative algorithm[8] to obtain the inverse matrix elements. We note that not all of the

corrections could be done in this way, since the technique is limited by the resolution of the detectors.

The design of MAGNEX has been optimised for precision nuclear spectroscopy up to about mass 40; detection of higher masses will be possible with some reduction in final energy resolution because of straggling effects in the detection system. Operation at very forward angles is anticipated; however, at 0° itself, the position-sensitive start detector (PSD) would generally have to be removed, resulting in some compromise in either angular resolution or acceptance. Mass resolution might be recovered if, e.g., a coincidence timing detector were installed in the region of the target.

Acknowledgments

We wish to thank G. Fremont, P. Gangnant, J.F. Libin, W. Mittig, P. Roussel-Chomaz, H. Savajols, C. Spitaels (GANIL, France) and A.V. Belozyorov, O. Malishev, M. Tchelnokov, A. Yeremin (JINR, Russia), who have collaborated with the detector development for MAGNEX.

References

1. G. Ciavola et al., *Nucl. Phys.* A **616**, 69c (1997).
2. A. Cunsolo et al., *Nucl. Instrum. Methods* A in press (2001).
3. A. Cunsolo et al. in *9th Intl. conf. on nuclear reaction mechanisms*, Varenna, 2000, ed. E. Gadioli (Publications of the Univ. of Milan, Sup. No. 115) p. 661.
4. F. Meot and S. Valero, ZGOUBI v. 4, DAPNIA/SEA-97-13 (1997).
5. J.S. Winfield, W.N. Catford and N. Orr, *Nucl. Instrum. Methods* A **396**, 147 (1997)
6. H. Lenske and G. Schreider, *Eur.Phys. J.* A **2**, 41 (1998).
7. J. Camplan and R. Meunier, *Nucl. Instrum. Methods* A **186**, 445 (1981).
8. M. Berz et al., *Phys. Rev.* C **47**, 537 (1993).

WHAT WE HAVE LEARNT FROM STUDY OF THE DIRECT REACTIONS WITH 25 MEV/N ^{6}HE AND ^{8}HE SECONDARY BEAMS ON HYDROGEN AN HELIUM TARGETS?

R. WOLSKI[1)], A.S. FOMICHEV, A.M. RODIN, S.V. STEPANTSOV, S.I. SIDORCHUK, G.M. TER-AKOPIAN, AND YU.TS. OGANESSIAN

Flerov Laboratory of Nuclear Reactions, JINR, Dubna, Russia,

[1)]Institute of Nuclear Physics, Cracow, Poland

e-mail: wolski@lnr.jinr.ru

The study of ^{6}He and ^{8}He elastic scattering on a helium target and 2-neutron transfer reactions on protons are reported. The elastic scattering data were discussed in terms of presence of the backward angle enhancement caused by structural effects. The transfer reactions appeared to be an important source of information on neutron correlations in these nuclei.

1 Introduction

Among exotic nuclei ^{6}He nucleus is the lightest system with 2-neutron halo. Notable by its large N/Z =3 value the ^{8}He nucleus is considered as a candidate for a neutron skin nucleus. The direct reactions have been widely used in the past for study properties of nuclei involved. A program to study the direct reaction with radioactive ^{6}He and ^{8}He beams has been undertaken by Flerov Laboratory of Nuclear Reactions (JINR, Dubna). In the present paper we discuss experiments aimed to study the elastic scattering and transfer reactions of these nuclei on helium and hydrogen targets.

2 Experimental technique

The beams of ^{6}He and ^{8}He ions have been obtained by bombarding a thick berylium target with primary beams of ^{7}Li, ^{11}B or ^{13}C at energies of a few tens MeV/n. The particles of interest were selected from the main bulk of other products using the doubly achromatic magnetic separator ACCULINNA [1]. The average beam intensities amounted up to 10^{5} of ^{6}He particles or 4×10^{3} of ^{8}He per second in long-term irradiation. The separator ion optics restricted the energy spread of the secondary beam to a value of 5% (FWHM). Two MWPC provided data of a track for each bombarding particle. The nuclear reactions products were identified by two position sensitive, large solid angle telescopes, see [2]. The helium target was made as a gas cell coupled to a cryogenic generator. At a helium gas pressure of 11atm the effective target thickness was 2.4×10^{21} atoms per cm^{2}. Polyethylene foils served as a hydrogen target. Typically, the content of hydrogen atoms in these targets was $(1.5\text{-}3.0) \times 10^{21}$ per cm^{2}. Differential cross sections for elastic scattering and transfer

reactions were obtained by averaging statistic over 5^0 angular bins in CM frame and estimating and estimating the effective solid angle for each bin.

3 Discussion of the Selected Results

The angular distributions for elastic scattering have been measured for the ^{6}He +^{4}He and ^{8}He +^{4}He systems at forward scattering angles and an attempt was undertaken to measure the cross-section at backward angles. The resulting differential cross section for the elastic scattering of 151MeV ^{6}He nuclei from ^{4}He has been reported in [3]. A one-step, 2n-neutron exchange process could account for the obtained distinct rise in the cross section at the CM backward angles. DWBA calculations were performed [3,4] with the use of a model wave function for ^{6}He derived from one given by the solution of the exact three-body problem [5]. The wave function assuming the presence of two prominent configurations – the "di-neutron" and "cigar"-like ones showed a good agreement with the experimental data obtained for the backward CM angles. This result indicates a dominance of the "di-neutron" configuration in this process. In an alternative, coupled channels approach [6], these data were described without the 2n-transfer process explicitly included. For the case of the ^{8}He +^{4}He scattering, in spite of rather low (around 1 μb/sr) cross-section limits attained at backward angles, the enhancement originated from 4n-transfer process was not observed, see Fig.1. The unobserved enhancement was estimated by DWBA calculations with the code FRESCO [7]. It was assumed that one-step as well as two-step transitions contribute. The calculated potential scattering with coherent inclusion of the one-step 4n-transfer is shown in Fig.1 as curve 2. The two-step processes were treated as several sequential transfers of 2 pairs of neutrons through the two intermediate ^{6}He nuclei, both in their ground and the first excited states. For simplicity all absolute values of the relevant Spectr. Amplitudes (SA) were taken to be equal to 1.0, their signs were determined with the help of [8]. It was found that the

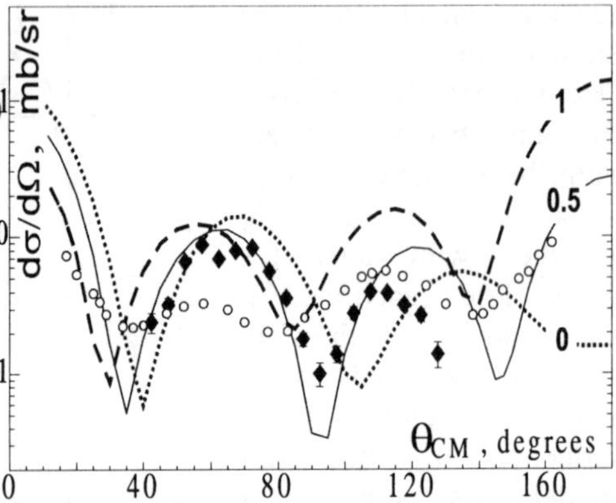

Fig.2: Solid symbols show our data on the cross section of ^{6}He+p→α+t reaction (E_{6He}=150 MeV). Open circles show similar data from Ref. [12] for p+^{6}Le→α+^{3}He reaction (E_p=25 MeV). Curves: DWBA calculations assuming 1 – the full shell-model spectroscopic amplitude for two triton clusters in ^{6}He, 0.5 – this amplitude was reduced by a factor of two, 0 – the two-triton clustering of ^{6}He neglected.

intermediate system of two excited ^{6}He nuclei generates the largest contribution to the backward-angle enhancement.

To further look into the structure of ^{6}He and ^{8}He nuclei the 2n- transfer reactions on protons were studied. In the case of the ^{6}He + p collision the two: 2n and triton transfers contribute coherently. Fig. 2 shows our data [2] together with data for ^{6}Li(p,^{3}He)α reaction [9]. The analysis of the ^{6}He data [2], enabled us to make a conclusion on a reduction of the shell model SA for t+t clustering in ^{6}He.

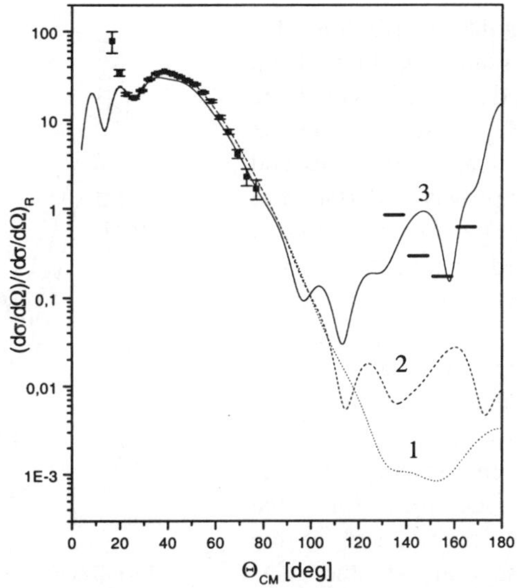

Fig.1. Experimental data for the elastic scattering of ^{8}He on ^{4}He. Symbols in the back hemisphere indicate upper limits only. Curves are for the calculations: curve 1 – OM, 2 – one-step transfer, 3 – two step transfer processes

For ^{8}He+p collision the 2n transfer leading to the ^{6}He ground and excited ^{6}He(2$^+$) states are shown in Fig.3. The one-step 2n and sequential 2-step (though the ground state of ^{7}He) transfers were calculated by the DWBA code [7]. To calculate SA for the relevant transitions by the method of Ref.[8] one should know the componetns of the wave functions (WF) in LS coupling for all states in question. Unlike ^{6}He(0$^+$) and ^{6}He(2$^+$) the wave functions for ^{8}He one is less known. In the case presented in Fig.3 SA were calculated assuming almost pure S (90%) wave function for the ^{8}Heg.s. The ratio of spectroscopic factors (SF) was 1.4 in favor of ^{6}He(2$^+$) channel. One can see in fig.3 that the calculation which reproduces the angular distribution of the one-step transition leading to ^{6}He(2$^+$) underestimates the 2n transition to ^{6}He(0$^+$) (dashed lines). It means that the ratio of SF is overestimated in present calculations. The inclusion ^{5}H cluster transfer which contributes to the ^{6}He(0$^+$) channel improves the agreement with the data at larger scattering angles only, see thick solid line in Fig.3. In order to investigate how much the two-step processes could change the picture, the sequential 2n transfer through the ^{7}He ground state was calculated. The resulting coherent sum of a one-step and two-step transfers is shown in Fig.3 with thin solid lines. One can see that the underestimation of ^{6}He(0$^+$) transition in respect to ^{6}He(2$^+$) one remains apparent.

It is generally accepted that unlike a single SF, which is highly model dependent, a ratio of two SF for transitions feeding the same residual nucleus is much more stable. We believe that the performed analysis of the data clearly recommend the relevant SF ratio to be rather in vicinity of 1.0.

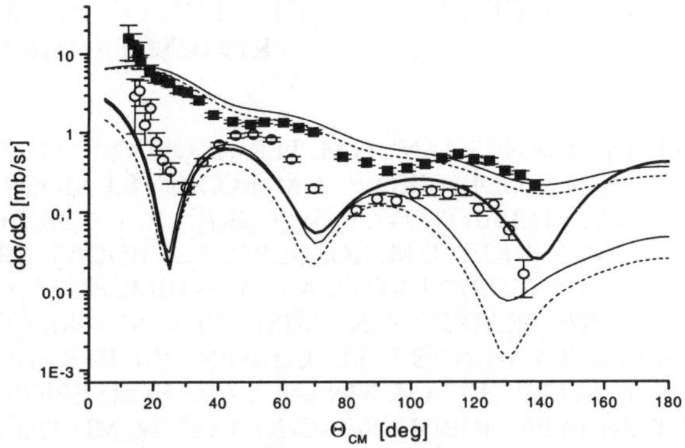

Fig.3. Experimental data (symbols) and DWBA calculations (lines) of differential cross-sections for 2n transfer reactions channels: $^{8}He+p \rightarrow {}^{6}He(0^{+})+t$ (circles), $^{8}He+p \rightarrow {}^{6}He(2^{+})+t$ (stars). The 1-step transfer calculations are shown by dashed lines, 2-step transfer, coherently added, by thin solid lines. The inclusion of 1-step transfer of ^{5}H cluster is shown by a thick solid line, see the text.

In this context it is worth to mention that for the pure LS coupling, what means S(100%), the shell model agebra of [8] yields the SF ratio to be quite close to 1.0, whereas for the pure jj coupling the ratio reaches value of 2.4.

Therefore the present discussion of experimental results suggests a week jj coupling of valence neutrons in ^{8}He. This important conclusion is at variance with the conjecture of $(p_{3/2})^4$ subshell closure for this nucleus.

The work was partially supported by the Russian Basic Research Foundation (grant No. 96-02- 17382) and INTAS grant No. 97-1668.

References

1. A.M. Rodin et al., *Nucl. Inst. Meth.* **B 126** (1997) 236.
2. R. Wolski et al., *Phys. Lett.* **B 467** (1999) 8.
3. G.M.Ter-Akopian et al., *Phys. Lett.* **B 426** (1998) 251.
4. Yu.Ts. Oganessian, V.I. Zagrebaev and J. Vaagen, *Phys. Rev. Lett.* **82** (1999) 4996, and *Phys. Rev.* **C 60** (1999) 044605.
5. M.V. Zhukov et al., *Phys. Rep.*, **v. 231** (1993) 151.
6. K. Rusek and K.W. Kemper, *Phys. Rev.* **C61** (2000) 044608
7. I.J. Thompson, *Com. Phys. Rep.* **7** (1988) 167
8. Yu.F. Smirnov and Yu.M. Thuvil'sky, *Phys. Rev.* **C15** (1976)
9. K. Schenk, at al., *Phys. Lett.* **B 52** (1974) 3

HYDROGEN-4 AND HYDROGEN-5 FROM TRANSFER REACTIONS INDUCED BY A 57.5-MEV TRITON BEAM ON DEUTERIUM AND TRITIUM TARGETS

YU.TS. OGANESSIAN[1], G.M. TER-AKOPIAN[1], D.D. BOGDANOV[1], A.S. FOMICHEV[1], M.S. GOLOVKOV[1], A.M. RODIN[1], S.I. SIDORCHUK[1], R.S. SLEPNEV[1], S.V. STEPANTSOV[1], R. WOLSKI[1,2], V.A. GORSHKOV[1], M.L. CHELNOKOV[1], M.G. ITKIS[1], E.M. KOZULIN[1], A.A. BOGATCHEV[1], N.A. KONDRATIEV[1], I.V. KORZYUKOV[1], A.A. YUKHIMCHUK[3], V.V. PEREVOZCHIKOV[3], YU.I. VINOGRADOV[3], S.K. GRISHECHKIN[3], A.M. DEMIN[3], S.V. ZLATOUSTOVSKY[3], A.V. KURYAKIN[3], S.V. FIL'CHAGIN[3], R.I. IL'KAEV[3], F. HANAPE[4], T. MATERNA[4], L. STUTTGE[5], A.H. NINANE[6], A.A. KORSHENINNIKOV[7], E.YU. NIKOLSKI[7], I.TANIHATA[7], P. ROUSSEL-CHOMAZ[8], W. MITTIG[8], N. ALAMANOS[9], V. LAPOUX[9], E.C. POLLACCO[9], and L. NALPAS[9]

[1] *Flerov Laboratory of Nuclear Reactions, JINR, Dubna, Russia,*
[2] *Institute of Nuclear Physics, Cracow, Poland,*
[3] *Russian Federal Nuclear Center -- All-Russian Scientific Research Institute of Experimental Physics, Sarov, 607190 Russia*
[4] *Universite Libre de Bruxelles, B-1050 Bruxelles, Belgium*
[5] *Institut de Recherches Subatomiques, F-67037 Strasbourg Cedex, France*
[6] *University of Louvain, Nuclear Physics Department Ch. du cyclotron, 2-1348 Louvain-la-Neuve, Belgium*
[7] *RIKEN, Hirosawa 2-1, Wako, Saitama 351-0198, Japan*
[8] *GANIL, BP 5027, F-14076 Caen Cedex 5, France*
[9] *DSM/DAPNIA/SPhN, CEA Saclay, F-91191, Gif/Yvette Cedex, France*

e-mail: wolski@lnr.jinr.ru

A state of ^{4}H with $E_{res}=3.22 \pm 0.15$ MeV and $\Gamma_{obs}=3.33 \pm 0.25$ MeV is obtained in t+d reaction. A valuable part of proton spectra observed in t+t reaction from ptn coincidence events is due to ^{5}H. The ^{5}H spectrum shows up a narrow maximum at about 2.5 MeV above the tnn decay threshold followed by a wide structure situated at 4-7 MeV.

Hydrogen-4 and hydrogen-5 are the lightest, so far little known nuclear objects where theory can try its growing ability to tackle the few-body problem. The most recent examples are papers [1,2] where the ^{5}H nucleus was investigated within a three-body model. Recent experimental work [3] dedicated to ^{4}H contains numerous references to earlier papers. Additional results about ^{4}H were reported in Ref. [4]. Papers [5,6] presenting the most recent experiment where ^{5}H was obtained give a complete list of references to the earlier attempts aimed at this nucleus.

To further investigate the resonance states of ^{4}H and ^{5}H we used transfer reactions induced by a 57.5-MeV triton beam on a liquid (solid) tritium (deuterium) target described in Ref. [7]. A 0.4-mm thick target cell had 10-μ thick stainless steel entrance and exit windows. For the sake of environmental safety the cell was embedded in a small protective volume also supplied with 10-μ stainless steel windows.

The modified achromatic beam line of separator ACCULINNA [8] was used to deliver a 58-MeV triton beam accelerated by the U-400M cyclotron to a shielded room where the reaction chamber housing the tritium target and charged particle detector telescopes was installed. This beam line was used also to cut the angular and energy divergence of the primary triton beam to FWHM values of 7 mrad and 0.3 MeV, respectively. Finally a triton beam with a typical intensity of 3×10^7 s^{-1} was focused in a halo-free 5-mm spot on the tritium target. The values of the beam energy, 57.5 MeV in the medium target plane, and the beam energy spread, 330 keV, were obtained in independent ways: from the separator tuning and by measuring tritons elastically scattered from tritium, hydrogen and iron.

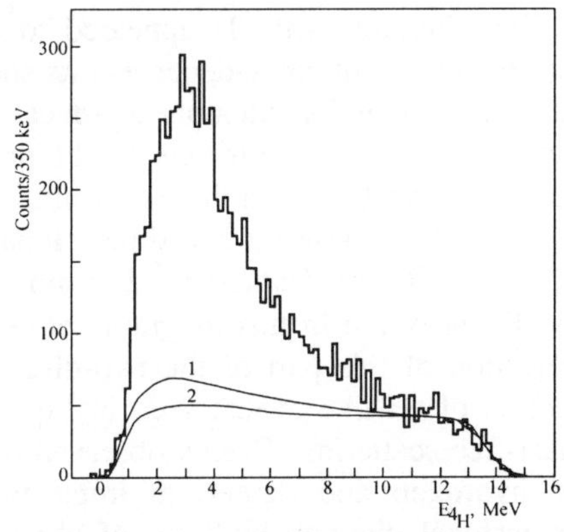

Figure 1. See text for explanations.

We used two Si detector telescopes for charged reaction products and 41 scintillation modules of the time-of-flight neutron spectrometer DEMON [9] for the neutron detection. The first telescope was set to detect protons emitted in an angular range of $18^\circ \le \theta_{lab} \le 32^\circ$, whereas the second one covered an angular range of $-15^\circ \ge \theta_{lab} \ge -39^\circ$ being optimized to to tritons originating from the ^{4}H (^{5}H) decay. The array of DEMON modules was installed just behind the triton telescope at a distance of 2.5 m from the target. The instrumental resolution for the searched ^{4}H and ^{5}H resonance states was about 450 keV.

As it is known from Ref. [3], the quasi-free scattering of projectile tritons on protons bound in deuteron nuclei makes a valuable contribution in the proton spectra obtained in the d(t,pt)n reaction. We reduced this contribution to a negligible level by selecting events with tritons emitted at $\theta_{lab} \ge 26^\circ$. Also, the triton selection at $\theta_{lab} \le 32^\circ$ reduced the effect of the n-p final-state interaction. The thus obtained energy spectrum of ^{4}H is shown in Fig. 1. This spectrum corrected for the energy dependent detection efficiency was fitted with a Breit-Wigner formula combined with simulations made for the phase space distribution and n-p final-state interaction (see curves 1 and 2 in Fig. 1). The best fit produced the following parameters for the resonance state of ^{4}H: the resonance position $E_{res}=3.22\pm0.15$ MeV and the resonance width $\Gamma_{obs}=3.33\pm0.25$ MeV. (An interaction radius of 4 fm was supposed to evaluate the penetrability factor.

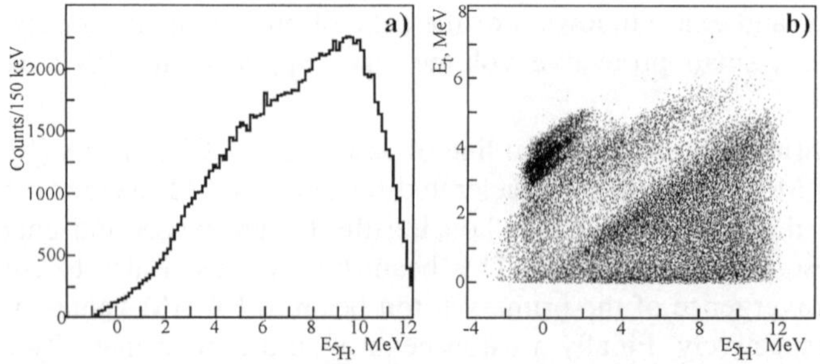

Figure 2. See text for explanations.

The situation with ^{5}H appeared to be more complex. The spectrum obtained with the use of pt coincidence events shown in Fig. 2a (the t(t,pt)nn case) does not leave any room for attempts to reveal a resonance-like structure. The quasi-free scattering of projectile tritons on the protons bound in tritium target nuclei makes the main contribution in this spectrum. One can see the effect of this mechanism in Fig. 2b, which shows a two-dimensional diagram in coordinates: the triton energy (E_t) in the ^{5}H rest frame *vs.* $^5\text{H} \rightarrow$ tnn decay energy (E_{5H}). Events associated with the ^{5}H decay are in this diagram below $E_t=(2/5)E_{5H}$ line. Shown in Fig. 2a is the projection of this part of the two-dimensional spectrum on the E_{5H} axis. The big fraction of ^{5}H-like events seen in Fig. 2b at about $E_t=(2/5)E_{5H}$ line is due to the quasi-free scattering. Events observed outside of the ^{5}H locus are produced due to the hydrogen and deuterium impurities of the target tritium. A similar two-dimensional diagram built for of ptn coincidence events (the t(t,ptn)n case) was entirely free of the quasi-free scattering. Also, events caused by hydrogen and deuterium presenting in the target were totally eliminated from the spectrum. The ^{5}H-spectrum obtained from the t(t,ptn)n data is shown as a histogram in Fig. 3a. Curves 1-4 in Fig. 3a show, respectively, simulation made for the phase space distribution, and distributions associated with different combinations of final state interactions: n-n interaction alone, n-t interaction alone (the ^{4}H resonance), and n-t and p-n interactions together. Neither of the curves or their combinations can be responsible for the relatively narrow maximum seen at about 2.5 MeV in Fig. 3a. This is also valid for the broad structure that shows up between E_{5H}=4 and 7 MeV. Assuming the incoherent summation, the "worst" case for the survival of these structures would be the subtraction of the sum of curves 2 and 3 from the spectrum. The rest of the spectrum obtained after this subtraction and corrected for our detection efficiency is shown in Fig. 3b. The obtained result shows, without doubts, a relatively narrow maximum, which evidently is the manifestation of a ^{5}H-resonance state. The structure appearing to the right of this maximum is associated, to all appearance, also with this nucleus.

In summary, we introduced the liquid/solid tritium target into the practice of experiments made with radioactive beams and investigated transfer reactions d(t,pt)n, t(t,pt)nn and t(t,ptn)n at a triton beam energy of 57.5 MeV. Parameters of

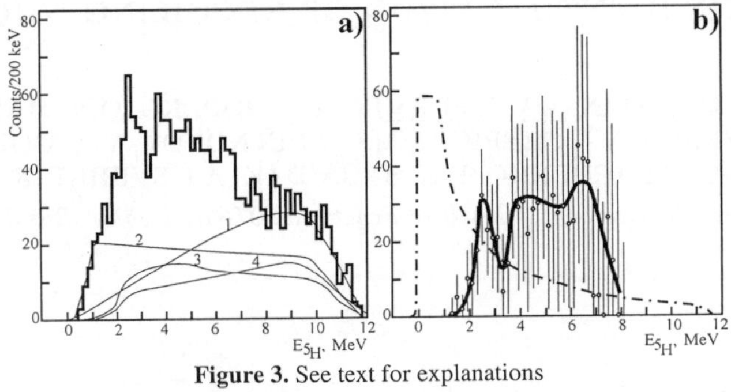

Figure 3. See text for explanations

the ^{4}H resonance state obtained in this work (E_{res}=3.22 ± 0.15 MeV, Γ_{obs}=3.33 ± 0.25 MeV) are close to those reported in a number of earlier papers (see quotation in Ref. [3]). Having smaller error bars, our data are in accord with the results of Ref. [3]. Evidently, after completing the analysis of other type data obtained for d(t,pn)t, t(t,dt)n and t(t,dn)t events we will be able to further reduce our errors. We offered convincing proofs that the ^{5}H system shows up in the proton spectra collected from ptn coincidence events detected in the t+ t reaction. The spectrum shown in Fig. 3b appears to be surprisingly close to the one predicted for ^{5}H in theory paper [1]. Notable is the fact that the position of the maximum seen in this spectrum is shifted to a higher energy as compared to the ^{5}H resonance reported in Refs. [5,6]. It might be that this shift, as well as the small width of the maximum obtained in the present work, are due to distortion caused by interference effects. This point must be the analyzed in a more detail. Also, the analysis of data on the t(t,pnn)t events will be quite useful for achieving a further progress in the ^{5}H nucleus study.

The partial support of this work by RFBR grant No. 99-02-17968 is acknowledged.

References

1. N.B. Shul'gina et al., *Phys. ReV. C* **62** 014312 (2000).
2. P. Descouvemont and A. Kharbach, *Plys. Rev. C* **63** 027001 (2001).
3. S. Blagus et al., *Phys. Rev. C* **44** 325-328 (1991).
4. D.V. Aleksandrov et al., *JETP Lett.* **62** 18-20 (1995).
5. M.S. Golovkov et al., *Physics of Atomic Nuclei*, **64** 1244-1248 (2001).
6. A.A. Korsheninnikov et al., will be published in *Phys. Rev. Lett.*
7. A.A. Yukhimchuk, et al., JINR Report P13-2001-25, Dubna, 2001; will be submitted to *Nucl. Inst. Meth.*
8. A.M. Rodin et al., *Nucl. Inst. Meth.* **B126** 236-241 (1997).
9. I. Tilquin et al., *Nucl. Inst. Meth.* **A365** 446-461 (1995).

KINEMATIC SEPARATOR AND MAGNETIC ANALYZER FOR THE IDENTIFICATION OF RECOILING NUCLEI.

YU. TS. OGANESSIAN, A.V. YEREMIN, A.G. POPEKO, O.N. MALYSHEV, A.V. BELOZEROV, V.I. CHEPIGIN, M.L. CHELNOKOV, V.A. GORSHKOV, A.P. KABACHENKO, R.N. SAGAIDAK, A.I. SVIRIKHIN.

Flerov Laboratory of Nuclear Reactions, JINR, Dubna, Russia

Introduction

A method for the investigation of consistent α decays, the so called α–α correlation analysis, has long been employed for the identification of new radionuclides. It has already been used in the works to discover and study the decay properties of elements from 102 (α–recoil milking experiments [1]) to 105 (delayed α correlation method [2,3]). Later this method was developed and successfully used for the identification and study of decay properties of elements 107 – 112 with a modern experimental set–up and detection module (position sensitive detectors array) [4]. This method is based on the fact that a decay chain starting from unknown isotope should be ended in the known region of isotopes with known decay properties. In the case if the statistics allow one to be sure that no members in the decay chains were missed, it is possible, starting from the known nuclei to go back to the beginning of the chain and to make an assignment, what definite isotope of what element was synthesized.

In the case when the neutron rich isotope ^{48}Ca is used as a bombarding beam we have found ourselves in a completely unknown region where all decay chains are started and finished with isotopes having unknown decay properties. According to the calculations [5,6] the decay chains started from neutron rich (N=171–175) isotopes of elements 112 – 114 after a few α decays should be terminated by spontaneous fission in the region of elements 104 – 110. As a first step in attempt to solve the problem with the newly synthesized isotope identification we plan a number of experiments, each experiment being a basis for the following one. One of the possibilities was to use targets which differ by an α particle. It means that the isotopes which can be produced in a subsequent experiment should decay to the isotopes which were obtained in the previous experiments. These two–step experiments could be the following ones: $^{48}Ca + {}^{232}\text{Th} \rightarrow {}^{280}110^*$ and $^{48}\text{Ca} + {}^{236}\text{U} \rightarrow {}^{284}112^*$, $^{48}\text{Ca} + {}^{238}\text{U} \rightarrow$

$^{286}112^*$ and ^{48}Ca + ^{242}Pu → $^{290}114^*$, ^{48}Ca + ^{244}Pu → $^{292}114^*$ and ^{48}Ca + ^{248}Cm → $^{296}116^*$.

In the first part of the experimental program the reactions ^{48}Ca + ^{206,208}Pb → $^{254,256-x}$No + xn were investigated. Detailed analysis of the results of these experiments will be published elsewhere, see e.g. [7]. Using the HIVAP code [8], the excitation energies for the maxima of the 3n and 4n channels for the reactions with ^{232}Th, ^{238}U and ^{242}Pu targets, respectively, were calculated. The corresponding beam energies for the reactions in the middle of the target thickness were 228±2 and 236±2 MeV for the ^{232}Th target, 231±2 and 238±2 MeV for the ^{238}U target, 234±2 and 245±2 MeV for the ^{242}Pu target [9].

Experiments, performed at the separator VASSILISSA using ^{48}Ca beams, and obtained results are listed in Table 1.

Table 1. Summary of experimental conditions for the irradiations with ^{48}Ca.

Reaction studied	Energy MeV	Beam time	Beam dose	Nuclide detected	Cross section
^{48}Ca+^{238}U	231	25	3.5x10^{18}	^{283}112	5^{+6}_{-3} pb
^{48}Ca+^{238}U	238	14	2.2x10^{18}	no	lim. 7 pb
^{48}Ca+^{232}Th	238	15	1.8x10^{18}	no	lim. 9 pb
^{48}Ca+^{232}Th	228	28	4.6x10^{18}	^{277}110	$2^{+4.6}_{-1.7}$ pb
^{48}Ca+^{242}Pu	235	32	7.5x10^{18}	^{287}114	$2.5^{+3.3}_{-1.6}$ pb

Experimental set up

The electrostatic separator VASSILISSA [10] with an upgraded detector system [11] was used in the experiments on the synthesis of the isotopes ^{283}112 and ^{287}114 using ^{238}U and ^{242}Pu targets [12,13].

An additional (in some cases perhaps the main) possibility for the distinguishing of the isotopes produced in complete fusion reactions from multinucleon transfer reactions products and identification of new nuclides is the method of measuring the atomic mass number of the ERs, synthesized during the experiment. In the case if the mass resolution of the experimental set up reaches the value of less than 0.5 % (for the heavy nuclei with masses in the region of 270 - 290 amu) one could make a direct identification of the obtained isotope on the basis of its mass measurement. But such a mass resolution needs rather big magnetic systems having deflection angles of $\geq 90^o$. Another possibility is the use of more simple and compact systems which allow one to have the mass resolution at the level of 1.5 - 3 %. For the mass region 270 - 290 amu it leads to the accuracy of 3 - 6 amu. In this case one can establish belonging of newly synthesized nuclide to the region of superheavy nuclei formed from compound nuclei as a result of complete fusion reaction between the heavy ion and the target nucleus.

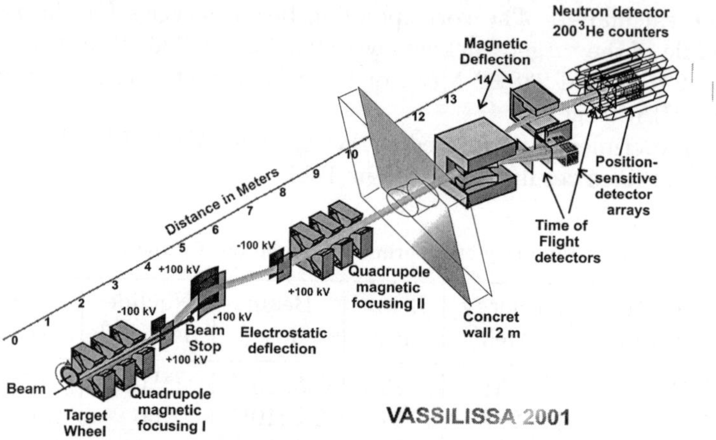

Figure 1. A schematic view of the upgraded separator VASSILISSA.

With the aim of continuation of the experiments on the synthesis and study of decay properties of superheavy nuclei the separator VASSILISSA was upgraded. For that purpose a new dipole magnet, having a deflection angle of 37.5 degrees, was installed behind the separator VASSILISSA replacing the old 8° magnet (see Fig. 1.). The new magnet will provide an additional suppression of unwanted reaction products by a factor of about 100 and a possibility to have the mass resolution at the level of 1.5 -2 % for heavy nuclei with masses of about 300 amu. Its bending radius is 1.08 m, the effective length is 69.8 cm and the vertical gap is 114 mm. The entrance and exit boundaries of the magnetic dipole have the values of 46° and -43°, respectively. To reduce chromatic abberations the entrance and exit boundaries of the pole have a radius curvature of +0.5 m and -0.5 m.

For the detection of heavy ER's at the focal plane of the new dipole magnet, a new detector system having a 32 strip detector assembly 60×120 mm^2 in size and surrounded by backward detectors was also developed (see Fig. 2.). It is planned to retain the old detector system [11] including the 8°

dipole magnet behind new the 37° magnet, thus organizing a second detector area. Due to the compact configuration of this detector system it is possible to surround it by ^{3}He counters, thus creating the neutron detector with a high (about 50 %) detection efficiency. For this purpose it will be necessary to build around the detector chamber an array with 200 ^{3}He counters (see Fig. 1.).

Figure 2. The new focal plane detector system.

Calculations show that the separator has a focal plane inclination of $\approx 24^o$, the mass (energy) dispersion D_x=3.9mm/% and the horizontal magnification M_x=1. Measurements showed good stability ($\sim$0.1%) of the effective length of the dipole up to the magnitude of the magnetic field H = 1.15 T. The results of the tests with the ^{226}Ra α source showed that the energy dispersion of the new magnet was about 3.8 mm/% and the resolution was about 2 %. (see Fig. 3.).

Experiment

In March 2001 the VASSILISSA separator with the new dipole magnet was tested. The analysis the evaporation residues produced in complete fusion reactions with ^{40}Ar bombarding ions were used in the analysis. The ^{40}Ar beam was delivered from the U-400 cyclotron. The energy range of the accelerated ions varied from 190 to 210 MeV, the test ^{164}Dy and ^{208}Pb targets were used.

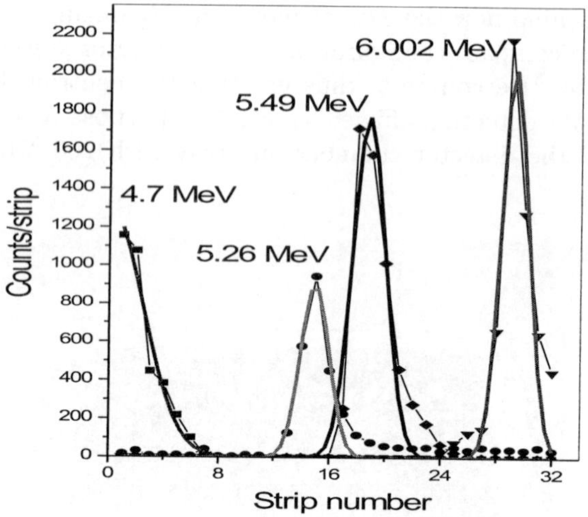

Figure 3. The distribution of the ^{226}Ra α source α peaks with different energies at the focal plane of the separator. Obtained energy resolution is about 2 %, dispersion - 3.8 mm/%

Using a specific ion optical regime of the separator it was possible to see at the focal plane of the separator the charge distribution of scattered ^{40}Ar ions (see Fig. 4). These data were used for the calibration of TOF - Energy spectra together with the relation between the strip number (deflection angle) and Bρ (magnetic rigidity) of ER's.

In Fig. 5. the experimental results obtained with the use of the ^{164}Dy target are presented. The isotope ^{197}Po was chosen and its distribution was plotted along the strip number of the focal plane detector, i.e. the deflection angle in the dipole magnet. The obtained structure in the ^{197}Po distribution allowed us to extract different charge states of ER's implanted into the focal plane detectors.

The time resolution of time–of–flight system of the separator was about 0.7 ns, which corresponds to the value of about 1 % (for slow evaporation residues). For each ER, passing through the separator and detected at the focal plane, the time–of–flight (speed "v" of ER) and position of implantation / strip number (magnetic rigidity "Bρ" of ER) were measured. Using the high resolution ion optical mode of the separator it became possible to separate different charge states for the Po ER's. At the two dimensional spectrum

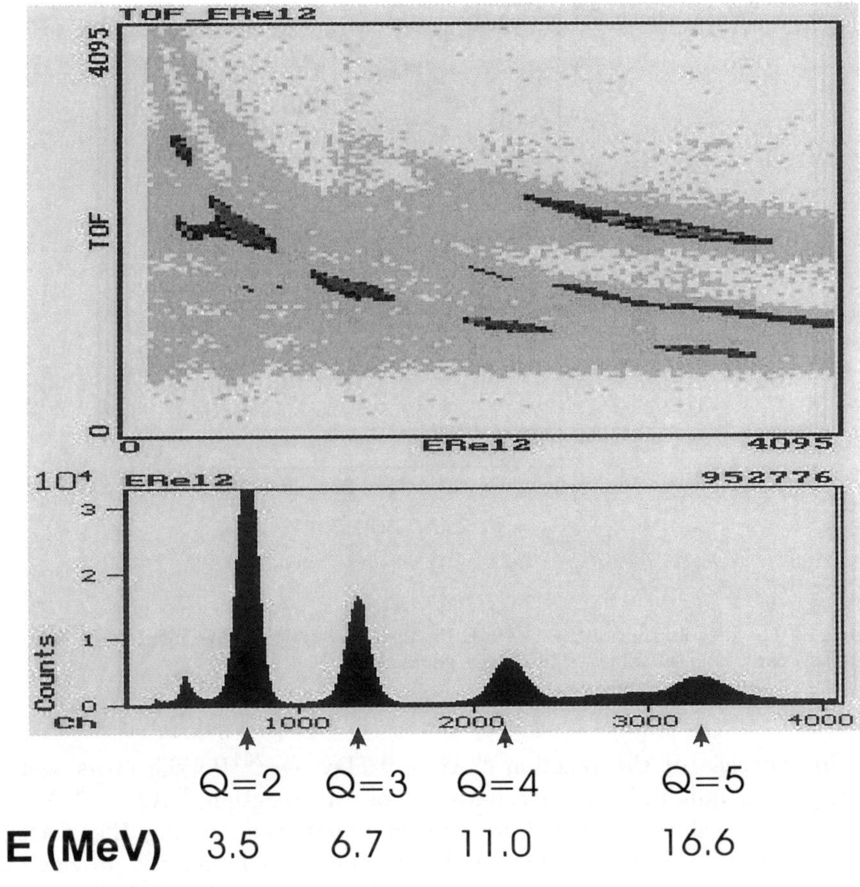

Figure 4. The TOF - Energy spectra, registered at the focal plane of the separator. The charge distribution of scattered ions is clearly seen.

Energy - Strip three ER's peaks corresponding to different charge states were clearly seen.

It was possible to calculate the mass of the ER's, detected at the focal plane. For different charge states (Q = 17, 18, 19) the result was A = 198.1 ± 1, 197.8 ± 1, 198.4 ± 1, respectively. Taking into account that the calculations were made for the isotope ^{197}Po, formed in the complete fusion reaction ^{164}Dy(^{40}Ar,7n)^{197}Po, the obtained mass resolution could be estimated as about 2 %.

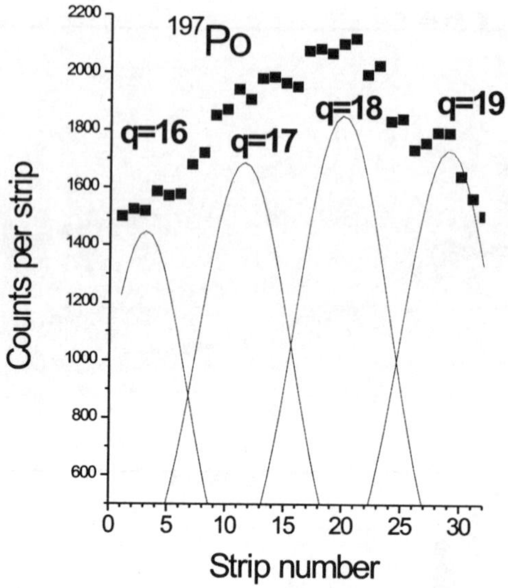

Figure 5. The distribution of the ^{197}Po ER's along the strips of the focal plane detector. Squares - experimental values, solid lines - gaussian fit.

In the case of the reaction ^{40}Ar + ^{164}Dy $\rightarrow$ ^{204}Po the cross section values reach hundreds of microbarns. For the reaction ^{40}Ar + ^{208}Pb $\rightarrow$ ^{248}Fm these values are equal to tens of nanobarns. Using the reaction ^{208}Pb(^{40}Ar,2n)^{246}Fm we estimated that the new dipole magnet provided an additional suppression factor of about 10 for scattered ions. Very clean α spectra were obtained. From the comparison with literature data [14] we could estimate that transmission efficiency for ^{246}Fm ER's was about 20 %. The obtained mass resolution for ^{246}Fm ER's, calculated according to the measured TOF and strip number, ranged from 244.5 to 247.2 that corresponded to the accuracy of about 2 %.

Discussion

The development of the experimental technique with the use of electrostatic deflectors leads to the creation of the new class of kinematic separators – mass spectrometers of recoiling nuclei (Recoil Mass Spectrometers). In these experimental set–ups the dipole magnet with a deflection angle of 25 – 30 degrees is placed between the electrostatic condensers (deflectors), i.e. the scheme E–D–E is used (E – electrostatic deflector, D - dipole magnet). It allows one to reach a rather good mass resolution $\Delta M/M \approx 1/300$ for the heavy products of nuclear reactions together with rather high suppression factors for the background products (better than 10^7 for scattered beam ions). One of the first experimental set–up of this type was a recoil mass spectrometer created at Rochester [15].

Later a number of experimental set–ups having the familiar scheme were created at different nuclear centers such as Legnaro (Italy)[16], Osaka (Japan)[17], Oak–Ridge (USA)[18], Argonne (USA)[19], Tokai (Japan)[20] and New-Dehli (India)[21]. All these set–ups have the same scheme E–D–E and differ only in the number of focusing quadrupole lenses in front and behind the spectrometer itself. All these set–ups can provide the mass resolution $\Delta M/M \approx 1/300$ for nuclear reaction products in the mass region around 200. A typical range of the formation cross section varied from a few millibarns to a few nanobarns and depended strongly on the value of suppression factors for the background products. Recently only two spectrometers have been used for the study of formation cross sections and decay properties of reaction products with heavy ions in the mass region of about and heavier than 200 amu. FMA in Argonne [19] was used for investigating proton radioactivity of neutron deficient isotopes of Pb and Bi [22], and JAERI–RMS in Tokai [20] for investigating decay properties of Sg isotopes in the reaction $^{30}Si + {}^{238}U$ [23]. But in the latter case the set–up was used not in the mass resolution mode.

Recently the first test experiments have been performed with the new 37° dipole magnet installed behind the recoil separator VASSILISSA. Obtained results are very promising, it is possible now to define masses of the synthesized ER's with an accuracy of 5 - 6 mass units. This provides an additional reliability of the identification in the experiments aimed at the synthesis of superheavy nuclei in complete fusion reactions between transactinide targets and ^{48}Ca accelerated beams.

With the use of the upgraded separator VASSILISSA we plan to continue the experiments aimed at the synthesis of superheavy nuclei in the vicinity of predicted spherical shells in complete fusion reactions between $^{34,36}S$, ^{48}Ca ions and ^{232}Th, $^{236,238}U$ and $^{24,244}Pu$ targets. Odd–Z isotopes which could be obtained in reactions with ^{237}Np and ^{243}Am targets may have even longer half–lives than those of even–Z elements 112 and 114. After the upgrade of the separator the search for long correlations (up to few hours) become possible.

As a second step, in order to increase the identification ability, a possibility of installing additional quadrupole lenses in front of and behind the 37 degree dipole magnet has been analyzed. Ion optical calculations show that in this case it could be possible to obtain the mass resolution for heavy nuclei at the level of 1% or even better.

Acknowledgments

This work was performed partially under the financial support of the Russian Foundation for Basic Research, contract N 99–02–16447 and INTAS, contract N 99–1344.

References

1. B.A. Zager et. al., Atom. Energia **20**, (1966) 230.
2. A. Ghiorso et. al., Phys. Rev. Lett. **22**, (1969) 1317.
3. V.A. Druin et. al., Yadernaya Fizika **13**, (1971) 251.
4. S. Hofmann, Rep. Prog. Phys. **61**, (1998) 639.
5. R. Smolańczuk, Phys. Rev. C **56**, (1997) 812.
6. P. Möller et. al., Atomic Data and Nuclear Data Tables **59**, (1995) 185.
7. A.V. Yeremin et. al., JINR Rapid Oommunications N6[92]-98, (1998) 21.
8. W. Reisdorf, Z. Phys. A **300**, (1981) 227.
9. Yu.Ts. Oganessian et. al., in: *Proc. of the Int. Conf. on Nuclear Physics "Shells 50"*, Dubna, Russia, 21–24 April, 1999, p. 167.
10. A.V. Yeremin et. al., Nucl. Instr. and Meth. B **126**, (1997) 329.
11. A.V. Yeremin et. al., Nucl. Instr. and Meth. A **440**, (2000) 86.
12. Yu.Ts. Oganessian et. al., Eur. Phys. J. A **5**, (1999) 63.
13. Yu.Ts. Oganessian et. al., Nature **400**, (1999) 242.
14. G. Münzenberg et. al., Z. Phys. A **302**, (1981) 7.
15. T.M. Cormier., Ann. Rev. Nucl. Part. Sci., **37**, (1987), 537.
16. P. Spolaore et. al., Nucl. Instr. and Meth. A **238**, (1985), 381.
17. S. Morinobu et. al., Nucl. Instr. and Meth. B **70**, (1992), 331.
18. J.D. Cole et. al., Nucl. Instr. and Meth. B **70**, (1992), 343.
19. C.N. Davids et. al., Nucl. Instr. and Meth. B **70**, (1992), 358.
20. H. Ikezoe et. al., Nucl. Instr. and Meth. A **376**, (1996), 420.
21. A.K. Sinha et. al., Nucl. Instr. and Meth. A **339**, (1994), 543.
22. G.L. Poli et. al., Phys. Rev. C **63**, 044304.
23. H. Ikezoe et. al., Eur. Phys. J. A **2**, (1998), 379.

DECAY TIME CHARACTERISTICS OF THE HEAVY EXCITED NUCLEI

D.O. EREMENKO, O.V. FOTINA, S.YU. PLATONOV AND O.A. YUMINOV

D.V. Skobeltsyn Institute of Nuclear Physics, M.V. Lomonosov Moscow State University, 119899 Moscow, Russia
E-mail:yuminov@p5-lnr.npi.msu.su

G. GIARDINA AND A. LAMBERTO

Istituto Nazionale di Fisica Nucleare,Sezione di Catania, and Dipartimento di Fisica dell' Universita', Messina, Italy
E-mail:giardina@nucleo.unime.it

F. MALAGUTI

Istituto Nazionale di Fisica Nucleare,Sezione di Bologna, and Dipartimento di Fisica della Università, Bologna, Italy
E-mail:malaguti@bologna.infn.it

The crystal blocking technique has been used to measure the fission decay times in the ^{nat}Pt + ^{28}Si(140-170 MeV), ^{232}Th + p(6-8 MeV), ^{232}Th + ^{3}He(20–24 MeV) and ^{232}Th + α(20-31 MeV) reactions. The experimental data are analyzed in the frames of the statistical theory of nuclear reactions with allowance for the lifetimes of the excited states of the first and second potential wells of the double-humped fission barrier and taking into account the phenomenon of shell effect damping with increase of nuclear temperature.

As it was shown in [1] the double-humped structure of fission barrier can have a strong influence on the time characteristics of nuclear decay. The fission decay time is due by the lifetimes of excited states in first and second potential wells. The measurements of the induced fission time allow us to obtain information on nuclear characteristics in the strongly deformed states. Moreover, the experimental investigation about *the energy dependence of the shell effects and fission barrier structure* is possible measuring the energy dependence of the fission decay time.

In this paper, the crystal blocking technique was used to measure fission decay times in the ^{28}Si(140-170 MeV)+^{nat}Pt, p(6-8 MeV)+^{232}Th, ^{3}He(20-24 MeV)+ ^{232}Th, α(20-31 MeV)+ ^{232}Th reactions. Experimental data have been compared with statistical calculations that take into account the lifetimes of excited states in both potential wells. The aim was to find evidence of additional time delays in the fission channel, induced by the lifetime of the excited states in the second well, and *to check whether the fission barrier keeps or loses its double-humped structure at the excitation energies considered here*.

Measurements on the silicon beam were done with Tandem XTU accelerator at LNL Laboratories (Italy) using thin ^{nat}Pt single crystals and a detector telescope made of a multiwire gas proportional chamber followed by an axial Bragg chamber for registration of the reaction products. Experiments on the light particle beams were performed with the cyclotron U-120 of Moscow State University (Russia) using a thick $^{232}ThO_2$ single crystals as a target. Angular distributions of fission fragments were measured by means of glass track detectors placed in the direction of the <111> crystallographic axis, which forms an angle of 160^0 with respect to the momentum vector of the bombarding particles. Other experimental details are described in [2]. The obtained fission decay times versus initial excitation energies of the uranium-like nuclei presented in Fig. 1.

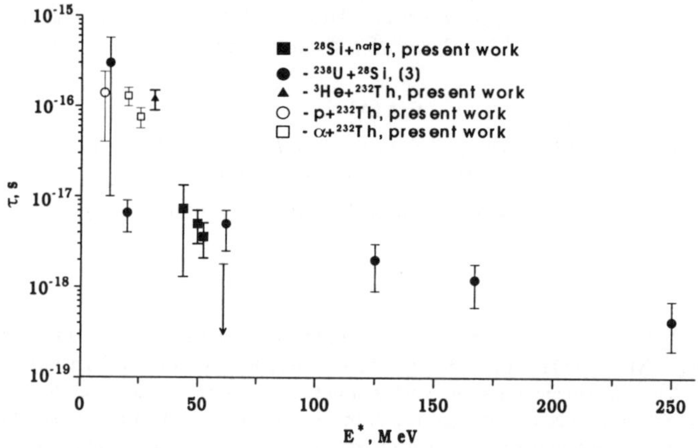

Figure 1. The fission decay times of uranium-like nuclei vs. initial excitation energy.

Our experimental decay times range from 10^{-15} s to 10^{-18} s and are in agreement with the data of [3], only the values obtained in the $^{232}Th + ^{3}He$ reaction are in excess. Probably this excess is connected with the influence of the cold nuclei produced in the direct channels – (^{3}He, df), (^{3}He, αf) - of the reaction under study. The fission decay times in the $^{28}Si + ^{nat}Pt$ reaction (see Fig. 2) are much longer than what expected from standard statistical calculations for the lifetimes of initial compound nuclei produced in the reaction under study. As one can see, a simple calculation in the framework of the liquid drop model, and taking into account only lifetimes of excited states under equilibrium deformation, underestimates experimental data by approximately three orders of magnitude. Introducing the double-humped fission barrier model with allowance for the lifetimes of excited states in the second and first wells [1] essentially improves the fit to experimental data. The calculations neglecting the damping of shell effects with increasing of nuclear temperature overestimate the experimental values.

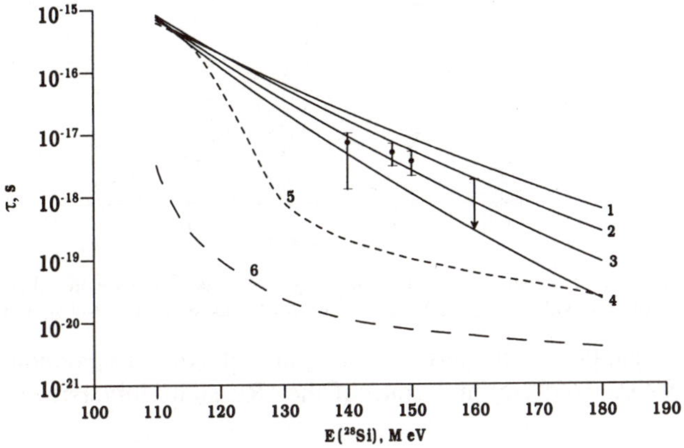

Figure 2. The fission decay times in the ^{28}Si + ^{196}Pt reaction at different beam energies. Dots are experimental data. Dashed lines do not contain effects of neutron evaporation cascade. In detail, 5 and 6 (short and long dashed curves) allow for (5) and not (6) delay times introduced by the excited states in the second well. The solid curves are the results of calculations allowing for the delay due to excited states in the first and second well and also for *all possible fission chances*. The solid curves were calculated taking into account *the temperature dependence of shell corrections* with different values of the temperature parameter T_0: 8 (1); 2.00 MeV (2); 1.85 MeV (3); and 1.75 MeV (4), respectively.

To take into account the decreasing shell correction part (δW) of the fission barrier parameters with increasing nuclear temperature we use $\delta W(T) = \delta W(T=0)F(T)$, where F(T) is the universal damping function of Fermi type: $F(T)=(1 + \exp\{(T-T_0)/d\})^{-1}$, where d = 0.2 MeV is the rate of disappearance of the shell corrections with temperature, and T_0 is an adjustable parameter. The best fit to experimental data was achieved using T_0 = 1.75 MeV.

An important property of *the shell structure* is that its influence on nuclear processes diminishes in highly excited nuclei and finally disappears at a certain temperature which was found to be $t^* = h\Omega/2\pi = 1.5\ 2$ MeV, where $h\Omega$ is the intershell energy spacing [4]. Theoretical considerations by A. Bohr and Mottelson [5] also predict that shell effects should disappear at similar temperatures. In addition, our results are very similar to the data of [6], where a semi-empirical analysis was done of temperature and spin dependence of shell corrections. However our data contradict some previous results (see, for example, [7]), that suggested a much faster decrease and from a smaller starting energy. For more details see Fig3. In conclusion, we have shown that fission barriers keep their double-humped structure in the excitation energy range considered here, and that the lifetimes of transitional excited states in the second potential well essentially contribute to the total observed decay time in the fission channel. This means that *shell corrections are important up to temperature 1.7 - 1.8 MeV.*

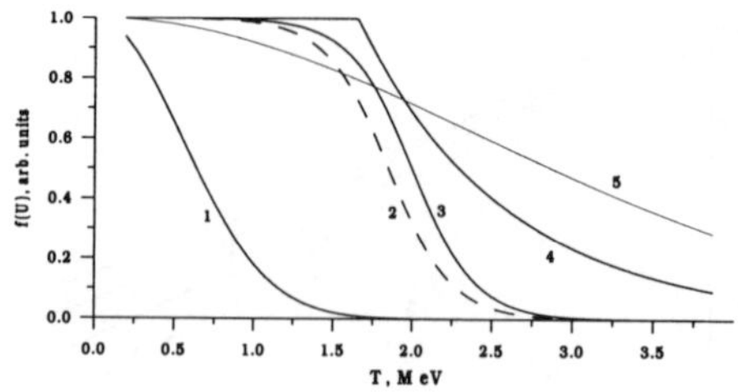

Figure 3. Different types of the shell correction damping functions. Curves were taken from: 1 – [7]; 2 and 3 – Fermi function with value T_0 equal 1.75 and 2.00 MeV, respectively; 4 – [6]; and 5 – [5].

This work was supported in part by the grant of Russian Foundation of Basic Research (N° 99-02-16992-a) and grant of the "Russian Universities" Program (N° 02-02-11).

References

1. Yuminov O. A., Platonov S. Yu., Fotina O. V., Eremenko D. O., Malaguti F., Giardina G. and Lamberto A., Statistical description of the decay of nuclei having two classes of excited states. *J. Phys. G* **21** (1995) pp. 1243-1254.
2. Yuminov O. A., Platonov S. Yu., Eremenko D. O., Fotina O. V., Fuschini E., Malaguti F., Giardina G., Ruggeri R., Sturiale R., Moroni A., Fioretto E., Ricci R.A., Vannucci L. and Vannini G., Investigation of shell effects for heavy fissionable nuclei by the blocking technique. *Nucl. Instr. and Meth. B* **164-165** (2000) pp. 973-977.
3. Morjean M., Chevallier M., Cohen C., Dauvergne D., Dural J., Galin J., Goldenbaum F., Jacquet D., Kirsch R., Lienard E., Lott B., Péghaire A., Périer Y., Poizat J.C., Prevot G., Rémillieux J., Schmaus D. and Toulemonde M., A direct determination of fission lifetimes as a function of excitation energy by the blocking technique. In *VI School-Seminar On Heavy Ion Physics*, ed. by Oganessian Yu. Ts. and Kalpakchieva R. (Dubna, Russia, 1997) pp. 683-687.
4. Strutinsky V. M., Macroscopic and microscopic aspects in nuclear fission. *Nucl. Phys. A* **502** (1989) pp. 67c-84c.
5. Bohr A. and Mottelson B. R. Nuclear Structure, vols I and II (Benjamin Press, New York, 1975).
6. D'Arrigo A., Giardina G., Herman M., Ignatyuk A. V. and Taccone A., Semi-empirical determination of the shell correction temperature and spin dependence by means of nuclear fission. *J. Phys. G* ***20*** (1994) pp. 365-376.
7. Ignatyuk A. V., Smirenkin G. N. and Tishin A.S., Phenomenological description of the energy dependence of level density parameter. *Yad. Fiz.* **21** (1975) pp. 485-490 (*Sov. J. Nucl. Phys.* **21** (1975) pp. 255-260).

NUCLEON COLLECTIVIZATION AS THE MAIN MECHANISM OF THE FUSION - FISSION DYNAMICS

V.I. ZAGREBAEV
Flerov Laboratory of Nuclear Reaction, JINR, Dubna, Moscow Region, Russia

A whole process of super-heavy nucleus formation is divided usually into three reaction stages even if connected with each other but treated and calculated separately: (1) overcoming the Coulomb barrier and approaching the point of contact $R_{cont} = R_1 + R_2$, (2) formation of the compound mono-nucleus, (3) decay ("cooling") of the compound nucleus. Different theoretical approaches are used for analyzing all the three reaction stages. However, the dynamics of the intermediate stage of the compound nucleus formation is the most vague. It is due the fact that in the fusion of light and medium nuclei, in which the fissility of the compound nucleus is not very high, the colliding nuclei having overcome the Coulomb barrier form a compound nucleus with a probability $P_{CN} \approx 1$. Thus, this reaction stage does not influence the yield of ER at all. However, in the fusion of heavy nuclei it is the fission channels (normal and quasi-fission) that substantially determine the dynamics of the whole process; the P_{CN} value can be much smaller than unit, while its accurate calculation is very difficult. Moreover, today there are no consensus for the mechanism of the compound nucleus formation itself, and quite different, sometimes opposite in their physics sense, models are used for its description.

The following quite natural mechanism can be proposed for the compound nucleus formation in competition with the quasi-fission process [1].

(1) Down to the instant of touch the nuclei keep their individualities and the potential energy of their interaction is defined in a usual manner. The point of contact R_{cont} can be defined as the sum of nuclear radii which is smaller by $1 \div 3$ fm than the radius of the Coulomb barrier and, thus, the nuclei have to overcome this barrier to reach it.

(2) In the point of contact the nuclei begin to lose their individualities due to an increasing number of common nucleons ΔA, here $A_1 + A_2 + \Delta A = A_{CN}$ (configuration (b) in Fig. 1). Interaction of two touching nuclei A_1 and A_2 weakens with increasing the number of common nucleons ΔA, and their specific binding energies approach a specific binding energy of the compound nucleus. Collectivized nucleons move in the whole volume occupied by the two nuclei and have the average over A_1 and A_2 specific binding energy.

(3) Thus, the process of compound nucleus formation in competition with quasi-fission occurs in the space (A_1, A_2), here the compound nucleus is finally formed when two fragments A_1 and A_2 go in its volume, i.e., at $R(A_1) +$

$R(A_2) = R_{CN}$ or at $A_1^{1/3} + A_2^{1/3} = A_{CN}^{1/3}$ (configuration (c) in Fig. 1).

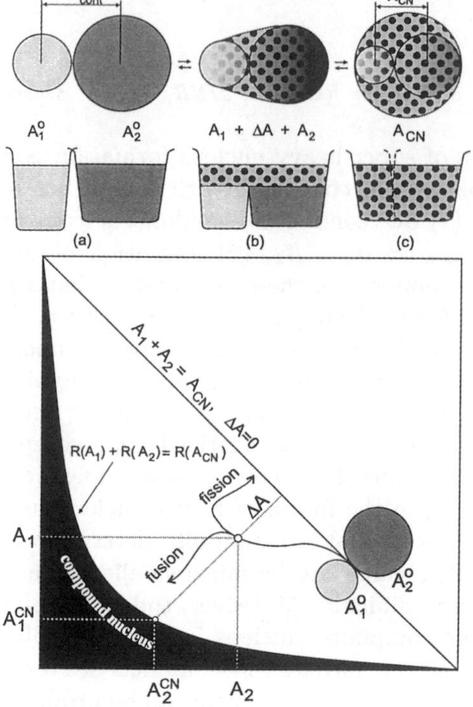

Figure 1: Schematic view of the process of compound nucleus formation and fission in the space of A_1, A_2 and ΔA, i.e., the number of nucleons in the projectile-like nucleus, target-like nucleus, and collectivized nucleons, here $A_1 + A_2 + \Delta A = A_{CN}$.

The topographical landscape of the driving potential $V(Z_1, Z_2; \beta_2 = 0)$ for the case of formation of the compound nucleus $^{296}116$ is shown in Fig. 2 (details of its calculation can be found in[1]). One can see that the shell structure, clearly revealing itself in the contact of two nuclei, i.e. at the borderline $A_1 + A_2 = A_{CN}$ (Fig. 2a), is also retained at $\Delta A \neq 0$ (see, e.g., the deep minima in the regions of $Z_{1,2} = 50$ and $Z_{1,2} = 82$ in Fig. 2b). As the calculations show, in the synthesis of the nucleus $^{296}116$ in the reaction ^{48}Ca+^{248}Cm, on the way from the contact point to the compound nucleus formation the system decays with a large probability into the quasi-fission channels (mainly asymmetric: Se+Pb, Kr+Hg and also about symmetric: Sn+Dy, Te+Gd) - solid arrow lines in Fig. 2b. Only a small part of the incoming flux reaches a compound nucleus configuration (dashed arrow line).

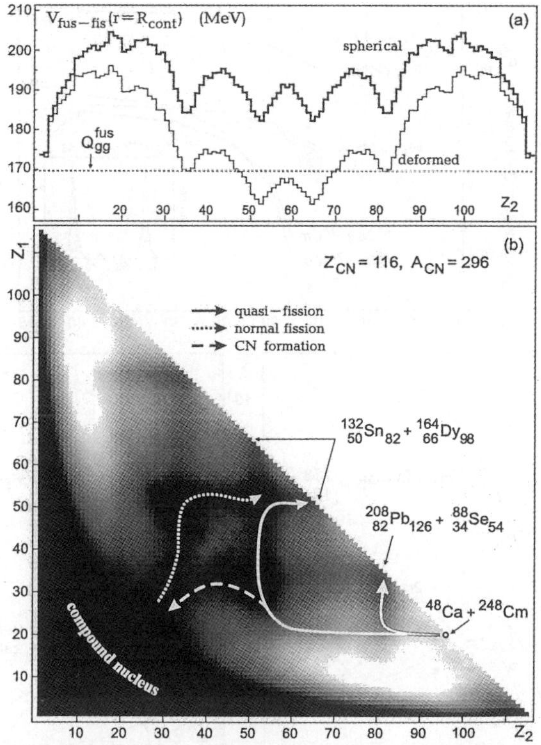

Figure 2: The driving potential $V_{fus-fis}(Z_1, Z_2)$ of the nuclear system consisting of 116 protons and 180 neutrons. (a) Potential energy of two touching nuclei at $A_1 + A_2 = A_{CN}$, $\Delta A = 0$, i.e., along the diagonal of the lower figure. (b) Topographical landscape of the driving potential on the plane (Z_1, Z_2) (zero deformation). The dark regions correspond to the lower potential energies.

Using the derived driving potential and solving the corresponding master equation we obtain the probability of the compound nucleus formation P_{CN} along with mass distribution of quasi-fission fragments. Results of such calculations are shown in Fig. 3a, while the corresponding ER cross sections are shown in Fig. 3b in comparison with the experimental data. Satisfactory agreement with experimental data (also for other reactions not shown here) allows one to conclude that the proposed mechanism of the fusion-fission dynamics is quite adequate and may be used for calculation of the probabilities of the compound nucleus formation and mass distributions of fission fragments.

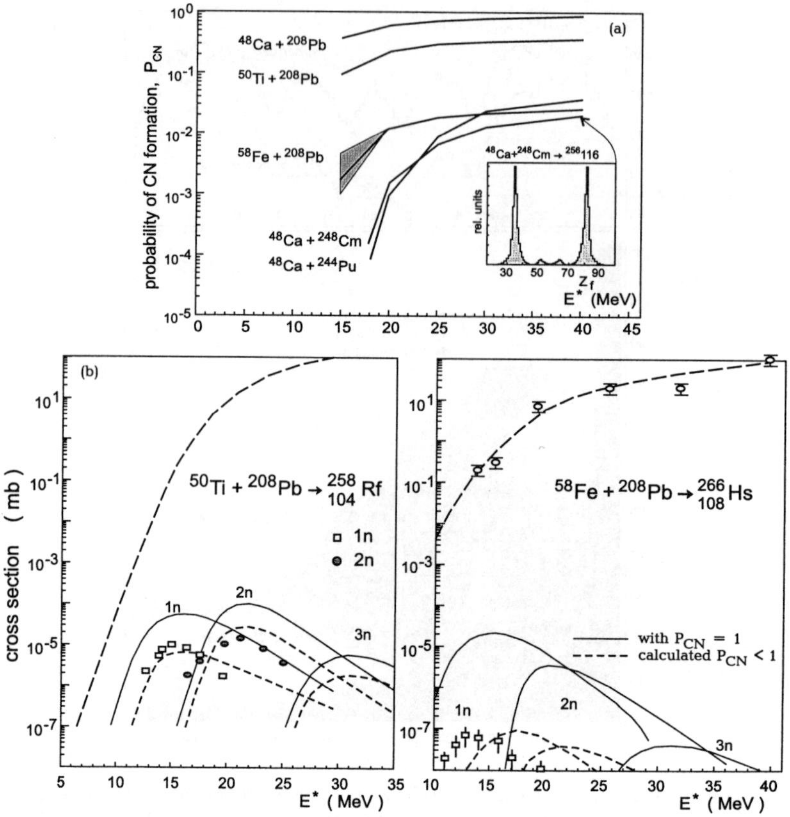

Figure 3: (a) The probability of the compound nucleus formation for the near-barrier fusion reactions. In the inset the corresponding charge distribution of quasi-fission fragments is shown for the $^{48}Ca + ^{248}Cm$ fusion reaction at $E^* = 40$ MeV (linear scale, relative units). (b) The capture cross section (long dashed curves) and formation cross sections for evaporation residues in the $^{50}Ti+^{208}Pb$ and $^{58}Fe+^{208}Pb$ fusion reactions. Experimental data for the capture and ER cross sections are from [2] and [3] correspondingly.

References

1. Zagrebaev, V.I., Phys.Rev. C **64** (2001), in print.
2. Itkis, M.G., et al., in *Fusion Dynamics at the Extremes*, Dubna, 2000, edited by Yu. Oganessian and V. Zagrebaev (WS, Singapore, 2001), p.93.
3. Hofmann, S., Rep.Prog.Phys. **61**, 639 (1998).

AUTHOR INDEX

[A]
Abbondanno U. 112
Achouri N.L. 30
Agodi C. 18
Alamanos N. 257,372
Alba R. .18
Alderighi M. 267
Alkhazov E.D. 224
Amorini F. 26,93,322
Amar N. 142,157,257
Andrighetto A. 197
Andronenko A. 224
Angélique J.C. 30,257
Angulo C. 93
Anne R. 257
Anzalone A. 267
Arickx F. 34
Aritomo Y. 38
Astier A. 322
Attallah F. 77
Auger G. 257
Azaiez F. 322
Azhari A. 216,340

[B]
Badalá A. 42
Baiborodin D. 85,322
Ban G. 176
Baran V. 68,267
Barnà R. 267
Bartel J. 300
Baumann P. 30
Baumann T. 77
Becker F. 60,257
Belleguic M. 322
Bellia G. 18
Belozerov A.V. 376
Bem P. 340
Benlliure J. 77
Benoit B. 56,275,326
Berceanu I. 267
Bertuglia I. 220
Bienczak K. 322
Biswas D.C. 305
Blagus S. 220
Blank B. .60
Blicharska J. 267
Blumenfeld Y. 46
Bogdanov D.D. 372
Bogatchev A.A. 56,142,146, 157,275,372
Bogovac M. 220
Boiano A. 305
Bonasera A. 112,253,267
Boninelli S. 64
Borcea C. 30,60,322
Borderie B. 267
Borge M.J.G. 77
Bouchet A. 224
Bouchez E. 60
Bougault R. 267
Bourgeois C. 322
Brenning H. 322
Broeckhove J. 34

Brondi A. 197,305
Bruno M. 18,112,267
Burjan V. 216,340
Buta A. 30,60

[C]
Caamaño M. 193
Cabaret L. 162
Cappuzzello F. 64,364
Cardella G. 26,93,267
Carstoiu F. 216,340
Casini G. 112
Catford W.N. 30
Cavallaro S. 112,267
Cejpek J. 340
Chbihi A. 267
Chelnokov M.L. 372,376
Chepigin V.I. 293,376
Cherubini S. 93
Chiari M. 112
Chizhov A.Yu. 157
Choudhury R.K. 305
Chubarian G. 157
Chulkov L. 77
Cinausero M. 197,305
Clark H.L. 340
Colonna M. 18,68,267
Colonna N. 18
Coniglione R. 18
Cortina-Gil D. 77
Cosmano A. 112
Cottereau E. 162
Courtin S. 30
Crawford J.E. 162
Cullen D.M. 322
Cunsolo A. 64,364

[D]
D'Agostino M. 18,112,267
Darius G. 176
Daugas J.M. 30
Davinson T. 93
Dax A. 336
Dayras R. 257,267
De Filippo E. 267
de France G. 60
De Góes Brennand E. 326
Delahaye P. 176
Deloncle I. 322
Del Zoppo A. 18
Demin A.M. 372
de Oliveira F. 30,60,257
de Oliveira Santos F. 322
De Pasquale D. 267
Descouvemont P. 81
Desesquelles P. 142,157,326
Dessagne P. 30
De Tourreil R. 257
Di Pietro A. 26,93
Di Toro M. 68,267
Dlouhý D.Z. 30,85,322
Dobrovolsky A.V. 224
Dombrádi Zs. 322
Donadille L. 326
Donzaud C. 322
Dorvaux O. 142,157,326
Dragulescu E. 322
Drouart A. 257
Drozdov V.A. 89
Duprat J. 322
Durand D. 176

[E]
Egelhof P. 224,336
Elliott J.B. 202
Emsallem A. 60
Eremenko D.O. 89,180,385
Essabaa S. 162

[F]

Fabris D. 197,305
Fazio G. 234
Fernandez-Vazquez J. 77
Fiandri M.L. 18,112
Figuera P. 26,93,220
Fil'chagin S.V. 372
Finocchiaro P. 18
Fioretto E. 197,305
Fomichev A.S. 368,372
Fontbonne J.M. 257
Forssén C. 77
Fortier S. 64
Foti A. 64,364
Fotina O.V. 89,180,385
Fraile L.M. 77
Fritz S. 224
Fritzsche S. 336

[G]

Gaelens M. 350
Gagarski A.M. 107
Gagliardi C.A. 216,340
Gall B. 157
Gauthier J.M. 157
Gavrilon G.E. 224
Geissel H. 77,224
Gelbke C.K. 97
Gelli N. 197,305
Geltenbort P. 350
Genevey J. 60
Geraci E. 267
Gerl J. 77,193
Gherghescu R.A. 289
Giardina G. . . 56,89,142,146,157,
180,234,275,293,385
Gillibert A. 257
Giustolisi F. 267
Goeminne G. 350
Golovkov M.S. 372
Gönnenwein F. 107
Gorbachenko O.M. 271
Gorshkov V.A. 372,376
Gorska M. 322
Gramegna F. 18,112
Grawe H. 322
Greco V. 68
Grévy S. 30,157,257,322
Grishechkin S.K. 372
Gross C. 224
Grzeszczuk A. 267
Guazzoni P. 267
Guerreau D. 257
Guillemaud-Mueller D. . . . 30,322
Guinet D. 267

[H]

Hadeler R. 30
Hagemann G. 322
Hanappe F. 56,142,146,157,
234,257,275,326,372
Hannachi F. 60
Hauschild K. 60
Heenen P.-H. 122
Hellström M. 193
Herskind B. 322
Heßberger F.P. 293
Hofmann S. 132,293
Hofmann F. 166
Horn R. 162
Huber G. 162
Hue R. 257
Hurstel A. 60

[I]

Iacono Manno M. 267
Il'kaev R.I. 372
Iori I. 18
Irnich H. 224
ISOLDE Collaboration 162

ISOMER Collaboration 193
Italiano A. 267
Ithashi K. 77
Itkis I.M. 56,142,146,157, 275,326
Itkis M.G. .. 56,142,146,157,275, 326,372

[J]
Jandel M. 56,146,157,275
Janik R. 77
Jesinger P. 107
Jonson B. 77

[K]
Kabachenko A.P. 293,376
Kapoor S.S. 305
Karlsson S. 77
Kavatsyuk M.O. 271
Keil C.M. 166
Khanzadeev A.V. 224
Kiener J. 322
Kliman J. ... 56,142,146,157,275
Kluge H.-J. 336
Knyajeva G.N. .. 56,142,146,157, 275
Knipper A. 30
Kondratiev N.A. 56,142,146, 157,275,326,372
Korolev G.A. 224
Korsheninnikov A.A. 372
Korten W. 60
Korzyukov I.V. ..56,142,146,157, 275,372
Köster U. 350
Kowalski S. 267
Kozhuharov C. 336
Kozulin E.M. ... 56,142,146,157, 275,326,372
Kraus G. 224
Kroha V. 216,340
Krupa L. 56,142,146,157,275
Kühl T. 336
Kuryakin A.V. 372

[L]
Lalleman A.S. 257
Lamberto A. 234,293,385
Lanchais A. 112,267
Lanzalone G. 26,267
Lanzanò G. 267
Lapoux V. 372
La Rana G. 197,305
Lassen J. 162
Lattuada M. 220
Lazzaro A. 64,364
Le Blanc F. 162
Lecesne N. 257
Lecolley F.R. 30
Lecouey J.L. 30
Le Coz Y. 60
Lee J.K.P. 162
Leenhardt S. 322
Legou T. 257
Lemmon R. 322
Le Neindre N. 267
Lenske H. 64,166
Lewitowicz M. 30,60,257,322
Li S. 267
Liatard E. 326
Lichtenthaler R. 257
Liénard E. 30,176,257
Lobodenko A.A. 224
Loiselet M. 350
Lo Nigro S. 267
Loukachine K. 18
Lu J. 26,93
Lucarelli F. 197,305
Lucas R. 60
Lui Y.-W. 340

Lukyanov S.M. 322
Lunardon M. 197,305

[M]
Mahmud H. 93
Maiolino C. 18,267
Majka Z. 267
Malaguti F. 180,385
Malyshev O.N. 293,376
Mandal S. 77
Margagliotti G.M. 112
Margagliotti G.V. 18
Markenroth K. 77
Marqués Moreno F.M. 183
Maruyama T. 253
Marx D. 336
Mastinu P.F. 18,112
Matea I. 60
Materna T. 56,142,146,157, 275,326,372
Mauger F. 176
Mayet P. 193,322
Mazzocco M. 197
Meister S. 77
Melita A.L. 64,364
Miehé C. 30
Migneco E. 18
Milazzo P.M. 18,112
Milin M. 93,220
Miljanic D. 220
Mineva M.N. 193
Mittig W. 257,372
Mocko X. 77
Moody K.J. 56,146
Moretto L.G. 202
Moro R. 197,305
Moroni A. 18,112
Mouze G. 213
Mrázek J. 30,85
Mukhamedzhanov A.M. 216,340
Muntian I. 314
Münzenberg G. 224
Musumarra A. 26,93,220
Mutterer M. 107,224,336

[N]
Nalpas L. 372
Nannini A. 112
Nasirov A.K. 234,293
Naviliat O. 176
Nayak B.K. 305
Nebbia G. 197,305
Negoita F. 30,60
Nesterov A. 34
Nesvizhevsky V. 107
Neumaier S.R. 224,336
Nikolski E.Yu. 372
Ninane A. 93,157,372
Ninov V. 293
Nociforo C. 64,364
Nörtershäuser W. 336
Novak J. 340

[O]
Obert J. 162
Oganessian Yu.Ts. 1,56,142, 146,157,275,293, 368,372,376
Ohtsubo T. 77
Olivo P. 180
Olkhovsky V.S. 244
Oms J. 162
Ordine A. 112,305
Orr N.A. 30,257
Orrigo S. 64
Ouchrif A. 162
Ozawa A. 77,248

[P]
Paduszynski T. 267

Pagano A. 267
Palamara R. 234
Pantelica D. 60
Pantalica D. 322
Papa M. 26,93,253,267
Patrizii L. 180
Pellegriti M.G. 93,220
Penionzhkevich D.Y. 30
Penionzhkevich Yu.-E. 322
Perevozchikov V.V. 372
Pesente S. 305
Péter J. 30,142,157,257
Petizon L. 322
Petrov G.A. 107
Petrovici M. 267
Pfützner M. 193
Phair L. 202
Piasecki E. 267
Piattelli P. 18
Pietri S. 30
Pinard J. 162
Pinston J. 60
Pirrone S. 26,267
Piskor S. 340
Pizzone R.G. 220
Platonov S.Yu. 89,180,385
Plujko V.A. 271
Podolyák Zs. 193
Pokrovski I.V. 56,142,146,157,275
Poirier E. 30
Politi G. 257,267
Pollacco E.C. 372
Pomorski K. 300
Ponomarenko V.A. . . 56,146,157, 275
Pop A. 267
Popeko A.G. 293,376
Porquet M.-G. 322
Porto F. 267
Pougheon F. 322
Prete G.F. 197,305
Pribora V. 77
Prokhorova E.V. 56,142,146, 157,275

[R]

Raabe R. 93
Radivojevic Z. 157
Rahkila P. 60
Redon N. 322
Regan P.H. 193
Rendić D. 220
Rejmund M. 60
Ricci R.A. 213
Riisager K. 77
Rivet M.F. 267,279
Rizzi V. 305
Rizzo F. 26,93
Rodin A.M. 368,372
Rolfs C. 220
Romano S. 220
Rosato E. 267
Roussel-Chomaz P. 372
Roussière B. 162
Rowley N. . . . 56,142,146,157,275
Royer G. 289
Rubchenya V.A. 157
Ruggeri R. 234
Rui R. 18
Ruiz C. 93
Rusanov A.Ya. 56,146,275

[S]

Sagaidak R.N. 293,376
Saint-Laurent M.G. 257,322
Samant A. 305
Sambataro S. 267
Santonocito D. 18
Sapienza P. 18
Sattarov A. 216,340

Sauvage J. 162
Saxena A. 305
Scalia A. 297
Scarpaci J.A. 322
Schäfer T. 224
Scheidenberger C. 224
Scheit H. 77
Schlegel Ch. 193
Schmitt Ch. 142,157,300,326
Schneider G. 77
Schrieder G. 77
Schwab W. 224
Sechi G. 267
Seliverston D.M. 224
Shchepunov V. 364
Shetty D.V. 305
Shotter A.C. 93
Sidorchuk S.I. 368,372
Signorini C. 197
Simeckova E. 340
Simion V. 267
Sitar B. 77
Siwek-Wilczynska K. . . . 310,326, 354
Slepnev R.S. 372
Sletten G. 322
Sobiczewski A. 314
Soic N. 93,220
Soramel F. 197
Sorlin O. 322
Sosin Z. 257
Sperduto M.L. 267
Spitaleri C. 220
Stanoiu M. 30,60,322
Steckmeyer J.C. 257,267
Stepantsov S.V. 368,372
Stodel Ch. 157,257,293
Stolz A. 77
Strmen P. 77
Stroe L. 197
Stuttgé L. . . . 56,142,146,157,234, 275,326,372
Sümmerer K. 77
Surowiec A. 300
Sutera C. 267
Suzuki T. 224
Svirikhin A.I. 376
Szarka X. 77

[T]

Taccone A. 89,234
Tang X. 216,340
Tanihata I. 336,372
Tarasov O. 30,322
Tassan-Got L. 64
Ter-Akopian G.M. 368,372
Theisen Ch. 60,322
Thiamová G. 85
Tillier J. 257
Timis C. 30
Timofeev N.A. 224
Togo V. 180
Tomaselli M. 336
Trache L. 216,340
Tribble R.E. 216,340
Trifirò A. 267
Trimarchi M. 267
Trzaska W.H. 107,157
Tudisco S. 26,93
Typel S. 220

[U]

Uguzzoni A. 180

[V]

Van Hove P. 176
Vannini G. 18,112,267
Vannucci L. 112
Vanroose W. 34
Vardaci E. 197,305

Vasilevsky V. 34
Verney D. 162
Viesti G. 197,305
Vigilante M. 267
Villari A.C.C. 257
Vincour J. 340
Vinodkumar A.M. 176
Vinogradov Yu.I. 372
Vitturi A. 197
Vorobyov A.A. 224
Voskresenski V.M. . . 56,142,146, 157,275

[W]
Wagemans C. 350
Wagemans J. 350
Walker P.M. 193
Walter G. 30
Wan S. 77
Wang H. 336
Weick H. 77
Weissman L. 93
Wieleczko J.P. 257,267
Wieloch A. 257
Wilczyński J. 267,326,354
Winfield J.S. 64,364
Wolski R. 368,372
Wolter H.H. 220
Wozniak G.J. 202
Wu H. 267

[X]
Xiao Z. 267

[Y]
Yatsoura V.I. 224
Yeremin A.V. 293,376
Yukhimchuk A.A. 372
Yuminov O.A. 89,180,385

[Z]
Zadro M. 220
Zagrebaev V.I. 389
Zbiri K. 289
Zetta L. 267
Zimmer O. 107
Zipper W. 267
Zlatoustovsky S.V. 372